Theory and Applications of Acoustic Waves

Theory and Applications of Acoustic Waves

Edited by **Sonny Lin**

CLANRYE
INTERNATIONAL

New Jersey

Published by Clanrye International,
55 Van Reypen Street,
Jersey City, NJ 07306, USA
www.clanryeinternational.com

Theory and Applications of Acoustic Waves
Edited by Sonny Lin

International Standard Book Number: 978-1-63240-491-6 (Hardback)

Printed in the United States of America.

Contents

Preface

This book discusses various approaches of the active research going on in the field of acoustic waves. The concept of acoustic wave is an extensive one; it materializes in a variety of mediums, from solids to plasmas, and at varying length and time scales which range from sub-micrometric layers in microdevices to seismic waves in the Sun's interior. Theoretical attempts lead to a comprehensive understanding of this phenomenon. Such waves are useful in exploring the characteristics of a variety of structures, be it inorganic layers and/or other bio-structures. These waves can also function as a tool to influence matter, from the gentle evaporation of bio-molecules to be evaluated, to the phase conversions inferred by intense shock waves. Furthermore, a whole class of widespread micro tools, inclusive of filters and sensors, depends upon the behavior of these waves propagating in fine layers. The book highlights acoustic waves as manipulative tools and as microdevices.

After months of intensive research and writing, this book is the end result of all who devoted their time and efforts in the initiation and progress of this book. It will surely be a source of reference in enhancing the required knowledge of the new developments in the area. During the course of developing this book, certain measures such as accuracy, authenticity and research focused analytical studies were given preference in order to produce a comprehensive book in the area of study.

This book would not have been possible without the efforts of the authors and the publisher. I extend my sincere thanks to them. Secondly, I express my gratitude to my family and well-wishers. And most importantly, I thank my students for constantly expressing their willingness and curiosity in enhancing their knowledge in the field, which encourages me to take up further research projects for the advancement of the area.

Editor

Part 1

Acoustic Waves as Manipulative Tools

Molecular Desorption by Laser–Driven Acoustic Waves: Analytical Applications and Physical Mechanisms

Alexander Zinovev, Igor Veryovkin and Michael Pellin
Argonne National Laboratory
USA

1. Introduction

Analytical mass-spectrometry (MS) is a powerful, widely-used tool for materials analysis, helping to make progress in materials and environmental sciences, chemistry, biology, astrophysics, etc (Dass 2007). Often the sample to be studied (analyte) is a solid requiring: a) volatilization/desorption of the analyte atoms/molecules and b) their consequent conversion to the charged particles (ionization) prior to mass analysis. The last two decades have seen revolutionary advances in these techniques (Dass 2007) and the use of direct laser irradiation to achieve volatilization is one of the wide-spread methods (Lubman 1990) These pulsed laser-based techniques for the desorption/emission of the atoms, molecules and ions from the surface of solids has benefitted from fundamental study of the process beginning with the invention of the lasers (Honig and Woolston 1963) . A short laser pulse hitting a solid absorbing surface delivers high energy in a small volume inducing a variety of state changes. One consequence is the evaporation/desorption of surface atoms and molecules could be used for further analysis by MS technique. However, the increasing use of MS methods in analytical chemistry of organic and biomolecules revealed that this direct desorption process had significant drawbacks for the analysis of molecular solids. Most importantly, the high energy density produced during irradiation results in not only surface heating but also in excitation of internal vibrational and electronic states of desorbed molecules leading to their partial or even complete fragmentation (Lubman 1990). This difficulty was overcome for many samples by the development of Matrix Assisted Laser Desorption and Ionization (MALDI), which by imbedding the analyte in a specialized UV absorbing molecular solid (the "matrix") allows UV lasers to both desorb and ionize large organic and biomolecules without significant fragmentation (Cole 2010). Because MALDI combines both of the needed initial processes (desorption and ionization) it very quickly following the pioneering publication (Karas, Bachmann et al. 1985) became a key analytical tool. MALDI is now one of principle research tools in proteomics (Cole 2010) and its discovery was recognized with the Nobel Prize in chemistry in 2002.

Despite the success of the MALDI technique current active areas of research include quantification and analysis in the low mass region. Application of MALDI to analyte quantification while possible requires careful attention to matrix/analyte sample preparation, a detailed understanding of the crystallization process with regard to the analyte, and careful many spot analyses (to find the sample signal average which often varies by orders of magnitude as a function of laser position). (Duncan, Roder et al. 2011)

This desire to find a discriminative, sensitive and more easily quantifiable alternative to MALDI has lead us to re-examine another molecular desorption method that doesn't require the use of a matrix. The first observation of this method of molecular desorption (later called Laser Induced Acoustic Desorption, LIAD) belongs (to the best of our knowledge) to B. Lindner (Lindner and Seydel 1985). This observation has been followed by several studies including (Golovlev, Allman et al. 1997) where the abbreviation LIAD was introduced and by (Perez, Ramirez-Arizmendi et al. 2000) who applied the technique to several classes of MS problems.

LIAD has rather simple experimental layout: an analyte is deposited onto the front surface of thin metal foil (the substrate), which is irradiated from the back (i.e. the side opposite to both the analyte and the mass spectrometer) by a pulsed laser beam with power density insufficient to pierce the foil. Such irradiation results in the volatilization of the analyte. The volatilization is largely in the form of neutral molecules that can be utilized for further MS analysis using an appropriate post-ionization method. The method seems to be relatively insensitive to the sample preparation method. Commonly the sample preparation requires only evaporating a drop containing a few nano-moles of the analyte. Remarkably little desorption induced fragmentation is seen when a suitable "soft" ionization method such as VUV photoionization can be found. It is also useful to note that while the molecular signal depends on drive laser intensity the fragmentation observed varies only weakly.

The advantages of LIAD have been demonstrated by many researchers; however, the mechanism is not well understood. The first desorption mechanism was proposed (Golovlev, Allman et al. 1997). It was supposed that because the metal is opaque and completely blocks the direct interaction between the drive laser light and adsorbed, front-side molecules, the only possible way of energy transfer is the mechanical. In this model, the interaction of laser pulse with metal foil backside resulted in formation of acoustic waves that move through the foil inducing a front surface oscillation motion. The molecules that are sitting on this surface desorb due to a simple "shake-off" mechanism similar to those that we use to remove dust particles from our clothes by shaking it.

A difficulty with this model arises when considering the relatively strong surface binding energy experienced even by physisorbed molecules. In order to be efficiently desorbed from the surface, surface molecules need to achieve initial kinetic energies exceeding their surface binding energies (typically in the range of 0.05 - 0.5 eV for physically adsorbed molecules (Adamson and Gast 1997)). This corresponds to velocities of a few hundred m/s for molecules with masses of a few hundred atomic mass units. Unfortunately, acoustic vibrations have mass transfer velocities much lower than the speed of sound, and in elastic deformation mode, this velocity does not exceed a few m/s (Landau and Lifshits 1987). While laser-driven acoustic wave generation in metals is very well studied problem (Hutchins 1985), the physics of their generation in metal foils is crucial to understanding the need for development of new desorption mechanisms. In the next paragraphs we will give a brief theoretical overview and will present our experimental results on laser-driven acoustic waves in thin metal foils.

2. Laser-driven acoustic waves in thin metal foils

The interaction of pulse laser beam with metal surface is very complex phenomenon but our specific interest is in formation of the acoustic waves in irradiated material. To generate an acoustic wave a time dependent stress needs to be applied to the solid. A laser pulse is an excellent tool to generate this kind of the stress. There are two principal mechanisms of laser-induced stress formation in the solids: a) thermal stress, resulted from the non-uniform heating of the irradiated surface by the laser beam; b) mechanical stress due to mechanical impulse transferred from the leaving plasma plume formed on the surface during laser ablation. The parameters of the acoustic waves generated for these two processes are slightly different and will be discussed in details later. Here we will use acoustic wave theory but one should note that its' use is applicable only when the magnitude of the applied stress is small in comparison with the Young's modulus of the material. Large applied stresses can cause the development of the shock waves a phenomenon with different characteristics than acoustic waves (Menikoff 2007). Shock waves have been also hypothesized to be the driver of the LIAD phenomenon, and, as such cannot be entirely excluded from consideration, especially in some extreme cases. Nevertheless, during the last decade, laser-driven acoustic waves emerged in the literature as the "prime suspect" in the LIAD case.

2.1 Acoustic waves in metal foils

The general governing equation for the generation of elastic waves in solids can be derived by combining the equation of motion and the Hooke's law. In general case it is a differential tensor equation, which interconnects the stress tensor, applied to the body, the displacement of the body's elemental volumes (strain), and their elastic properties. In order to analyze the data in detail, the appropriate stress tensor needs to be determined, and the corresponding set of partial differential equations for strain and stress must be resolved (Pollard 1977). One can simplify this analysis by taking into account specifics of the experiments. For thin foils with $h/R_0 \ll 1$ (where R_0 is the radius of the target foil, and h is its thickness), a round thin plate approximation can be applied to describe and analyze this problem (Smith 2000). The rise of the strain due to laser heating and the consequent development of the plasma plume can be considered as an external driving force. Generally speaking, the thin plate equation is a differential equation of the forth order,

$$\left(\frac{\partial^2}{\partial r^2} + \frac{1}{r} \cdot \frac{\partial}{\partial r} + \frac{1}{r^2} \cdot \frac{\partial^2}{\partial \theta^2} \right)^2 \xi + v_L^{-2} \cdot \left(2 \cdot \alpha \frac{\partial \xi}{\partial t} - \frac{\partial^2 \xi}{\partial t^2} \right) = F(r,t) \tag{1}$$

where $\xi = \xi(r,t)$ is the surface displacement in the z direction (perpendicular to the sample surface), $F(r,t)$ is the external driving force caused by the laser irradiation, v_L is a parameter depending on the material density, ρ, thickness, h, and flexural rigidity, D. Furthermore,

$$v_L = \sqrt{D / \rho \cdot h} \ , \text{ and } , \ D = (N \cdot h^3 / 12 \cdot (1 - \varepsilon^2)) \tag{2}$$

where ε is the Poisson ratio, N is Young's modulus and α is the oscillation decay constant. A vibrating thin plate is one of the most frequently analyzed mathematical problems, and its detailed analysis can be found elsewhere (McLachlan 1951; Smith 2000). In the discussion given below, we will remain in the framework of the analysis given by (Smith 2000).

Assuming separate solutions for the radial and tangential terms, Eq. (1) can be converted into a system of three differential equations of the second order. Being primarily interested in the foil displacement in normal (z) direction to the surface at the epicenter ($r=0$) and assuming that the external driving force $F(t)$ is a pulse function lasting a specific time τ, we can separate the variables in the Eq. (2) for $t>>\tau$. Under the assumption of harmonic motion for all modes, the surface displacement can be expressed as $\xi(r,t)=\xi_1(r)\cdot e^{i\omega_{m,n}t-\alpha\cdot t}$, where $\omega_{m,n}$ is the vibration frequency. The general governing equation can then be written in the following form:

$$\left(\frac{\partial^2}{\partial r^2}+\frac{1}{r}\cdot\frac{\partial}{\partial r}-\frac{n^2}{r^2}\pm k_{n,m}^2\right)\xi_1=0 \tag{3}$$

where n is an integer number, $k_{n,m}=\omega_{n,m}/v_L$ and v_L is the wave velocity as in Eq.(2). Under the assumption that the oscillations are harmonic and decay exponentially, the following solution for the surface displacement can be obtained (McLachlan 1951):

$$\xi_n(r,t)=\xi_0\cdot\left(J_n(k_{n,m}r)+\chi\cdot I_n(k_{n,m}r)\right)\cdot\exp(i\omega_n t)\cdot\exp(-\alpha\cdot t) \tag{4}$$

where J_n and I_n are Bessel functions and χ is a constant. The term describing the angular dependence of the oscillations is omitted in Eq.(4). The fact that foils used in LIAD experiments are typically glued or welded on their perimeter corresponds in our analysis to the situation when edges of the round plate are fixed (i.e. non-vibrating), and is described by the following boundary conditions:

$$\xi(R_0,t)=0,\frac{d\xi(R_0,t)}{dt}=0, \tag{5}$$

that lead to an equation, whose solutions $j_{n,m}$ have tabulated values (Smith 2000). The corresponding vibration frequencies, $\omega_{n,m}$, can be then expressed as

$$\omega_{n,m}=\sqrt{j_{n,m}^4\cdot\frac{v_L^2}{R_0^4}-\alpha^2}\approx\frac{j_{n,m}^2\cdot h}{R_0^2}\cdot\sqrt{\frac{N}{12\rho\cdot(1-\varepsilon^2)}} \tag{6}$$

Because values of α are small compared to the first term under the square root sign in Eq.(6), it is a reasonable assumption that the frequency is proportional to the square root of the ratio of Young's modulus, N, to the density of the foil material, ρ. As described by Eq. (6), in the steady-state regime (driving force $F(t)=0$ for $t>>\tau$) the frequency spectra and decay times of the oscillation will remain the same while the laser intensity is varied, and only the amplitude should change.

For short times, ($t<\tau$), the approach to solving Eq.(1) is to assume that the external force is a delta-function in space $F(r,t)=F(t)\delta(r)$ (point source), allowing the solution of Eq.1 to be expressed in the form

$$\xi(r,t)=g(r)\cdot\xi(t) \tag{7}$$

where $g(r)$ and $\xi(t)$ represent the spatial and the time dependencies of the final solution, respectively. This approximation should help to develop a clearer understanding of physical problems related to the laser generation of acoustic waves in thin foils.

Using Eq. (7), Eq. (1) can be split into two independent equations, with the equation for $\xi(t)$ having the form

$$\frac{d^2\xi}{dt^2} - 2 \cdot \alpha \cdot \frac{d\xi}{dt} + \omega_n^2 \cdot \xi = F(t) , \tag{8}$$

and the equation for $g(r)$ being a non-uniform Bessel equation with the $\delta(r)$- function in the right side of it.

The exact solution of Eq. (8) can be expressed as the convolution of the Green function of the problem (8) and real time shape of $F(t)$. The governing equation for the Green function will be Eq.(8) with δ-function in the right side. It can be easily derived by applying the Laplace transform to the Eq.(8), subsequently solving the obtained linear equation in the s-space and returning back to the time space with using the inverse Laplace transform. As a result of these procedures, one can finally obtain the following equation

$$\xi_n(t) = \int_0^t \frac{e^{-\alpha \cdot (t-\tau)}}{\sqrt{\omega_n^2 - \alpha^2}} \cdot \sin\left[\sqrt{\omega_n^2 - \alpha^2} \cdot (t-\tau)\right] \cdot F(\tau) d\tau , \tag{9}$$

and the complete solution of Eq.(8) can then be obtained as the sum of the components of Eq.(9) over n.

Thus, the generation of acoustic waves in thin foils can be described by Eq.(9) which strongly depends on driving force $F(t)$ after whose cease the vibration evolves into decaying harmonic oscillations (Eq.(4)) with frequencies defined by Eq.(6). It is apparent that maximal surface velocities can be achieved only at the initial stage of acoustic wave generation when $t<\tau$. The application of Eq.(9) to analysis of laser-driven acoustic vibrations is complicated by the lack of the exact knowledge of the time profile of the driving force $F(t)$. Depending on regime of the surface irradiation, this force may be of different origins and, accordingly, have strongly varying magnitudes and time profiles. The appropriate mechanisms will be discussed in the following section.

2.2 Generation of the acoustic waves by laser pulses
2.2.1 The action of laser pulse on the metal surface: heating and plasma generation
Metals subjected to pulsed laser irradiation absorb energy within a very thin surface layer (the skin-depth for most metals is less than 10^{-7} m) so that the temperature of the irradiated surface can rise extremely fast. For moderate laser intensities (below the plasma formation threshold) and a Gaussian-shaped laser beam, the maximum surface temperature can be estimated using a well-known expression (Prokhorov, Konov et al. 1990)

$$T_{max}(0) = 2.15 \frac{A I_{max} \tau^{1/2}}{(\pi c \rho K)^{1/2}} , \tag{10}$$

where $T_{max}(0)$ is the maximum surface temperature at $z=0$ (the z direction is orthogonal to the target surface), A is the laser radiation absorption coefficient, I_{max} is the peak laser power, τ is the laser pulse duration, and c, ρ, and K are the specific heat, density and thermal conductivity of the corresponding metal, respectively.

After cessation of the laser pulse, the adsorbed energy continues to diffuse into the bulk metal and along the metal surface, resulting in a temperature increase underneath the irradiated spot and outward from the spot along the surface. Temperature evolution at any moment, $t>\tau$, and for any position, $z>0$, proceeds according to the following equation (Prokhorov, Konov et al. 1990):

$$T(z,t) = \frac{2AI_{max}\gamma^{1/2}}{K} \cdot [t^{1/2}ierfc(\frac{z}{2(\gamma t)^{1/2}}) - (t-\tau)^{1/2} \cdot ierfc(\frac{z}{2[\gamma(t-\tau)]^{1/2}})] \tag{11}$$

where γ is the thermal diffusivity of the metal, which can be expressed as $\gamma = K / c\rho$. The function $ierfc(x)$ is given by

$$ierfc(x) = \pi^{-1/2}\{\exp(-x^2) - x(1 - erf(x))\} \tag{12}$$

where $erf(x) = \frac{2}{\sqrt{\pi}}\int_0^x \exp(-\xi^2)d\xi$

Eqs. (10) and (11) are the solutions of the one-dimensional heat diffusion equation and are valid only if the laser beam size, r_0, is significantly greater than both the foil thickness h and the thermal diffusion length l_{th} calculated as $l_{th}=(\gamma \cdot t)^{1/2}$.

The strong rise of the surface temperature given by Eq.(10) results in the surface melting and evaporation, as well as in plasma plume formation (ablation regime) (Miller and Haglund 1998). Despite the fact that laser plasma generation and evolution have been the focus of numerous studies, no general mechanisms exist that describe the plasma recoil pressure on the surface for a broad range of laser intensities (Phipps, Turner et al. 1988), due to the complexity of the phenomenon. For GW/cm² peak laser powers, hot and dense plasma is formed in the vicinity of the surface, which can screen the surface and prevent laser radiation from reaching it. In this case, the ablative pressure very weakly depends on the target material parameters (Phipps, Turner et al. 1988) and has a sub-linear dependence on laser intensity. In a semi-regulating, one-dimensional plasma model (which can be applied to our case as a simplified, first-order approximation), this equation is written, as follows (Gospodyn, Sardarli et al. 2002) :

$$P_{a,max} \approx 7.26 \cdot 10^8 \cdot I^{3/4} \cdot (\lambda \cdot \sqrt{\tau})^{-1/4} , \tag{13}$$

where I is expressed in GW/cm², λ in microns, τ in nanoseconds and $P_{a,max}$ in Pa. While Eq. (13) was derived for an aluminum target in vacuum and for a supercritical plasma density, it exhibits only a weak dependence on the atomic mass, A, of the material irradiated ($A^{-1/8}$) (Gospodyn, Sardarli et al. 2002) and may be applicable to specific experiments only as an upper limit estimate. For lower laser intensities (<1 GW/cm²), the plasma plume transmittance strongly varies with laser intensity (Song and Xu 1997), depending upon the plasma density and temperature. In this case the evaporated surface material is ionized only partially and the total mass of the evaporated atomic cloud are exponentially increasing with the surface temperature and, hence with the laser intensity (Murray and Wagner 1999).

2.2.2 Thermal and plasma driven acoustic waves in metal foils

The temperature rise, as heat is transported into the solid causes linear thermal expansion resulting in the development of thermoelastic acoustic waves in the irradiated metal. Eqs. (10) and (11) can be used to determine the driving force which produces the waves. In accordance with the general theory of thermal stresses in thin plates (Boley and Weiner 1960), a non-uniform heating of the surface is equivalent to a *negative loading pressure* and may be expressed as

$$P_T = -\frac{N \cdot \eta}{1-\varepsilon} \cdot \nabla_r^2 \left[\int_{-h/2}^{h/2} T(r,z) \cdot z\,dz \right], \qquad (14)$$

where the temperature distribution over z is described by the Eq. (11) and η is the linear thermal expansion coefficient.

If the loading force is negative (i.e., directed backwards, towards the heating laser beam, Eq. 14), it is not surprising that an initial depression observed in the foil surface is opposite to the heating laser beam. Similar results have been reported in the literature (Scruby 1987) for thicker metal samples where the thin plate approximation was not applicable. Maximum amplitudes and shapes of the observed depression vary for different metals and are defined by both the temperature profile (Eq. (11)) and by elastic properties of the material.

In the case of plasma formation, the situation becomes more complex. The amplitude of the driving force can be estimated using an expression similar to Eq.(13), whereas the time profile of generated stress pulse is the subject of experimental study (Krehl, Schwirzke et al. 1975). Laser plasmas formation and their interaction with the surface is very complex and multi-variable problem, which can only be solved in the framework of some model assumption (Mora 1982). This is why direct experimental studies of laser-driven surface vibrations should be an essential part of any acoustic wave related desorption phenomena.

2.2.3 Experimental observations of laser-generated acoustic waves in thin foils

Experimental studies of acoustic waves in solids due to pulsed laser irradiation have started with the advent of such lasers (White 1963). A great collection of experimental results and theoretical analyses of acoustic wave generation in solids driven by laser pulses has been accumulated since (Hutchins 1985), and these studies continue at present (Xu, Feng et al. 2008). Regrettably, there is a very limited data set, which could be used to interpret of LIAD experiments. To prove (or disapprove) the "shake-off" hypothesis of molecular desorption, direct measurements of thin foil surface velocities in back-side irradiation geometry are required. The scarcity of such data motivated us to setup a series of our own experiments aiming at measurements of thin foils vibrations under typical LIAD conditions. Experimental approaches to this problem are well known and described in the literature (Scruby and Wadley 1978; Royer and Dieulesaint 2000). Nevertheless, we will briefly describe below our system, in order to create a better stage for presentation and discussion of the original results.

2.2.3.1 Experimental technique: optical and electrical methods

One of the most popular and widely used methods to studies of acoustic waves is based on non-contact optical measurements. Figure 1a shows the experimental setup for measurements of surface displacement using interferometry-based approach. A He-Ne laser (1) (Melles-Griot, 543 nm, 0.5 mW) was used as the light source for a Michelson interferometer. It consisted of a beam splitter (5), an etalon and steering mirrors (3, 4), a focusing lens (6), a target (7), an imaging lens (9), an aperture (13), a focusing lens (14) and a photomultiplier (15). The target was back-irradiated by a pulsed laser (12) through a fused silica lens (8). Laser beam parameters were measured by intersecting the laser flux with two partially reflecting (8%) quartz plates (10, 11) directing reflected beams onto a fast photodiode (16) and an energy meter (17), respectively. A wedge-type optical attenuator (18) was used to balance the Michelson interferometer shoulders and to increase the contrast of the resulting interference pattern. Focusing lenses (6) and (8) were mounted on three-axis translation stages. In this arrangement, both the acoustic-wave generating and diagnostic laser beams could be independently focused and translated to different points on the target surface. A lens (9) formed the magnified interference pattern in the plane of an aperture (13) whose diameter was selected to be equal to the width of the dark band of the lowest interference order. A second lens (14) was used to collect the light, which passed through the aperture (13), and to direct it to the photomultiplier (PMT, 15).

The anode of the PMT was terminated with a 50 Ohm load to allow optical signals with rise times as short as 5 ns to be measured. This capability was confirmed by demonstrating that the shape of a 12 ns, NdYAG laser pulse was identical when measured by using this detection system and by a high-speed avalanche photodiode. The measurements bandwidth was limited by the PicoScope 3206 oscilloscope (200 MHz bandwidth and 200 Ms/s sampling rate), which was used for signal acquisition. The digitized PMT signal was transmitted to a PC via USB port and stored for further processing. The oscilloscope was triggered by a pulse from the fast photodiode (17). The measured lag between the trigger signal and the PMT signal was less than 40 ns. The maximum signal amplitude, corresponding to the peak-to-valley ratio of the interference pattern, was 80 mV. The minimum detectable signal was 5 mV at signal-to-noise ratio of about 3, which corresponds to a surface displacement of approximately 25 nm. However, because of the strong electrical noise generated by the Q-switch of the laser, the smallest surface displacement detectable in this series of experiments was about 40 nm.

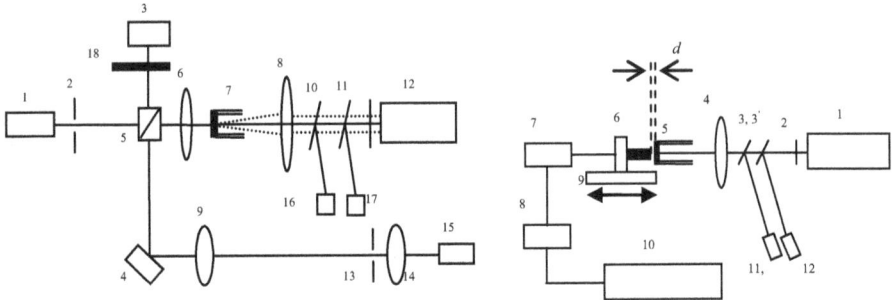

Fig. 1(a,b). Schematic drawing of the experimental setup for laser-driven acoustic wave studies. (a) Interferometry method; (b) Capacitance method

The same target irradiation scheme as described above was also used for the capacitance transducer measurements (Fig.1b). In contrast to the previous approach, a metal pin was placed in front of the target. This pin served as the second plate of a capacitor whose first plate was the target. These two capacitor electrodes were separated by a small gap d, typically about 100 μm. Both the target and the pin were fixed in optical mounts that allowed alignment in the plane of the sample surface. In addition, the mount of the pin was placed on a translation stage (9), driven by a picomotor, which could move the target (with the precision of 1 μm) in the direction orthogonal to the target surface. The pin had a diameter of 3 mm and its end was polished flat. In general, the design of this detector is similar to that described in Ref.19. The pin was connected to the input of a miniature charge amplifier (7) powered by a constant (20 mA) direct current (DC) supply (8) and connected to the oscilloscope (10). The bandwidth of the charge amplifier was about 2 MHz, which provided a signal rise time of less than 400 ns. To increase the overall sensitivity of the detector, a positive bias potential of 100 V was applied to the target.

The sensitivity of the transducer to surface displacement can easily be expressed in terms of a planar capacitor

$$\Delta q = \frac{-\varepsilon V S}{d^2} \Delta d \tag{15}$$

where Δq is the change in the electric charge, V is the applied voltage, S is the surface area of the pin tip, d is the width of the gap between the electrodes and Δd is the change of the width. The sensitivity of the charge preamplifier was 10 mV/pC, which corresponded to a minimal detectable signal of about 5 mV (thus yielding reasonably good signal-to-noise ratio). With an applied bias potential of -100 V, the estimated detection limit of the transducer-based sensor was about 5 nm.

2.2.3.2 Experimental results: displacement and surface velocity of thin foils

Foils with various thicknesses (from 12.5 μm up to 100 μm) made from different materials were used in our experiments. Materials were selected to span the range from soft metals (Au, Al and Ni) to refractory metals (W, Mo and Ta) and semiconductors (Si). For each experiment, the front surface of the sample was mechanically polished to roughness of less than 0.250 μm (RMS). After polishing, the foils were glued with silver epoxy to the rim of a hollow quartz cylinder (8 mm outside diameter, 8 mm height, and 0.5 mm wall thickness). The epoxy was cured for 2 hours in an oven at temperature of 100° C. Due to the differences between thermal expansion coefficients of the foil materials and quartz, the foil stretched over the top of the quartz cylinder once the assembly cooled to room temperature. The tension was not very strong (according to our estimates, the total radial force did not exceed 1 N), and therefore, the foils in our experiments may be considered as supported at the edges. The silver epoxy also provided a conductive path between the sample and the instrument by placing a silver epoxy track along the quartz cylinder side.

Lasers generating both ultraviolet (UV) and infrared (IR) light were used for target irradiation. The UV light was generated by an ArF excimer laser (EX10-300, GAM, Inc.) having a wavelength 193 nm and a pulse duration of 15 ns. The output pulse energy could be varied from 0.4 to 4 mJ by using neutral density optical filters. The laser radiation was focused onto the backside of the target (opposite to the surface displacement sensors) using a fused silica lens with a focusing distance of f=300 mm. The irradiated spot on the target had rectangular dimensions of 100×500 μm, which corresponded to a UV laser power density in the range of 50–500 MW/cm². For IR irradiation of the target, a Q-switched Nd:YAG laser (Continuum) was used. This IR light had a wavelength of 1064 nm, a 12 ns pulse duration and an output energy in the range of 1–15 mJ/pulse. The IR laser beam had a Gaussian profile and produced a spot on the target surface with a nominal diameter of 500 μm, corresponding to peak power density of 40–600 MW/cm².

Waveforms representing foil oscillations were measured over the time range from 5 μs to 5 ms. For times much greater than the laser pulse duration (t>>τ), a decaying quasi-harmonic oscillations were observed for all materials. The measured time dependence of the displacement of the foils irradiated by laser pulses with different intensities exhibited qualitatively similar behavior, although amplitudes, frequencies and decay times varied. These results suggest that each foil behave as a mechanical system able to oscillate in a free-running mode after the external force is removed. Fast Fourier Transform (FFT) analysis applied to the measured data has shown that the frequency spectra consisted of discrete lines (modes) appearing in the range of 10–100 kHz.

In contrast to the steady-state regime ($t \gg \tau$) when different foils oscillated very similarly, the initial moment ($t < \tau$) of the evolving oscillation was distinctly different for each foil. Fig. 2 presents the time dependence of the surface displacement for different metals at low irradiation intensities (50 MW/cm²) for "early" times in the range up to 50 μs. For all measurements at low intensities, the initial displacement is found to be negative, indicating that the surface is first depressed (i.e. towards to the driving laser beam and, correspondingly, away from the detector). Increasing the laser intensity leads to the initial surface vibration waveform displacement changing from negative to positive. This is due to the recoil pulse which occurs when material is ablated from the irradiated surface due to plasma formation. Fig. 3 shows the time dependences of the displacement of Ta foil surface for different laser irradiation intensities. The plasma formation threshold for this Ta foil was ~220 MW/cm², as determined by the observation of the plasma plume glow in a separate experiment. The FFT analysis of the experimental data at higher laser intensities revealed that the frequency spectrum of the foil oscillations in the non-steady-state regime contains much higher frequency components than found for the steady-state case. After a few tens of microseconds, the high frequency components disappear as the foil oscillation become harmonic as described by Eq.(4).

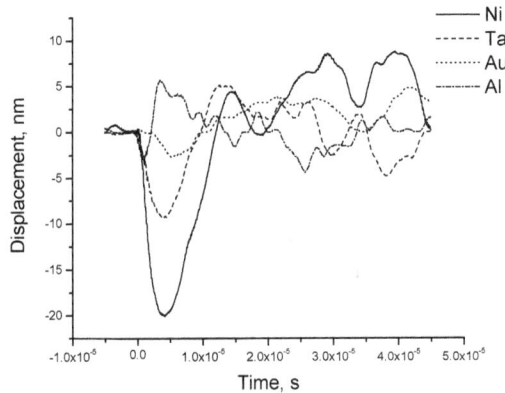

Fig. 2. Time dependent displacement for different metal foils at low laser intensities (50 MW/cm²

The obtained results are in good agreement with time dependencies of surface displacement measured under slightly different experimental conditions (Hutchins 1985). This fact clearly demonstrates that laser-driven acoustic waves in thin metal foils have no experimental peculiarities distinguishing them from well-known acoustic wave mechanisms. This result allowed us to calculate surface velocities using the measured surface displacements (Fig. 4). As one can see from Fig.4, these velocities are indeed in the range of meters per second.

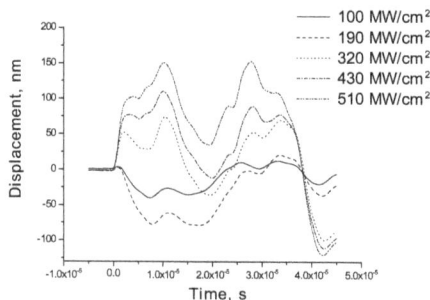

Fig. 3. Time dependent displacement of Ta foil (12.5 μm thick) at different laser intensities

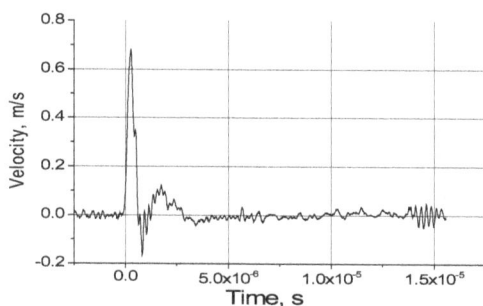

Fig. 4. Time dependence of Ta foil (12.5 mm thick) surface velocity at 400 MW/cm² driving laser intensity

This picture clearly confirms the statement above that the mass transfer velocity (or surface displacement velocity, in terms of our experiment) is much slower than the speed of sound in metals. In its turn, this result supports our hypothesis that the vibrational motion of the foil surface cannot serve as direct cause of molecular desorption, and that *the real physical mechanism of LIAD is not as simple as mechanical shake-off.*

3. Desorption of the molecules from back-irradiated thin metal foils

3.1 Laser desorption in modern MS: methods and applications

As shown in the previous sections, laser induced desorption phenomena play important role in modern MS. The primary role of the laser beam there is to deliver high energy density into some (typically small) volume of the analyte. Due to the local overheating this volume is volatilized forming hot and dense vapor plume, which might be partially ionized. This ionization phenomenon can be considered as a great advantage of laser desorption because there is no need for an additional ionization step, so that the desorbed ions may be directly analyzed by a mass-spectrometer. At the same time, this can be a significant drawback, because, due to collisions in the plume, organic molecules may fragment to the point that their mass analysis becomes meaningless (Miller and Haglund 1998). While using UV lasers in MS analyses of organic materials often produced encouraging results, it is well recognized in the literature that "general mechanism that is applicable to all organic solids at all UV wavelengths does not exist"(Srinivasan and Braren 1989). The introduction of MALDI gave strong indication that many problems, associated with laser desorption MS might have been solved. However despite popularity of MALDI in MS analyses of proteins, lipids and many other organics (Schiller, Suss et al. 2007), this method cannot be considered as universal because it requires to identify efficient matrix substances for different organic species, and often to develop specialized sample preparation protocols. And, regrettably, MALDI MS cannot be used to directly characterize mixtures of unknown molecules. This is why the search for more versatile and universal methods of molecular desorption/ionization is still on in the analytical mass spectrometry community. From this perspective, the ability of LIAD to volatilize different kinds of organic molecules without noticeable (or, very often, without any) fragmentation has attracted strong interest among researchers.

3.2 Laser-induced acoustic desorption

As described above, the acronym LIAD was suggested in the work conducted by the Chen's team (Golovlev, Allman et al. 1997) where the hypothesis about the acoustic wave nature of the desorption process was expressed. However, LIAD remained just an interesting observation until Kentamaa's team of researchers from Purdue University (Perez, Ramirez-Arizmendi et al. 2000) took on it and demonstrated successful applications of LIAD for the MS analysis of different organic species like cytosine, guanine, thimidine and some others. This work was followed by the series of studies where the applicability of LIAD to the MS analysis of a wide range of organic samples has been demonstrated. The LIAD volatilization method was successfully coupled with Fourier transform ion cyclotron resonance mass spectrometer and alanylglycine (Reid, Tichy et al. 2001), saturated hydrocarbons (Campbell, Crawford et al. 2004), polyethylene (Campbell, Fiddler et al. 2005) and even petroleum distillates (Crawford, Campbell et al. 2005) were analyzed. The great advantages of this technique were the ability to efficiently volatilize various organics and the simplicity of its incorporation into different classes of MS instruments, such as Linear Quadrupole Ion Trap MS (Habicht, Amundson et al. 2010) and Time-of-Flight MS (Zinovev, Veryovkin et al. 2007). Various approaches for the ionization of the desorbed molecules were successfully employed, among them: electron impact and chemical ionization (Crawford, Campbell et al. 2005), single-photon ionization (SPI) (Zinovev, Veryovkin et al. 2007), as well as Elecro- Spray Ionization (ESI) (Cheng, Huang et al. 2009). Moreover, the ability of the LIAD process to non-destructively eject from solid surfaces not only single molecules but also larger intact biological particles, such as viruses (Peng, Yang et al. 2006) and 1 μm size tungsten particles (Menezes, Takayama et al. 2005) have been demonstrated.

In our opinion, the wider spread of LIAD among analytical MS applications is now limited by the lack of an adequate theoretical concept able to explain the existing observations and to predict optimal experimental conditions for future measurements. The mechanical "shake-off" model was proposed only as a qualitative explanation of observed desorption process, and as such was never used to obtain any quantitative agreement between the observable LIAD parameters and the generated acoustic waves. Moreover, to date, there was no work published in the literature, which would be devoted to systematic studies of physical parameters of the molecules desorbed by LIAD. Since we have attempted such as study, we feel it would be beneficial for the research community if we describe here briefly our own experimental methods and experimental results on LIAD.

3.3 Energy and velocity distributions of desorbed molecules

Because the dominant fraction of the desorbed flux in LIAD are neutral molecules, it is very important to select an appropriate ionization method for the molecules as well as the type of their mass analysis technique. Single-photon ionization (SPI) is well suited for characterization of this phenomenon (Pellin, Calaway et al. 2001) because of its ability to efficiently ionize the desorbing flux with minimal fragmentation. SPI occurs following absorption of a single photon whose energy exceeds the ionization potential (IP) of the molecule of interest, creating a cation. For many molecules, particularly those with aromatic rings to stabilize the cation, fragmentation from photoionization is minimized, and thus the state of the initial LIAD flux can, in principle, be revealed(Lipson and Shi 2002). Currently, the shortest wavelength of commercially available energetic lasers suitable for SPI is 157 nm (F_2 laser), which corresponds to the photon energy of 7.9 eV. This energy limits the range of species that can be ionized by SPI to atoms and molecules with IPs less than 7.9 eV. Therefore, organic dyes were chosen as analytes for this LIAD study due to their low IPs, ability to form stable cations, and high photon absorption coefficients. We selected dyes that have IPs in the range from 5 eV to 7 eV and can be easily ionized by the F_2-laser radiation.

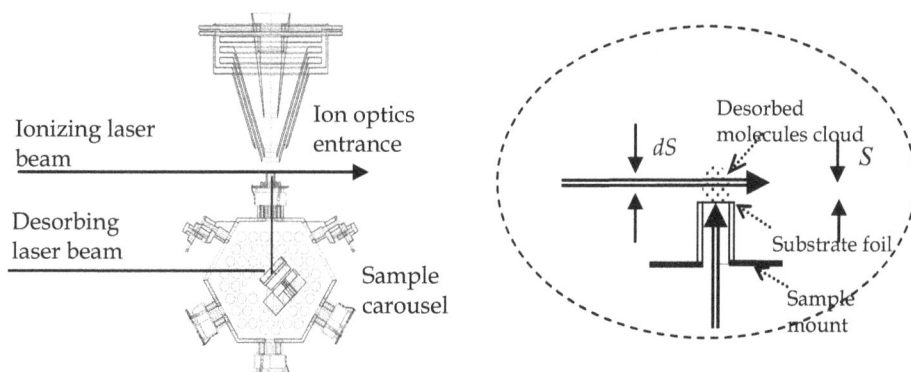

Fig. 5. Schematic drawing of target assembly and laser irradiation pathways. On the right the enlarged view of desorption/ionization scheme is displayed

A time-of-flight mass spectrometer (TOF MS) with a combined LIAD/SPI ion source was employed in our studies of the LIAD phenomenon. The experiments were conducted under ultra high vacuum conditions, with the residual gas pressure in the sample chamber less than 3×10^{-7} Pa. The schematic drawing of the target assembly in our instrument is shown in Fig. 5. The sample was mounted on one of six sample holders that were supported by a hexagonal carousel. This carousel was driven by an ultrahigh-vacuum compatible motion stage with closed-loop precision of better than 50 nm. The sample holders were secured on the carousel via three 30 mm long alumina ceramic insulators and connected (using vacuum feedthroughs) to a high-voltage pulser unit, which provided voltages necessary for the operating of TOF MS instrument. Each new sample was inserted into the UHV chamber through a vacuum loadlock. Using the motion stage, the sample was then positioned in the focal plane of the TOF MS source optics. The ion optics and operational principles of our instrument are described in more detail elsewhere (Veryovkin, Calaway et al. 2004). A dielectric mirror with 98% reflection at 248 nm was mounted in the center of the carousel, in order to deliver the laser beam to the back side of the sample. Note that we will use the convention that the front of the sample is the side facing the ion source and TOF, while the back side is the opposite. For desorption, an excimer KrF laser with wavelength 248 nm (EX10/300 GAM Laser Inc.) was used. The output energy of the laser pulse could be varied between 0.5 - 5 mJ by adjusting the laser discharge voltage and by an additional attenuation with a set of neutral optical filters. The driving laser beam was focused on the target back surface into a spot of rectangular shape ~200×800 μm^2 by the fused silica lens with focusing distance of 500 mm. The laser pulse duration was 7 ns, producing a peak power density on the irradiated surface ranging from 50 to 500 MW/cm^2. These laser intensities are close to those used in most of LIAD experiments (Perez, Ramirez-Arizmendi et al. 2000; Campbell, Crawford et al. 2004; Campbell, Fiddler et al. 2005; Crawford, Campbell et al. 2005) taking into account that the reflection coefficient in the UV is normally less than it is at visible wavelengths. Post-ionization of the desorbed molecules was performed with an F_2 laser, with output energy of 2 mJ/pulse, and pulse duration of 10 ns. The F_2 laser beam was focused just above the front target surface, with a waist of 400×2000 μm^2, using of a combination of MgF_2 spherical and cylindrical lenses. The F_2 laser radiation power density in the focal plane was ~10 MW/cm^2, which assured the saturation of the photoionization process for the investigated molecules (as verified by a laser power study).

For comparison with LIAD, direct laser desorption (LD) mass spectra were measured for the same samples, also using the F_2 laser for post-ionization. To this end, an N_2 laser (337 nm wavelength, 100 μJ/pulse energy and 7 ns pulse duration) was focused onto the target front surface using an in-vacuum Schwarzschild optical microscope (Veryovkin, Calaway et al. 2004). The beam spot size on the surface was about 50 μm in diameter.

The delay between the driving KrF (or N_2) laser pulses and the ionizing F_2 laser could be precisely controlled and varied from 0 to 1000 μs. The desorbed molecules that move away from the surface could therefore be ionized at a precisely defined moment in time and volume in space above the target after the desorption event, with the photoions then analyzed by the TOF MS. This approach allowed us to measure mass spectra for the (postionized) desorbed neutral molecules and determine their velocity distribution. Each mass spectrum was the sum of 128 individual acquired spectra. To prevent the rise of the average foil temperature due to adsorption of laser power, the repetition rate of the laser pulses was maintained at 8 Hz.

Foils from different materials with different thicknesses were used in the experiments. The foil preparation procedure was the same as described in paragraph 2.2.3.1. Before applying the analyte to the top surface of the foil, each substrate was cleaned in methanol-acetone solution (1:1) in an ultrasonic bath (10 minutes).

Organic dyes rhodamine B, fluorescein, methylanthracene (MA), coumarin-522 (N-Methyl-4-trifluoromethylpiperidino3,2-gcoumarin), and BBQ (4,4"-Bisbutyloctyloxy-p-quaterphenyl) were used as received (Eastman Kodak). The dyes were dissolved in methanol (for MA and BBQ, mixed xylenes were also used as solvent), and then the resulting solution (about 10^{-3} M) was used for sample preparation. One μl of the analyte solution was pipetted onto the foil surface, and then the quartz cylinder-foil assembly was spun at 4500 rpm for 30 seconds to coat the analyte uniformly over the surface. During spin-coating, a significant part of the solution (90% or more) was taken off the surface and, surface concentrations of the analyte could be estimated to be less than 0.5 nM/cm^2. After the sample preparation, the foil was introduced into the instrument via the loadlock for analysis.

In good agreement with the previously published results on LIAD, we detected strong and stable desorption signals from foils with thickness of 12.5 μm (Fig. 6). Thicker foils (25 μm) produced relatively weak signals for the range of acoustic wave driven laser intensities used in our experiments. For comprehensive experiments, a Ta foil with 12.5 μm thickness was chosen as optimal, not only because high desorption signals were detected from it, but also for its good mechanical strength, high melting point, and durability under powerful laser irradiation. The TOF mass spectra of different organic dye molecules desorbed from the front of the back-irradiated Ta foil surface and ionized by the 157 nm laser radiation are shown in Fig.6. This figure shows three major features, (1) all analytes display large parent molecular ion signals, (2) all spectra display a small number of peaks, with a few or none in the mass range below 100 Da., (3) the number of fragment ion peaks is specific to each molecular analyte. In order to characterize the desorption process in terms of the corresponding molecular fragmentation, a parameter ς, can be defined as a ratio of the sum of intensities of the fragment peaks A_f to the parent molecular peak intensity A_p, $\varsigma = \sum A_f / A_p$. As will be shown below, this parameter depends on laser intensities that drive the acoustic waves. For SPI , the parameter ς characterizes not only the peculiarities of the desorption process, but it also generally depends on the photoionization cross-section of the parent molecule and specifics of its photofragmentation such as possible decay channels and their activation energies.

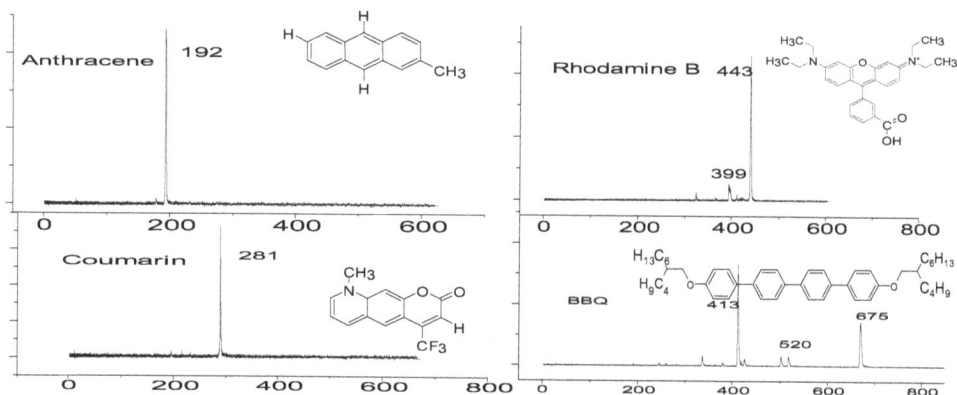

Fig. 6. Mass spectra of organic dyes in LIAD experiments

For rhodamine B, the parameter ς grew linearly with KrF laser intensity (Fig. 7), and a similar behavior was also detected for fluorescein, although values of ς were different ($\varsigma =0.8$ for fluorescein at 300 MW/cm2 of KrF laser intensity). On the contrary, the same experiments conducted for BBQ, surprisingly revealed that the parameter ς decreased with increasing desorption laser power. This discrepancy in ς parameter dependency may indicate that the relationship between these two observations is not a trivial one. This raises the question of whether the desorption and fragmentation phenomena are driven by the same fundamental process. We note that fragmentation is not intrinsic to LIAD. In our experiments, we observed some indications of fragmentation for only three analytes out of five. The analysis of existing data from the literature shows that in some cases fragmentation was observed even with soft ionization of the desorbed molecules whereas in other cases fragmentation was very small or completely absent. A detailed study of fragmentation in LIAD has not been done yet and the presented here results are the first attempt to quantify this characteristic of LIAD.

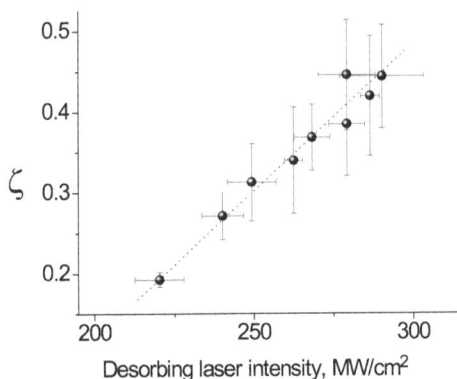

Fig. 7. Power dependence of fragmentation parameter

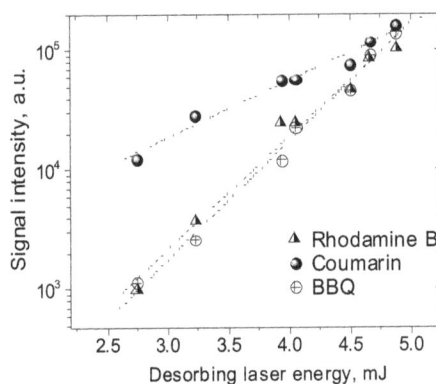

Fig. 8. Power dependence of total desorption yield for rhodamine B molecules

An important characteristic of any desorption phenomenon is its yield. This is why in order to understand the basic processes driving the phenomenon, one has to identify external parameters that have the strongest effect on the desorption yield and then measure a dependence of the yield for each parameter. In the case of LIAD, the dependence of key peak intensities in the mass spectra and the ς parameter on the driving laser intensity appears to be the most important for understanding this phenomenon. The overall desorption yield for all studied analytes strongly increased with desorption laser intensity (within our experimental range), displaying for most peaks approximately exponential dependency. Figure 8 demonstrates this dependence clearly. Plotted on a semi-logarithmic scale, these dependences appear linear, with different slopes for each analyte.

For the characterization of physical nature of the desorption processes, the knowledge of velocity distributions of desorbed neutral molecules is extremely important (Levis 1994). As mentioned above, time-of-flight mass spectra of such molecules can be measured using the laser post ionization (LPI) technique (Spengler, Bahr et al. 1988). This experimental arrangement makes it possible to determine the distribution of the desorbed neutral molecules over their translational velocities (in the direction normal to the substrate surface) by varying the time delay between desorbing and post-ionizing laser pulses. Raw experimental data in this case are the dependencies of the observed signal on the laser delay time, called below as *signal-vs-delay* dependencies. In order to be able to directly compare LIAD and LD results, the energies of the desorbing N_2 (LD) and KrF (LIAD) lasers were adjusted such that the output molecular ion signals reached about the same intensities for the same gains of the TOF MS detector.

In good agreement with the previous experiments (Perez, Ramirez-Arizmendi et al. 2000), dramatic differences between LIAD and LD in widths of the *signal-vs-delay* dependencies and in positions of their maximums have been observed in these measurements. The most tempting explanation for this experimental finding was that the mean velocities of desorbed molecules in LIAD and LD processes were very different. Unfortunately, no actual energy or velocity distributions of desorbed molecules in LIAD process have been measured experimentally and reported in the literature to date. For the first time, this gap in knowledge can now be filled by processing our experimental data from the *signal-vs-delay* dependencies and converting them into kinetic energy distributions corresponding to our experimental conditions.

Our experiment geometry (Fig.5) is rather common, and the detailed discussion of the method as well as the appropriate conversion equations can be found elsewhere (Young, Whitten et al. 1989; Balzer, Gerlach et al. 1997). Typical values of the distance S and the thickness of ionization volume dS were set to 3 and 0.4 mm, respectively. While, as mentioned above, the value of S could be varied in our experiments between 1 mm and 5 mm, in order to achieve optimal compromise between the energy/velocity resolution and the signal-to-noise ratio, most of our data were obtained at $S=3$ mm. At this distance, the relative velocity resolution was $dv/v=dS/S=0.4/3=0.13$ and, correspondingly, the energy resolution was $dE/E=0.18$. A typical velocity distribution obtained for rhodamine B is demonstrated on Fig.10. It is apparent that the desorbed molecules in LIAD are very slow. The average velocities of rhodamine B and BBQ molecules were found to be 59±12 m/s and 47±9 m/s, respectively. For methylanthracene and coumarin522 molecules, the measured average velocities were 76±20 m/s and 70±18 m/s. It is important to recognize that these LIAD-desorbed molecules are much slower than what could be expected assuming the thermal mechanism of LIAD. In order to fit these rhodamine B data with Maxwellian distributions (dashed curve on Fig.10), physically unrealistic low temperatures of about 100 K, were required. On the other hand, the measured velocities are much higher than that of the laser-induced acoustic motion of the foil surface in normal direction, which, measured in the same experimental arrangement, did not exceed 1 m/s. This observation caused serious doubts on the validity of the "shake-off" mechanism considered by many as the most likely cause of LIAD. The measurements of velocity distribution of species desorbed in LD geometry were used for direct comparison with LIAD, and as the proof of validity of our experimental procedure. A smaller insert plot on Fig.10 demonstrates the velocity distribution of desorbed rhodamine B main fragment (399 amu) obtained in the LD irradiation scheme. The mechanisms of the velocity distribution formation in the LD process are well-known and discussed in many reviews (Levis 1994). According to a commonly used procedure described in (Natzle, Padowitz et al. 1988), this distribution can be fit by a two-temperature bi-modal velocity distribution: dashed lines in the insert plot represent spectral components with different temperatures, and the solid line corresponds to their sum. Being in a good agreement with general LD regularities, these results also confirm the validity of the experimental procedure used for both LD and LIAD.

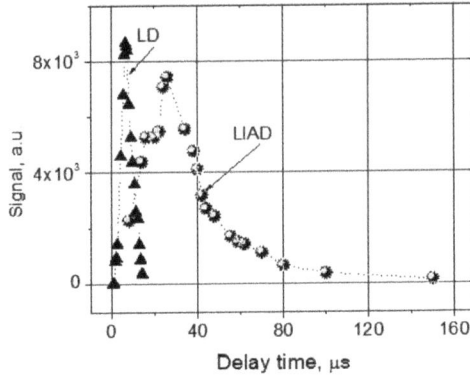

Fig. 9. The comparison of *signal-vs-delay* for LIAD and LD desorbed rhodamine B molecules

Velocity distributions were measured also at different fluences of the driving KrF laser. On Fig. 10, two distributions corresponding to laser fluences of 2.3 J/cm² and 3.4 J/cm² are plotted. Within the limits of accuracy of our measurements, no change of average velocity has been detected, which suggests that both thermal and "shake-off" models are not applicable for the explanation of LIAD process.

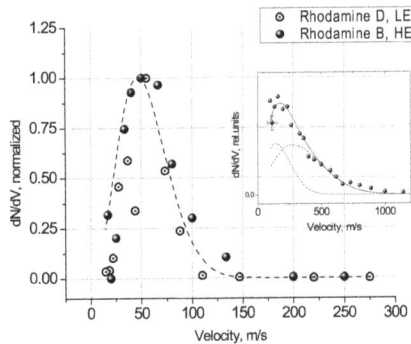

Fig. 10. Velocity distribution for rhodamine B molecules at different LIAD desorption laser intensities. On insert there is the same distribution for LD regime

In order to calculate the mean energies of desorbed molecules, we used the described above approach and took into account that the Jacobian of this variable transformation was given by S^2/t^3 (Balzer, Gerlach et al. 1997). The mean energies of molecules in LIAD experiments (as well as their mean velocities) showed no apparent trend with the increase of desorbing laser fluences keeping the average value at about 9 meV for rhodamine B, 9.5 meV for BBQ, 6.5meV for coumarine 522 and 7.5meV for methylanthracene. Typical energy distributions measured for these molecules are shown on Fig.11. Dashed line represents Maxwell distribution for rhodamine B corresponding to the 80 K temperature. This temperature was chosen to obtain the best fit at the distributions maximums. At the same time, high energy tails of the experimental energy distributions strongly deviate from the exponential law, which is apparent with the double logarithmic scale in Fig.11, and reveal behavior close to the power dependence.

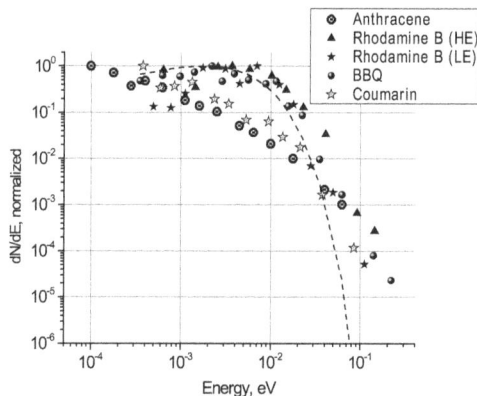

Fig. 11. Energy distribution of different organic dyes molecules. Dashed line represents equilibrium Boltzmann distribution at T=100 K

4. Is desorption process in LIAD really driven by acoustic waves?

4.1 "Shake-off "mechanism versus thermal origin of molecular desorption

The "shake-off" mechanism of molecular desorption in LIAD for a long time was considered as the only sensible explanation of the observed phenomenon. However, as was shown above, it contradicts both with general physical considerations and with the experimentally measured parameters of acoustic vibrations of the surface as well as with the observed energy and velocity spectra of desorbed molecules.

An important consequence of the backside irradiation is the heating of the front side. Considering this effect, we should keep in mind that this heating process in LIAD is distinctly different from the case of direct front-side laser irradiation (LD). In the latter case, due to the small value of skin depth in metals, the rate of the temperature rise is extremely high. In case of backside irradiation of the foils, the front surface temperature is governed by the heat conduction through the metal foil, which makes the heating rate much slower than that for the front side irradiation geometry. For one-dimensional heat conduction problem, the specific time of the temperature rise is defined by the heat propagation time $\tau = l^2 / \alpha$, where a is the thermal diffusivity of the metal and l is the foil thickness. For the foil thicknesses typical for LIAD (~10 µm) and the thermal diffusivity of the most metals $a\approx0.2$ – 0.7 cm^2·s^{-1}, τ has the values in the range of a few µs. Our numerical calculations using the heat conduction equation showed that for Ta foil with the thickness of 12.5 µm and the driving laser fluence 3.5 J/cm^2 (corresponding under conditions of our experiments to the peak laser intensity of about 500 MW/cm^2), the front surface temperature rise is 375 K, which reaches its maximum 1.75 µs after the laser desorbing pulse ceases. If experiments start with room temperatures, the peak foil surface temperatures then can reach 668÷673 K. Because the melting point of rhodamine B is just 438 K, the thermal origin of the LIAD process can, in fact, come into the focus of our consideration. However, other experimental results obtained in this work, make this mechanism very unlikely. These results are:

1. rather small fragmentation of molecules in LIAD, compared to that for LD;
2. distinct differences in velocity and energy distributions between LIAD and LD;
3. slow velocities and low kinetic energies of desorbed molecules compared to those required by the thermal mechanism;
4. the apparent independence of mean energies of the desorbed molecules in LIAD on the driving laser fluence (also observed by by Kenttamaa et al. in their recent work (Shea, Petzold et al. 2007).

Moreover, the mechanical "shake-off" mechanism is also in contradiction with the observation (4), because the amplitudes and velocities of laser generated acoustic waves should increase with the driving laser fluencies. Thus we can conclude that both mechanisms of the direct energy transfer (acoustic waves and heat conduction) cannot serve as the primary explanation of the LIAD phenomenon, and apparently more complicated processes are involved here.

4.2 Stress and strain of the foil surface due to the laser irradiation

It is a well-known fact that a film deposition on a substrate surface results in many cases in the residual mechanical stress (due to the lattice parameters mismatch between the film and the substrate) and thus in some excessive potential energy stored in the film. This stress can be produced by two ways: one is the growth stress and the other is the induced (or extrinsic) stress (Freund and Suresh 2003). An external impact, such as acoustic or thermal wave generated by the laser irradiation of the substrate, should initiate the reconstruction of the film, which can result in releasing this excess energy or, possibly, generating an additional extrinsic stress. In both cases, it can cause formation of cracks in the film so that intermolecular bonds at the edges of the cracks can break. As a result of this crack formation process, excited electronic states can form at the crack edges and induce desorption of the molecules.

Despite a great variety of mechanisms of stress formation and evolution do exist, a quantitative description of this process is possible only in limited cases even for simple adsorbent-adsorbate systems. Accurate modeling of the film behavior requires precise knowledge of the thermal and mechanical properties of the film material and their interaction with substrate surface. For most organic materials, such data are unavailable, and thus only limited estimates may be done, based on the little amount of data published in the literature (Bondi 1968). Moreover, the deposited films of organic materials tend to consist of a mixture of micro-crystals with different crystallographic orientation and, therefore, there are many grain boundaries, defects, dislocations in such films that could significantly change their local mechanical properties.

A simplified numerical estimate of the stress energy, related to the thermal mismatch of the substrate foil and the analyte film on its top could be done, as follows. An equi-biaxial stress resulting from acoustic vibration of the back-irradiated metal foil may be estimated using the following expression (Boley and Weiner 1960; Freund and Suresh 2003)

$$\sigma_m = \frac{E \cdot h_f}{R \cdot (1 - v)} \tag{16}$$

where E is the Young's modulus, v is the Poisson ratio, h_f is the thickness of organic film and R is the radius of curvature of the foil surface. From the other side, the stress, resulting from thermal mismatch has the following value (Boley and Weiner 1960)

$$\sigma_t = (\alpha_s - \alpha_f) \cdot \Delta T \cdot E / (1-v),$$
(17)

where a_s and a_f are thermal expansion coefficients of the substrate and the film respectively, ΔT is the temperature increase. There is not much information in the literature on thermo-mechanical parameters for molecular crystals but based on existing data for anthracene (Bondi 1968) we can estimate the order of the generated stress values. Under the assumption that $a_f = 2.8 \cdot 10^{-4}$ K^{-1} , $a_s = 6.3 \cdot 10^{-6}$ K^{-1}, $E = 13$ GPa, $v \approx 0.25$, and $\Delta T = 100$ K, we can obtain $\sigma_t = 485$ MPa. At the same time, the stress associated with acoustic vibrations is much lower: $\sigma_m \approx 1.7$ kPa (taking into account that maximal value of R is approximately 1 m, and the film thickness never exceeded 10^{-7} m,). This is a negligible value, in comparison with σ_t, which means that the thermal mismatch stress is the principal reason for cracks formation. The estimate of internal energy, stored in thermally strained organic films can be done with using of following expression (Boley and Weiner 1960)

$$G = \frac{E \cdot (1+v)^2}{2 \cdot (1-v^2)} \cdot (\alpha_s - \alpha_f)^2 \cdot \Delta T^2 \cdot 2 \cdot \pi \cdot r_0 \cdot h_f$$
(18)

The average energy per analyte molecule can easily be calculated

$$g_a = \frac{E \cdot (1+v)^2}{2 \cdot (1-v^2) \cdot \rho} \cdot \frac{(\alpha_s - \alpha_f)^2 \cdot \Delta T^2 \cdot M}{N_A}$$
(19)

Here M is the molar mass, ρ is the specific gravity and N_A is the Avogadro number. It is interesting to note that g_a does not depend on the analyte island size but strongly depends on the thermal and mechanical parameters. Again with the use of existing data for anthracene we can estimate the value of g_a, and for $\Delta T = 100$ K we will get $g_a = 0.025$ eV . This is not enough to break intermolecular bonds but when thermally induced stress exceeds a critical value, the film can start to fracture and the stored energy is released in a small volume in the vicinity of the stress cracks. Due to the strong spatial nonuniformity of thermo-mechanical properties of molecular crystals (Bondi 1968) the physical mechanisms involved in this process are very complicated in nature; therefore we can give only a qualitative picture of this phenomenon. Presumably, the cracks are formed along grain boundaries, defects and interfaces. The increase of the desorption laser intensity that causes a rise in ΔT and, in accordance with Eq. 18, an increase in energy G, results in the formation of additional cracks. Some part of this excess energy can be then converted into the increased free surface energy, other part – into electronic excitations, but because these processes are obviously non-adiabatic, the crack formation most likely will be accompanied by breaking intermolecular bonds and forming new desorption sites.

4.3 Proposed mechanisms of the molecules desorption

The desorption process itself appears to be the most obscure part of the LIAD phenomenon. The formation of the electronically excited states on the surface due their mechanical fracture is considered to be the main physical nature of triboemission (Nakayama, Suzuki et al. 1992), also known as "Kramer effect" (Oster, Yaskolko et al. 1999; Oster, Yaskolko et al.

2001). The essence of this effect is the emission of charged particles and photons initiated by surface distortion (in particular by mechanical deformation, scratching, bending, etc.) and do not connected with thermal excitation. The luminescence of thin metal discs, irradiated from back side by the laser pulses (Abramova, Shcherbakov et al. 1999; Abramova, Rusakov et al. 2000) can also serve as evidence of formation of excited electronic states by the laser-driven stress in thin foils.

The mechanism of molecular desorption due to such laser-generated stress has been proposed earlier by Vertes (Vertes and Levine 1990; Vertes 1991) for MALDI, and was based on the thermal stress generation in the layer of organic film deposited on solid substrates. One should notice, however, a clear differences in the physical conditions between these sample volatilization methods. For MALDI, the absorption of the laser pulse energy occurs in an optically and thermally dense film, which experiences thermal stress due to its non-uniform and fast heating. In the case of LIAD, the laser radiation is absorbed by the back side of the metal foil substrate, opposite from where the sample was deposited. The amount of energy transmitted through the metal foil to the analyte layer on the front site should be so strongly attenuated, compared to the direct (front side) irradiation, that it cannot directly be sufficient for desorbing molecules with velocities observed in our experiments. On the other hand, a typical average thickness of the analyte film in LIAD can be estimated to be on the order of several molecular layers. Because of this, the specific density of energy stored in each analyte island due to intrinsic stress can be high, and during laser irradiation of the back side of the foil, the laser irradiation can simply trigger the release of this energy and to induce molecular desorption event.

The other experimental fact that could help in the interpretation of LIAD phenomenon is the similarity of the energy and velocity spectra of desorbed neutral molecules in case of LIAD and Electron Stimulated Desorption (ESD) (Young, Whitten et al. 1989). The primary mechanism of ESD is supposed to be the formation of the repulsive states in the surface due to electron excitation.

Thus we could hypothesize that in case of LIAD, due to complete opaqueness of metal foil for laser radiation, the only channel for energy transfer is the formation of acoustic and thermal field by laser pulse impact. As shown above, the result can be the mechanical distortion of the analyte film followed by film cracking and delamination and the consequent formation of excited and repulsive states electronic for analyte molecules.

This will increase the number of desorption sites and finally the total number of desorbed molecules as observed in the nonlinear laser intensity dependence. But because the formation of the any individual cracks is defined only by the intermolecular bonding forces in the vicinity of the crack, the translational kinetic energy of desorbed molecules should still remain independent of driving laser intensity, also matching our observations and other recent LIAD publications.

5. Conclusion

Traditionally, the name LIAD combine all desorption phenomena taking place when an opaque target is irradiated from the back side, ignoring differences in experimental conditions. In our opinion, it is not correct. As it was demonstrated in this work, under some experimental conditions (most commonly used in many present studies), the physical origin of the observed desorption phenomenon is not (and could not be) connected with the acoustic waves generated in the foil and most likely is defined by the film stress and cracking due to thermal and mechanical mismatch of the analyte and substrate. Therefore, ironically, the acronym LIAD in this case does not correctly reflect the physical nature of the process.

From the other hand, we cannot exclude that some strong change of the experimental conditions can also change the relationship between various physically possible mechanisms of molecular desorption, similarly to the case of direct laser desorption, when the thermal mechanism dominates under wide range of conditions and makes other possible mechanisms undetectable. One can expect, for example, that strong increase of laser power density (10 GW/cm^2 and above) will cause the corresponding increase of the foil temperature and formation of hot and dense plasma plume near its surface facing the laser. In some such cases and strongly depending on the foil material properties, the material motion could evolve from elastic into plastic regime of deformation. The acoustic wave relation could not then apply, and the velocity of the surface linear motion may strongly increase. "Large and heavy" objects weakly bound to the substrate surface may be then kinematically removed from it ("shaken-off"). This could indeed serve as an explanation of observations from recent experiments where the desorption of intact viruses and biological cells were reported (Peng, Yang et al. 2006). However, this regime cannot be connected with generation of the acoustic waves but most likely corresponds to a physically different mode of shock-wave generation (Menikoff 2007). One cannot exclude that under some experimental LIAD conditions only these shock-wave induced phenomena can be responsible for molecular desorption, particularly in the experiments where emission of ions was detected. In one of such pioneering experiments (Golovlev, Allman et al. 1997), where the laser generated pressure pulse was apparently much stronger than that in our work, because of confined ablation conditions (Fabbro, Fournier et al. 1990), the emission of both electrons and ions was observed. This may have been connected with significant surface disruption at the microscale generated by shock waves.

From another standpoint, the backside irradiation of very thin films (about two or three hundred nanometers), which also could be called LIAD, has demonstrated domination of the thermal mechanism in the desorption process (Ehring, Costa et al. 1996). It is clear that the thermal equilibrium between front and back sides in such thin films can establish within the time interval of tens of nanoseconds, and the absolute temperature difference between front and back sides is negligibly small. Thus, at some laser power densities, the front side temperature could reach the melting temperature of the metal that would be enough to cause an efficient thermal desorption of the most organic molecules.

To conclude, the variety of desorption and emission phenomena observed on the front side of thin metal foils whose back sides are subjected to pulsed laser irradiation, combined under the general name of LIAD, could, in fact, have a number of different physical origins depending on specific experimental conditions. According to our observations conducted at moderate laser power densities (0.1 - 1 GW/cm^2) and foils thicknesses of about 5 – 20 µm, that appear to be the most commonly used conditions in LIAD experiments, the predominant desorption mechanism is connected with the reorganization of the deposited analyte film and the consequent breaking of molecular bonds on the edges of these cracks.

6. Acknowledgment

This work is supported by the U.S. Department of Energy, BES-Materials Sciences, under Contract DE-AC02-06CH11357, by UChicago Argonne, LLC.

7. References

Abramova, K. B., A. I. Rusakov, et al. (2000). Photon emission from metals under fast nondestructive loading. *J. Appl. Phys.* 87(6): 3132-3136.

Abramova, K. B., I. P. Shcherbakov, et al. (1999). Emission processes accompanying deformation and fracture of metals. *Phys.Solid State* 41(5): 761-762.

Adamson, A. W. and A. P. Gast (1997). Physical Chemistry of Surface. New York, John Wiley &Sons, Inc.

Balzer, F., R. Gerlach, et al. (1997). Photodesorption of Na atoms from rough Na surface. *J. Chem. Phys.* 106(19): 7995-8012.

Boley, B. A. and J. H. Weiner (1960). Theory of thermal stresses. New York,, Wiley.

Bondi, A. (1968). Physical properties of molecular crystals, liquids and gases. New York, London, Sydney, Hohn Wiley & Sons.

Campbell, J. L., K. E. Crawford, et al. (2004). Analysis of saturated hydrocarbons by using chemical ionization combined with laser-induced acoustic desorption/Fourier transform ion cyclotron resonance mass spectrometry. *Analytical Chemistry* 76(4): 959-963.

Campbell, J. L., M. N. Fiddler, et al. (2005). Analysis of polyethylene by using cyclopentadienyl cobalt chemical ionization combined with laser-induced acoustic desorption/Fourier transform ion cyclotron resonance mass spectrometry. *Analytical Chemistry* 77(13): 4020-4026.

Cheng, S. C., M. Z. Huang, et al. (2009). Thin-Layer Chromatography/Laser-Induced Acoustic Desorption/Electrospray Ionization Mass Spectrometry. *Analytical Chemistry* 81(22): 9274-9281.

Cole, R. B. (2010). Electrospray and MALDI mass spectrometry: fundamentals, instrumentation, practicalities, and biological applications. Hoboken, N.J., Wiley.

Crawford, K. E., J. L. Campbell, et al. (2005). Laser-induced acoustic desorption/Fourier transform ion cyclotron resonance mass spectrometry for petroleum distillate analysis. *Anal. Chem.* 77(24): 7916-7923.

Crawford, K. E., J. L. Campbell, et al. (2005). Laser-induced acoustic desorption/Fourier transform ion cyclotron resonance mass spectrometry for petroleum distillate analysis. *Analytical Chemistry* 77(24): 7916-7923.

Dass, C. (2007). Fundamentals of contemporary mass spectrometry. Hoboken, N.J., Wiley-Interscience.

Duncan, M. W., H. Roder, et al. (2011). Quantitative matrix-assisted laser desorption/ionization mass spectrometry. *Briefings in Functional Genomics* 7(5): 355-370.

Ehring, H., C. Costa, et al. (1996). Photochemical versus thermal mechanisms in matrix-assisted laser desorption/ionization probed by back side desorption. *Rapid Communications in Mass Spectrometry* 10(7): 821-824.

Fabbro, R., J. Fournier, et al. (1990). Physical Study of Laser-Produced Plasma in Confined Geometry. *Journal of Applied Physics* 68(2): 775-784.

Freund, L. B. and S. Suresh (2003). Thin Film Materials. Stress, Defect Formation and Surface Evolution. Cambridge, Cambridge University Press.

Golovlev, V. V., S. L. Allman, et al. (1997). Laser-induced acoustic desorption of electrons and ions. *Applied Physics Letters* 71(6): 852-854.

Gospodyn, J. P., A. Sardarli, et al. (2002). Ablative generation of surface acoustic waves in aluminum using ultraviolet laser pulses. *Journal of Applied Physics* 92(1): 564-571.

Habicht, S. C., L. M. Amundson, et al. (2010). Laser-Induced Acoustic Desorption Coupled with a Linear Quadrupole Ion Trap Mass Spectrometer. *Analytical Chemistry* 82(2): 608-614.

Honig, R. E. and J. R. Woolston (1963). Laser-induced emission of electrons, ions and neutral atoms from solid surfaces. *Applied Physics Letters* 2(7): 138-139.

Hutchins, D. A. (1985). Mechanisms of Pulsed photoacoustic generation. *Can. J. Phys.* 64: 1247-1263.

Karas, M., D. Bachmann, et al. (1985). Influence of the Wavelength in High-Irradiance Ultraviolet-Laser Desorption Mass-Spectrometry of Organic-Molecules. *Analytical Chemistry* 57(14): 2935-2939.

Krehl, P., F. Schwirzke, et al. (1975). Correlation of Stress-Wave Profiles and Dynamics of Plasma Produced by Laser Irradiation of Plane Solid Targets. *Journal of Applied Physics* 46(10): 4400-4406.

Landau, L. D. and E. M. Lifshits (1987). Fluid mechanics. New York, Pergamon.

Levis, R. J. (1994)."Laser-Desorption and Ejection of Biomolecules from the Condensed-Phase into the Gas-Phase. *Annu. Rev. Phys. Chem.* 45: 483-518.

Lindner, B. and U. Seydel (1985). Laser Desorption Mass-Spectrometry of Nonvolatiles under Shock-Wave Conditions. *Analytical Chemistry* 57(4): 895-899.

Lipson, R. H. and Y. J. Shi (2002). Ultraviolet Spectroscopy and UV lasers. New York-Basel, Marcel Deccer, Inc.,.

Lubman, D. M. (1990). Lasers and mass spectrometry. New York, Oxford University Press.

McLachlan, N. W. (1951). Theory of vibrations. New York, Dover Publications.

Menezes, V., K. Takayama, et al. (2005). Laser-ablation-assisted microparticle acceleration for drug delivery. *Applied Physics Letters* 87(16): -.

Menikoff, R. (2007). Empirical Equations of State for Solids. *Shock Wave Science and Technology Reference Library*. Y. Horie. Berlin, Springer. 2: v. <1-3>.

Miller, J. C. and R. F. Haglund (1998). Laser ablation and desorption. San Diego, Academic Press.

Mora, P. (1982). Theoretical-Model of Absorption of Laser-Light by a Plasma. *Physics of Fluids* 25(6): 1051-1056.

Murray, T. W. and J. W. Wagner (1999). Laser generation of acoustic waves in the ablative regime. *Journal of Applied Physics* 85(4): 2031-2040.

Nakayama, K., N. Suzuki, et al. (1992). Triboemission of Charged-Particles and Photons from Solid-Surfaces during Frictional Damage. *J. Phys. D: Appl. Phys.* 25(2): 303-308.

Natzle, W. C., D. Padowitz, et al. (1988). Ultraviolet-Laser Photodesorption of No from Condensed Films - Translational and Internal Energy-Distributions. *Journal of Chemical Physics* 88(12): 7975-7994.

Oster, L., V. Yaskolko, et al. (1999). Classification of exoelectron emission mechanisms. *Physica Status Solidi a-Applied Research* 174(2): 431-439.

Oster, L., V. Yaskolko, et al. (2001). The experimental criteria for distinguishing different types of exoelectron emission mechanisms. *Physica Status Solidi a-Applied Research* 187(2): 481-485.

Pellin, M. J., W. F. Calaway, et al. (2001). Laser Post Ionization for Quantative Elemental Analysis. *ToF-SIMS: Surface Analysis by Mass Spectrometry*. J. Vickerman and D. Briggs, Surface Spectra Ltd. and IM Publications 375.

Peng, W. P., Y. C. Yang, et al. (2006). Laser-induced acoustic desorption mass spectrometry of single bioparticles. *Angew. Chem., Int. Ed.* 45(9): 1423-1426.

Perez, J., L. E. Ramirez-Arizmendi, et al. (2000). Laser-induced acoustic desorption/chemical ionization in Fourier-transform ion cyclotron resonance mass spectrometry. *Int. J. Mass Spectrom.* 198(3): 173-188.

Phipps, C. R., T. P. Turner, et al. (1988). Impulse coupling to targets in vacuum by KrF, HF and CO_2 single-pulse lasers. *Journal of Applied Physics* 64(3): 1083-1096.

Pollard, H. F. (1977). Sound waves in solids. London, Pion.

Prokhorov, A. M., V. I. Konov, et al. (1990). Laser Heating of Metals. Bristol, Philadelphia, New York, Adam Higler.

Reid, G. E., S. E. Tichy, et al. (2001). N-terminal derivatization and fragmentation of neutral peptides via ion-molecule reactions with acylium ions: Toward gas-phase Edman degradation? *Journal of the American Chemical Society* 123(6): 1184-1192.

Royer, D. and E. Dieulesaint (2000). Elastic waves in solids. Berlin ; New York, Springer.

Schiller, J., R. Suss, et al. (2007). Maldi-Tof Ms in Lipidomics. *Frontiers in Bioscience* 12: 2568-2579.

Scruby, C. B. (1987). An introduction to acoustic emission. *Journal of Physics E (Scientific Instruments)* 20(8): 946-953.

Scruby, C. B. and H. N. G. Wadley (1978). Calibrated Capacitance Transducer for Detection of Acoustic-Emission. *Journal of Physics D-Applied Physics* 11(11): 1487-1494.

Shea, R. C., C. J. Petzold, et al. (2007). Experimental investigations of the internal energy of molecules evaporated via laser-induced acoustic desorption into a Fourier transform ion cyclotron resonance mass spectrometer. *Analytical Chemistry* 79(5): 1825-1832.

Smith, S. T. (2000). Flexures: elements of elastic mechanisms. Amsterdam, Gordon & Breach.

Song, K. H. and X. Xu (1997). Mechanisms of absorption in pulsed excimer laser-induced plasma. *Applied Physics a-Materials Science & Processing* 65(4-5): 477-485.

Spengler, B., U. Bahr, et al. (1988). Postionization of Laser-Desorbed Organic and Inorganic-Compounds in a Time of Flight Mass-Spectrometer. *Anal. Instrum.* 17(1-2): 173-193.

Srinivasan, R. and B. Braren (1989). Ultraviolet-Laser Ablation of Organic Polymers. *Chemical Reviews* 89(6): 1303-1316.

Vertes, A. (1991). Laser Desorption of Large Molecules: Mechanisms and Models. *Methods and Mechanisms for Producing Ions from Large Moleculres.* K. G. Standing and W. Ens. New York, Plenum Press.

Vertes, A. and R. D. Levine (1990). Sublimation Versus Fragmentation in Matrix-Assisted Laser Desorption. *Chem. Phys. Lett.* 171(4): 284-290.

Veryovkin, I. V., W. F. Calaway, et al. (2004). A new time of flight instrument for quantative surface analysis. *Nucl. Instrum. Methods Phys. Res., Sect. B* 219-220: 473.

White, R. M. (1963). Generation of elastic waves by transient surface heating *Journal of Applied Physics* 34(12): 3559-&.

Xu, B. Q., J. Feng, et al. (2008). Laser-generated thermoelastic acoustic sources and Lamb waves in anisotropic plates. *Applied Physics a-Materials Science & Processing* 91(1): 173-179.

Young, C. E., J. E. Whitten, et al. (1989). Electron-Stimulated Desorption of Neutrals from 6063 Aluminum - Velocity Distributions Detected by 193 Nm Non-Resonant Laser Ionization. *Surface and Interface Analysis* 14(10): 647-655.

Zinovev, A. V., I. V. Veryovkin, et al. (2007). Laser-driven acoustic desorption of organic molecules from back-irradiated solid foils. *Analytical Chemistry* 79(21): 8232-8241.

2

Use of Acoustic Waves for Pulsating Water Jet Generation

Josef Foldyna
Institute of Geonics of the ASCR, v. v. i., Ostrava
Czech Republic

1. Introduction

The technology of a high-speed water jet cutting and disintegration of various materials attained considerable growth during the last decades. Continuous high-speed water jets are currently used in many industrial applications such as cutting of various materials, cleaning and removal of surface layers. However, despite the impressive advances made recently in the field of water jetting, substantial attention of number of research teams throughout the world is still paid to the improvement of the performance of the technology, its adaptation to environmental requirements and making it more beneficial from the economic point of view.

An obvious method of the water jetting performance improvement is to generate jets at ultra-high pressures. The feasibility of cutting metals with pure water jets at pressures close to 690 MPa was investigated already in early nineties of the last century (Raghavan & Ting, 1991). Such a high pressure, however, induces extreme overtension of high-pressure parts of the cutting system which has adverse effect on their lifetime.

An alternate approach, as shown in this chapter, is to eliminate the need for such high pressures by pulsing the jet. It is well known that the collision of a high-velocity liquid mass with a solid generates short high-pressure transients which can cause serious damage to the surface and interior of the target material. The liquid impact on a solid surface consists of two main stages (see Fig. 1). During the first stage, the liquid behaves in a compressible manner generating the so-called "water-hammer" pressures. These high pressures are responsible for most of the damage resulting from liquid impact on the solid surface. The situation shortly after the initial impact of the liquid on the solid surface is illustrated in Fig. 2. After the release of the impact pressure, the second stage of the liquid impact begins. Once incompressible stream line flow is established, the pressure on the central axis falls to the much lower Bernoulli stagnation pressure that lasts for relatively long time.

The force distribution on liquid jet impact on the solid surface can be summarized as follows: initially a small central area of the first contact is compressed under a uniform pressure. The magnitude of the impact pressure p_i on the central axis is given by

$$p_i = \frac{v\rho_1 c_1 \rho_2 c_2}{\rho_1 c_1 + \rho_2 c_2} \tag{1}$$

where v is the impact velocity and ρ_1, ρ_2 and c_1, c_2 are the densities and the shock velocities in the liquid and the solid, respectively (de Haller, 1933).

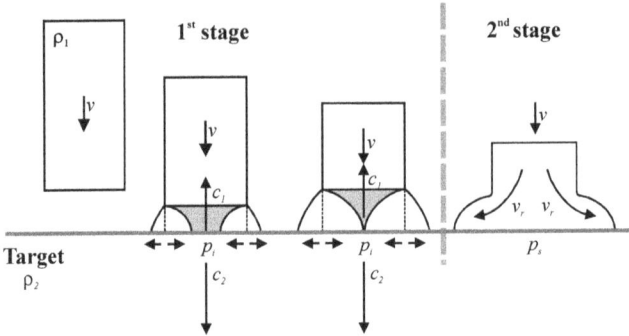

Fig. 1. Two stages of liquid impact on a solid target

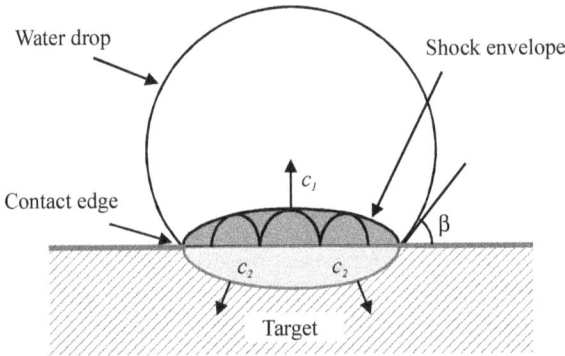

Fig. 2. Initial stage of impact between a water drop and a solid target with the contact edge moving faster than the shock velocity in the liquid. The liquid behind the shock envelope is compressed and the target beneath this area subjected to high pressure

The magnitude of the impact pressure is independent of the geometry of the drop (Thomas & Brunton, 1970), but the duration of the pressure is affected by the size and shape of the drop. For a sphere or cylinder the corresponding radius or half-width of the contact area R exposed to this pressure is given by

$$R = \frac{rv}{c_1} \tag{2}$$

where r is the radius of curvature of the drop or cylinder (liquid mass) in the region of contact (Bowden & Field, 1964).

The initial area of contact grows as the impact continues; there is very little reduction in pressure on the surface until appreciable outward flow begins. The outward flow of the liquid becomes possible when the limit of the compressible deformation of the liquid is exceeded. The limit is given by

$$\frac{v}{c_1} = \sin\beta \tag{3}$$

where β is the liquid/solid interface angle – see Fig. 2 (Hancox & Brunton, 1966).

At this stage there is a rapid fall in pressure along the periphery of contact. As the outward flow continues, the water-hammer compression at the centre of impact is relieved until the maximum pressure acting on the surface is the central stagnation pressure for the incompressible flow. The stagnation pressure is given by

$$p_s = \frac{1}{2}\rho_1 v^2 \tag{4}$$

When the liquid begins to flow away from the point of impact, there is evidence that the velocity of this tangential flow may be as much as five times the impact velocity (Thomas & Brunton, 1970). The velocity increase is thought to be connected with the shape of the head of the jet. It has been observed that an increase in velocity along the surface occurs only in cases where the jet head is inclined at an angle to the surface. Since spherical drops (and/or spherical heads of a train of pulses of pulsating jet) always provide a sloping interface to a plane solid surface it might be expected that high radial velocities will occur on impact. Therefore, there are additional shear forces associated with the high speed flow across the surface acting on the surface in addition to the normal forces. The shear forces acting on a roughened surface are large enough to cause local shear fractures, even in high strength materials (Hancox & Brunton, 1966).

Exploitation of above described effects associated with water droplet impingement on solids in a high-speed water jet cutting technology should lead to considerable improvement of its performance, better adaptation to more and more demanding environmental requirements, and consequently to more beneficial use of the technology also from the economical point of view. Generating sufficiently high pressure pulsations in pressure water upstream the nozzle exit enables to create a pulsating water jet that emerges from the nozzle as a continuous jet and it forms into a train of pulses at certain standoff distance from the nozzle exit. Such a pulsating jet produces all of the above mentioned effects associated with water droplet impingement on solids. In addition, the action of pulsating jet induces also fatigue stress in the target material due to the cyclic loading of the target surface. This further improves the efficiency of the pulsating liquid jet in comparison with the continuous one. Thus, destructive effects of the continuous high-speed water jet can be enhanced by the introduction of high-frequency pulsations in the jet, i.e. by generation of pulsating water jets.

Recently, a special method of the generation of the high-speed pulsating water jet was developed and tested extensively under laboratory conditions. The method is based on the generation of acoustic waves by the action of the acoustic transducer on the pressure liquid and their transmission via pressure system to the nozzle. The high-pressure system with integrated acoustic generator of pressure pulsations consists of cylindrical acoustic chamber connected to the liquid waveguide. The liquid waveguide is fitted with pressure liquid supply and equipped with the nozzle at the end. The acoustic actuator consisting of piezoelectric transducer and cylindrical waveguide is placed in the acoustic chamber (see Fig. 3). Pressure pulsations generated by acoustic actuator in acoustic chamber filled with pressure liquid are amplified by mechanical amplifier of pulsations and transferred by liquid waveguide to the nozzle. Liquid compressibility and tuning of the acoustic system are utilized for effective transfer of pulsating energy from the generator to the nozzle and/or nozzle system where pressure pulsations transform into velocity pulsations. The acoustic generator can be used for generation of both single and multiple pulsating water jets (e.g.

rotating) using commercially available cutting heads and jetting tools. Laboratory tests of the device based on the above mentioned method of the pulsating liquid jet generation proved that the performance of pulsating water jets in cutting of various materials is at least two times higher compared to that obtained using continuous ones under the same working conditions.

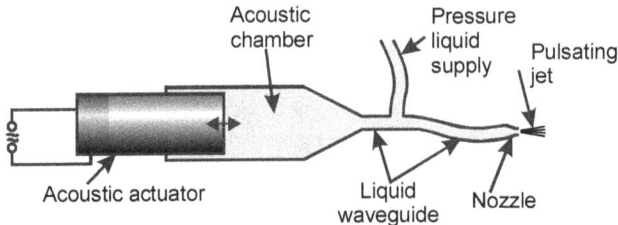

Fig. 3. Schematic drawing of the high-pressure system with integrated acoustic generator of pressure pulsations

However, further improvement of the apparatus for acoustic generation of pulsating liquid jet requires thorough study oriented at determination of fundamentals of the process of excitation and propagation of acoustic waves (and/or high-frequency pressure pulsations) in liquid via high-pressure system and their influence on forming and properties of pulsating liquid jet.

Problems related to the generation and propagation of pressure pulsations with frequency in the order of tens of kHz in liquid under pressure of tens of MPa and subsequent discharge of the liquid influenced by the pulsations through the orifice in the air (producing pulsating liquid jet with axial velocity in the order of hundreds meters per second) were not investigated in detail so far. Only partial information on this topic can be found in publications dealing with processes of a fuel injection for combustion in diesel engines (see e.g. Pianthong et al., 2003 or Tsai et al., 1999) and/or underwater acoustics (Wong & Zhu, 1995).

Therefore, the research on pulsating water jets was focused recently on the study of fundamentals of the process of excitation and propagation of acoustic waves (high-frequency pressure pulsations) in liquid via high-pressure system and their influence on forming and properties of pulsating liquid jet as well as on the visualization of the pulsating jets and testing of their effects on various materials. Results obtained in above mentioned areas so far are summarized in following sections.

2. Acoustic wave propagation in high-pressure system with integrated acoustic generator

The efficient transfer of the high-frequency pulsation energy in the high-pressure system to longer distances represents one of the basic assumptions for generation of highly effective pulsating water jets with required properties. To achieve that goal, the amplification of pressure pulsations propagating through the high-pressure system is necessary. The amplification can be accomplished by properly shaped liquid waveguide that is used for the pulsations transfer to the nozzle. In addition, maximum effects will be obtained if the entire high-pressure system from the acoustic generator to the nozzle is tuned in the resonance. To be able to study theoretically process of generating and propagation of pressure pulsations in the high-pressure system, both analytical and numerical models of the system with integrated acoustic generator were developed.

2.1 Analytical solution

The analytical solution of both pressure and flow oscillation waveforms in the conffuser-shaped tube with circular cross-section is based on linearized Navier-Stokes equations and wave equation for propagation of pressure wave. The wave equation incorporates both the standard kinematical viscosity and the kinematical second viscosity that is related to the liquid compressibility. Therefore, the irreversible stress tensor Π_{ij}, on the basis of which the wave equation is derived, can be written as follows:

$$\Pi_{ij} = 2\eta c_{ij} + \delta_{ij} \int_0^t \Theta(t-\tau)c_{kk}(\tau)d\tau \tag{5}$$

where the function Θ (dynamic second viscosity) is related to the voluminous memory, and c_{ij} represents the tensor of deformation velocity.

In the frequency domain (ω), equation (5) can be written in simplified form verified experimentally:

$$\Pi_{ij\omega} = 2\eta c_{ij\omega} + \delta_{ij}\frac{k}{\omega}c_{kk\omega} \tag{6}$$

whereby δ_{ij} represents Kronecker delta, and η dynamic viscosity. It is obvious from (6) that the dynamic second viscosity is frequency dependent. The kinematical second viscosity is then defined using following formula:

$$\xi = \frac{k}{\rho\omega} \tag{7}$$

where ρ represents liquid density.

2.1.1 Wave equation

If one considers linearized Navier-Stokes equations, the wave equation for pressure function can be written using the Laplace operator Δ in the following form:

$$\frac{\partial^2 p}{\partial t^2} - 2\gamma\frac{\partial}{\partial t}(\Delta p) - \int_0^t \rho^{-1}\Theta(t-\tau)\frac{\partial}{\partial\tau}(\Delta p)\,d\tau - v^2\Delta p = 0 \tag{8}$$

where γ is kinematical viscosity, p pressure, t time and v speed of sound in water, respectively.

If Laplace transformation for zero initial conditions is applied in (8), following equation can be obtained:

$$s^2\sigma - s[2\gamma + \xi(s)]\Delta\sigma - v^2\Delta\sigma = 0 \tag{9}$$

where s represents parameter of the Laplace transformation according to time ($\xi(s)$ is the Laplace function of the second kinematical viscosity), and, at the same time, following is valid:

$$L\{p(t)\} = \sigma(s) \tag{10}$$

If following expression is denoted κ:

$$\kappa^2 = -s^2[v^2 + (2\gamma + \xi)]^{-1} \tag{11}$$

then it can be written

$$\kappa^2\sigma + \Delta\sigma = 0 \tag{12}$$

In the frequency domain it is valid that $\psi(i\omega) = k/\omega$.
The solution of (12) can be performed by the implementation of spherical coordinate system (r, φ, v), see Fig. 4. Now, the wave equation can be written in the following form:

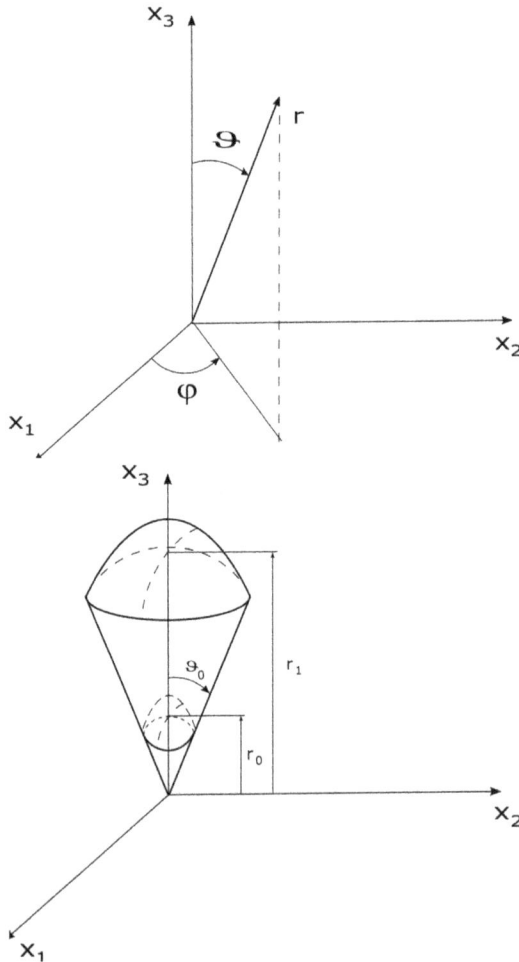

Fig. 4. Implementation of spherical coordinate system

$$\frac{\partial^2 \sigma}{\partial r^2} + \frac{2}{r}\frac{\partial \sigma}{\partial r} + \frac{1}{r^2}\frac{\partial^2 \sigma}{\partial \upsilon^2} + \frac{1}{r^2}\cotg\frac{\partial \sigma}{\partial \upsilon} + \frac{1}{r^2}\frac{1}{\sin^2\upsilon}\frac{\partial^2 \sigma}{\partial \varphi^2} + \kappa^2\sigma = 0 \tag{13}$$

Let's assume the solution of (13) as a product of functions:

$$\sigma = Z(r)W(\cos\upsilon)\Phi(\varphi) \tag{14}$$

Then, individual particular integrals can be expressed as follows:

$$\Phi_p = A_n\cos n\varphi + B_n\sin n\varphi \tag{15}$$

$$W_p(\cos\upsilon) = M_{nm}P_m^n(\cos\upsilon) + N_{nm}Q_m^n(\cos\upsilon) \tag{16}$$

where P, Q are special Legendre polynomials:

$$Z_p = \frac{1}{\sqrt{r}}[F_m J_D(\kappa r) + G_m Y_D(\kappa r)] \tag{17}$$

$$D = \frac{\sqrt{1 + 4m(m+1)}}{2} \tag{18}$$

2.1.2 Transfer matrix

The objective is to determine transfer matrix P that can be used in solving pressure and flow pulsations in hydraulic systems in conffuser-type tubes. For this purpose, it is convenient to introduce the mean velocity of the liquid c_r in a direction of r using following formula:

$$\tilde{c}_r = \frac{1}{2\upsilon_0}\int_0^{\upsilon_0} c_r(r,\varphi,\upsilon)d\upsilon \tag{19}$$

The solution can be simplified by the assumption that the flow is rotary symmetrical. It can be derived under the above mentioned assumption that functions P_m^n, Q_m^n will be streamlined to the following form: P_m^0, Q_m^0. Further, considering that the pressure function p varies only a little with respect to the angle υ, the following relation for the mean velocity c_r can be written based on Navier-Stokes equations:

$$\frac{\partial \tilde{c}_r}{\partial t} = -\frac{2\upsilon_0}{\rho}\frac{\partial p}{\partial r} - \frac{2\upsilon_0\xi}{\rho^2\upsilon^2}\frac{\partial^2 p}{\partial r\partial t} \tag{20}$$

The continuity equation and component c_r in Navier-Stokes equations expressed in the spherical coordinate system were used in the above mentioned derivation. Withal, effects of dynamic viscosity were neglected. If we will keep considering zero initial conditions, it can be written after the Laplace transformation (20):

$$sw_r = \alpha\frac{\partial \sigma}{\partial r}; L\{\tilde{c}_r(t)\} = w_r(s) \tag{21}$$

$$\alpha = -\frac{2v_0}{\rho}\left(1 + \frac{s\xi}{v^2}\right) \tag{22}$$

If all assumptions of the solution are considered, following can be written for Laplace images of both the pressure function and the velocity $\tilde{c}_r$ and with respect to (21):

$$\sigma = \frac{1}{\sqrt{r}}[FJ_{0.5}(\kappa r) + GY_{0.5}(\kappa r)] \tag{23}$$

$$m = 0; \quad D = \frac{1}{2} \tag{24}$$

$$w_r = \frac{\alpha}{s\sqrt{r}}\left[F\frac{\partial J_{0.5}(\kappa r)}{\partial \gamma} + G\frac{\partial Y_{0.5}}{\partial r}\right] \tag{25}$$

If we introduce for $r = r_0$ the state vector

$$\mathbf{u}^T = [w_r(r_0, s), \sigma(r_0, s)] \tag{26}$$

and for r the state vector

$$\mathbf{u}^T = [w_r(r, s), \sigma(r, s)] \tag{27}$$

the dependence in locations r and r_0 can be expressed by means of the transfer matrix:

$$\mathbf{u}(r, s) = \mathbf{P}\mathbf{u}(r_0, s) \tag{28}$$

Then, the matrix $\mathbf{P}$ will be derived from (26) and (27) by the elimination of integration constants F, G. If we designate:

$$\delta = \frac{\alpha}{r}\left[\frac{\partial J_{0.5}}{\partial r}(\kappa r_0)Y_0(\kappa r_0) - \frac{\partial Y_{0.5}}{\partial r}(\kappa r_0)J_{0.5}(\kappa r_0)\right] \tag{29}$$

following relation can be written for matrix $\mathbf{P}$:

$$\mathbf{P} = \frac{1}{\delta\sqrt{r_0 r}}\left\|\begin{matrix} \alpha\frac{\partial J_{0.5}}{\partial r}(\kappa r) & \alpha\frac{\partial Y_{0.5}}{\partial r}(\kappa r) \\ J_{0.5}(\kappa r) & Y_{0.5}(\kappa r) \end{matrix}\right\|\left\|\begin{matrix} Y_0(\kappa r_0) & -\alpha\frac{\partial Y_{0.5}(\kappa r_0)}{\partial r} \\ -J_{0.5}(\kappa r_0) & \alpha\frac{\partial J_{0.5}}{\partial r}(\kappa r_0) \end{matrix}\right\| \tag{30}$$

In the frequency domain, $s = i\omega$ is substituted.

Both pressure and flow pulsations of hydraulic systems with conffuser-shaped tubes can be solved on the basis of the transfer matrix (30). Individual elements of the transfer matrix are dependent on values of the speed of sound and the second viscosity. Values of both these quantities depend on the static pressure and the value of second viscosity depends also on the frequency. The values can be determined experimentally using the transfer matrix.

2.1.3 Application of the transfer matrix

The transfer matrix derived in the previous section can be used in solving transmission of pressure and flow pulsations in complex hydraulic systems. Such a system can consist of cylindrical and confusser-shaped sections; the system can also be bifurcated.

The advantages of use of the transfer matrixes for determination of both pressure and flow oscillation waveforms in the hydraulic system are illustrated on the model of the conffuser-shaped tube with the circular cross-section. The tube consists of series of coaxial cylinders with various diameters connected with cone frustums and filled with water at a pressure of 30 MPa. The acoustic generator of pressure pulsations located at the end of the largest diameter cylinder vibrates at the frequency of 20 kHz. The cylindrical nozzle is situated on the other end of the tube. The length of the largest diameter cylinder L_1 (and thus also total length of the tube L) can be changed (see Fig. 5).

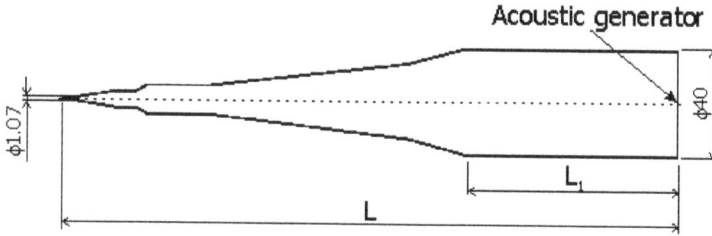

Fig. 5. Schematic drawing of the conffuser-shaped tube with circular cross-section

Effects of acoustic generator on the pressurized fluid in the tube are simulated by mean velocity:

$$\tilde{c}_r = c_{r0} e^{i\omega t}; w_{r0} = \frac{1}{s - i\omega} \tag{31}$$

The cylindrical nozzle at the end of the conffuser is represented by the linear hydraulic resistance, simulated by following equations:

$$p = \lambda c_r; \sigma = \lambda w_r \tag{32}$$

where λ represents the discharge coefficient.

Based on the transfer matrixes, a simulation model with the above described geometrical configuration was elaborated. Firstly, the natural frequency of the hydraulic system in question was determined using the simulation model. Next, the frequency was applied as oscillation frequency of the acoustic generator and propagation of a generated forced pressure waveform in the hydraulic system was investigated. Fig. 6 illustrates calculated forced pressure waveforms in the simulated geometry related to the phase angle.

It can be seen from the Fig. 6 that oscillations of the acoustic actuator generate a standing wave in the hydraulic system. The standing wave converts to the travelling wave at the area close to the nozzle exit. It is obvious from the presented solution that a properly designed conffuser-type tube can amplify the amplitude of pressure pulsations at the exit of the pulsating nozzle.

The presented analytical solution of the problem of the acoustic (pressure) wave propagation in the high-pressure system can be used for the determination of the optimal geometrical configuration of the high-pressure system to operate in the resonance mode. The solution can also be used in design of a transmission line for efficient transfer of high-frequency pulsation energy to longer distances in high-pressure systems for generation of pulsating high-speed water jets.

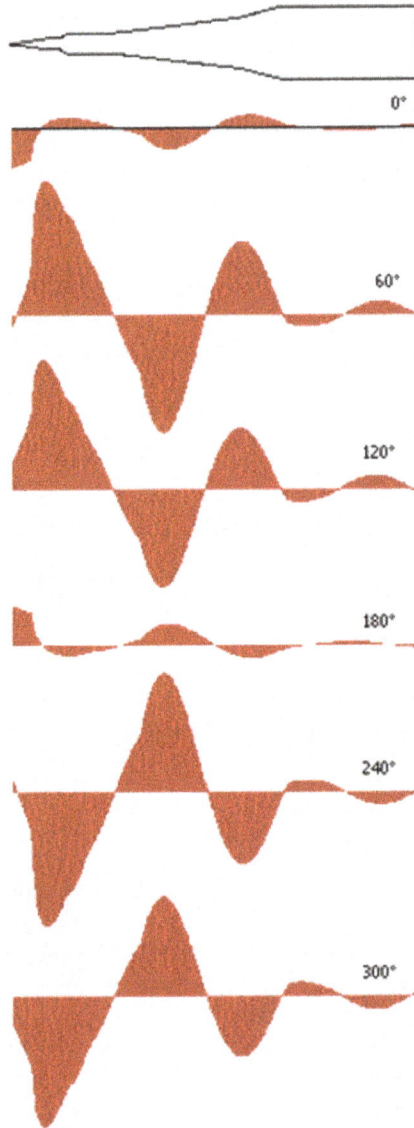

Fig. 6. Amplitudes of forced pressure waveforms in the simulated geometry related to the phase angle calculated from the analytical model

2.2 Numerical model

Computational Fluid Dynamics (CFD) models of selected geometrical configurations of the high-pressure system with integrated acoustic generator were created using CFD code ANSYS CFD to simulate numerically the influence of operating and configuration parameters of the acoustic generator and transmitting line on the generation and propagation of acoustic waves (pressure pulsations) in high-pressure system and properties of pulsating jet. The high-pressure system consisted of cylindrical acoustic chamber, liquid waveguide provided with high-pressure water supply and the nozzle. To simplify the model, acoustic actuator was substituted by vibrating wall of the acoustic chamber. The fluid flow in the model was solved as 3-D turbulent compressible unsteady flow of water. Water compressibility was taken into account in the numerical model using so called user defined function (UDF). The UDF covers calculations of both water density and speed of sound in water related to pressure:

$$\rho = \frac{\rho_{ref}}{1 - \dfrac{\Delta p}{K}} = \frac{\rho_{ref}}{1 - \dfrac{p - p_{op}}{K}} \tag{33}$$

$$a = 2.10^{-6}p + 1432 \tag{34}$$

where ρ is water density [kg.m-3], ρ_{ref} is reference water density under normal conditions (998,2 kg.m-3), p and p_{op} are real and operating pressures [Pa], K represents bulk modulus of water (2.2 . 10⁹ Pa) and a is speed of sound in water [m.s-1] determined experimentally.

The numerical simulation of a high-pressure system equipped with an acoustic generator was verified by the measurement of pressure pulsations in the high-pressure system upstream to the nozzle exit using dynamic pressure sensors. The pressure waveform in the numerical model was recorded at the same location as the pressure sensor was installed during the laboratory measurement. It was found out that numerical model provides information on the pressure waveform in high-pressure system that is in relatively very good agreement with experimental measurement. Comparison of results of numerical simulation and measurement also proved that the numerical model is able to simulate influence of geometry changes on the amplitude of dynamic pressure accurately and thus also to simulate pressure wave propagation and transmission in the high-pressure system.

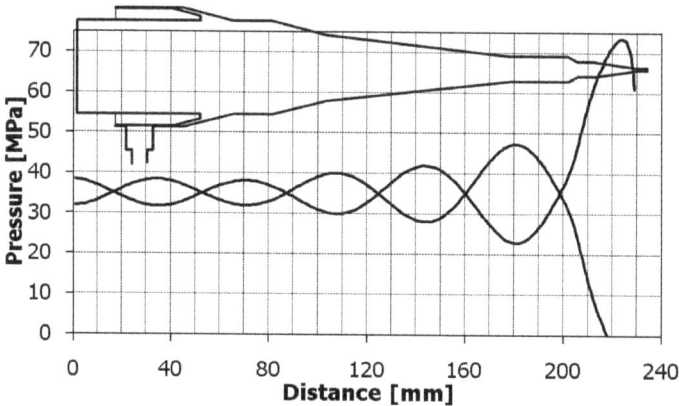

Fig. 7. Standing wave amplitudes along longitudinal axis of the high-pressure system

After the verification of plausibility of results of numerical simulation by the laboratory measurement, the model was used in studying of the process of propagation and transmission of acoustic waves in the high-pressure system from the acoustic actuator to the nozzle. An example of the behaviour of amplitudes of standing wave along the longitudinal axis of high-pressure system can be seen in Fig. 7. Figure 8 illustrates forced pressure waveforms in the simulated geometry related to the phase angle.

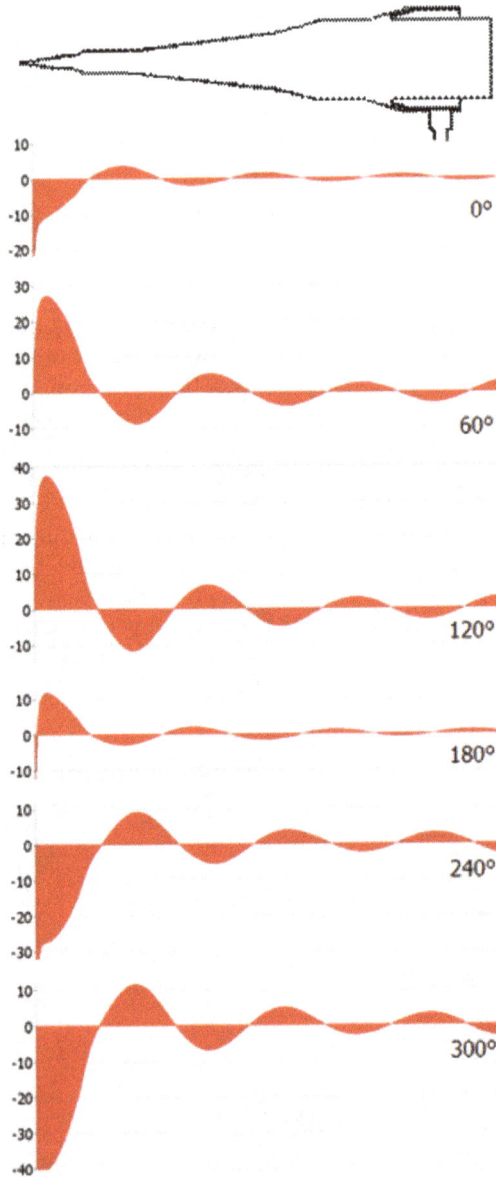

Fig. 8. Calculated forced pressure waveforms in the simulated geometry related to the phase angle recorded along the longitudinal axis of high-pressure system – from numerical model. (Scale indicates amplitude of dynamic pressure in MPa.)

Results obtained from the numerical simulation correspond to results obtained from the analytical one even if the numerical model used is not physically accurate in the close vicinity of the nozzle outlet where cavitation occurs. Cavitation model was not implemented in the numerical model with respect to the computational speed. Results of numerical modelling clearly indicate that the geometrical configuration of high-pressure system influences significantly propagation and transmission of pressure pulsations from the acoustic actuator to the nozzle. The amplitude of pressure waves increases towards the nozzle outlet due to the proper shaping of the liquid waveguide – its frustums act as mechanical amplifiers of the acoustic waves. At the same time, the amplitude of pressure pulsations close to the nozzle outlet (where it has crucial influence on the pulsating jet generation) changes significantly with respect to the geometrical configuration of the high-pressure system.

3. Visualization of pulsating jet

The use of visualization plays an important role in the study of behaviour of pulsating water jet. It enables not only the examination of characteristics of the jet such as mean velocity and break-up length of the pulsating jet but also to study the morphology and processes of formation of the pulsating jet and development of pulses in the jet. Furthermore, the visualization can be used to validate results obtained from numerical simulation of the process of generation of pulsating jets using CFD methods.

An original method of visualization of pulsating water jets based on the application of stroboscopic effect was elaborated for the above mentioned purposes. The method enables to obtain visual information not only on instantaneous structure of the pulsating jet but also on the mean structure of the jet. In addition, the stroboscopic effect allows observing process of formation of pulsating water jet by the naked eye. Special stroboscope for the pulsating jet visualization was developed where the frequency of stroboscope flashing is controlled by the frequency of pressure pulsations in the high-pressure system measured upstream from the nozzle exit. An example of the mean structure of pulsating jet with pulsating frequency of 20 kHz can be seen in Fig. 9. Exposure time of the photograph was 1/1000 s and the frequency of stroboscope flashing was about 20 kHz, therefore the figure represents superposition of 20 images of pulsating jet "frozen" by the stroboscope flashing.

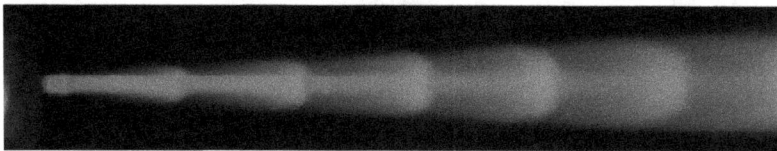

Fig. 9. The mean structure of the pulsating water jet generated at 30 MPa (illumination by the stroboscope)

An instantaneous structure of the pulsating water jet with the frequency of 20 kHz was studied using the high-speed camera LaVision VC-HighSpeedStar 5 equipped with image amplifier LaVision HighSpeed IRO. The record rate of the high-speed camera was 35 000 frames per second and the gate was set to 1μs. In addition, visualization of instantaneous structure of the pulsating water jet was also performed using Particle Image Velocimetry (PIV) system consisting of LaVision Imager Intense camera and New Wave Research laser,

Model Solo 120 with pulse duration 3-5 ns; the optical system was used to produce 1 mm thick sheet of light. An example of the pulsating jet visualized by high-speed camera can be seen in Fig. 10, the same jet visualized by PIV system is presented in Fig. 11.

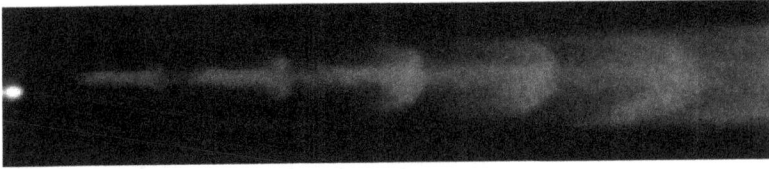

Fig. 10. The instantaneous structure of the pulsating water jet generated at 30 MPa (high-speed camera)

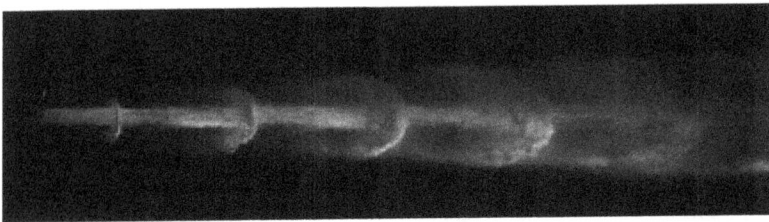

Fig. 11. The instantaneous structure of the pulsating water jet generated at 30 MPa (PIV system)

The fan (flat) pulsating water jet visualization was performed using the pulsed laser New Wave Research and digital camera Nikon D70s. Figure 12 shows the morphology of fan pulsating water jet generated at a pressure of 20 MPa.

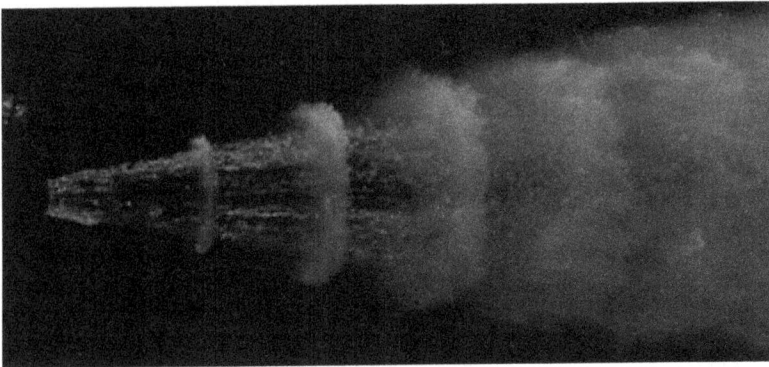

Fig. 12. The instantaneous structure of the fan pulsating water jet generated at 20 MPa (illumination by pulsed laser, camera Nikon D70s)

4. Effects of pulsating water jets on materials

Effects of pulsating water jets with the frequency of 20 kHz were tested on various types of materials, such as metals, rocks and concrete. Tested materials were exposed to the action of diverse types of jets: single round and fan pulsating jets as well as rotating pulsating jets. The effects of pulsating jets were evaluated in terms of cutting depth, rate of mass-loss or volume removal rate respectively and compared with the effects of continuous water jets under the same operating conditions.

Obtained results show clearly the supremacy of pulsating water jets over continuous ones in terms of their effects on material. Figures 13 to 15 illustrate the effects of various types of both pulsating and continuous jets on metal, rock and concrete samples. Differences in the surface structures created by pulsating and continuous water jets on individual materials are clearly visible in the above mentioned figures. An example of erosion effects of pulsating fan water jets (generated at various pressures) on aluminium samples at variable stand-off distance is presented in Figure 16.

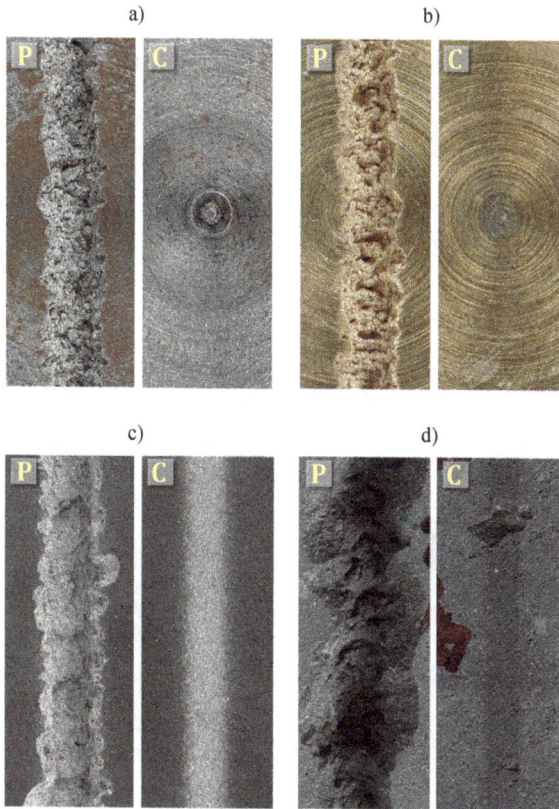

Fig. 13. Comparison of effects of pulsating (P) and continuous (C) water jets on samples of: a) mild steel (pressure 40 MPa, nozzle dia. 1.98 mm, traverse speed 0.03 m.min⁻¹, standoff distance 140 mm), b) brass (pressure 40 MPa, nozzle dia. 1.98 mm, traverse speed 0.03 m.min⁻¹, standoff distance 140 mm), c) duralumin (pressure 50 MPa, nozzle diameter 1.45 mm, traversing speed 0.05 m.min⁻¹, standoff distance 60 mm) and d) basalt (pressure 50 MPa, nozzle diameter 1.45 mm, traversing speed 1.0 m.min⁻¹, standoff distance 40 mm (P) and 20 mm (C))

Fig. 14. Comparison of effects of rotating pulsating (A) and rotating continuous (B) water jets on concrete sample (V_r – removed volume, pressure 30 MPa, nozzle diameter 2x1.47 mm, traversing speed 0.5 m.min^{-1}, standoff distance 40 mm (A) and 20 mm (B))

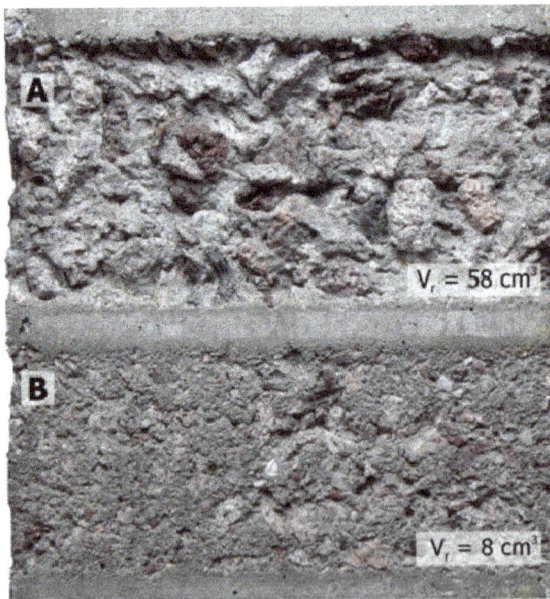

Fig. 15. Comparison of effects of fan pulsating (A) and fan continuous (B) water jets on concrete samples (V_r – removed volume, pressure 30 MPa, equivalent nozzle diameter 2.05 mm, traversing speed 0.2 m.min^{-1}, standoff distance 40 mm)

Results of the measurement of surface roughness characteristics on surfaces created by fan pulsating jets indicate that the characteristics are strongly influenced by both the standoff distance and the operating pressure. An example of the influence of a standoff distance on arithmetic mean roughness (SRa) and average maximum height roughness (SRz) can be seen in Figures 17 and 18, respectively. It should be pointed out that surface roughness (both SRa and SRz) produced by the pulsating fan water jet out of the range of "optimum" standoff distances (where the pulsating jet acts as a continuous jet) correspond to those produced by continuous jets reported by Kunaporn et al. (2009). On the other hand, the fan pulsating water jet produces surfaces with much higher values of the surface roughness (up to 20 times higher) within the "optimum" range of standoff distances (where the pulses are well-developed in the jet) compared to continuous jets.

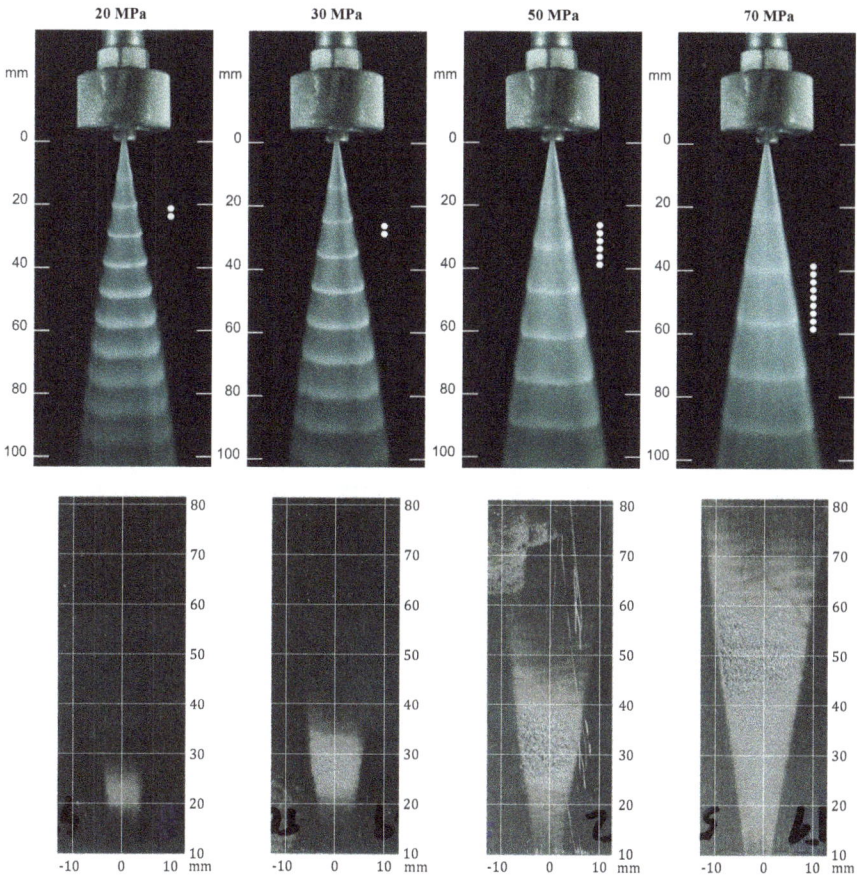

Fig. 16. Top: Pulsating fan water jets generated by nozzle with equivalent of diameter 1.10 mm and spraying angle of 10° at various operating pressures. Scale on the left side of photographs represents standoff distance in millimeters. Dots indicate the range of standoff distances where maximum erosion effects of pulsating fan water jet occur. Bottom: Erosion effects of the above pulsating fan water jets on duralumin samples. Scale on the right side of photographs indicates standoff distance in millimeters; scale on the bottom indicates width in millimeters

Results of the research of effects of pulsating jets on various materials obtained so far indicate that the pulsating jets can be used advantageously for the removal of surface layers of materials and/or "rough" cutting. However, further research will be necessary to be able to use the pulsating water jets in applications of precise cutting.

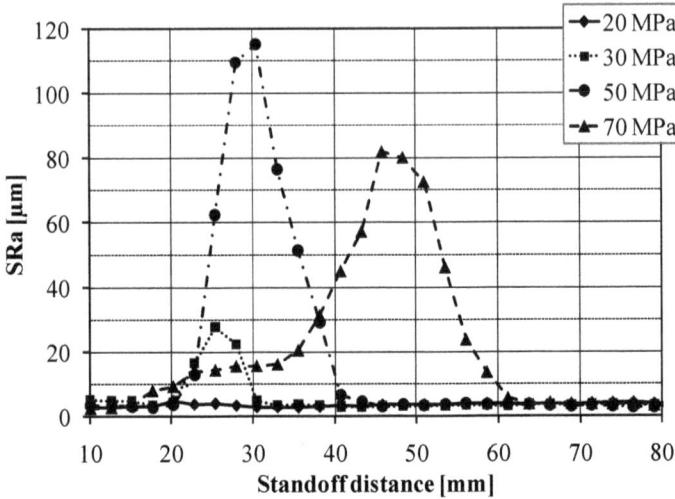

Fig. 17. Influence of a standoff distance on arithmetic mean roughness (SRa) of the surface created by the action of pulsating fan water jet generated by the fan jet nozzle with equivalent of diameter 1.10 mm and spraying angle of 10°

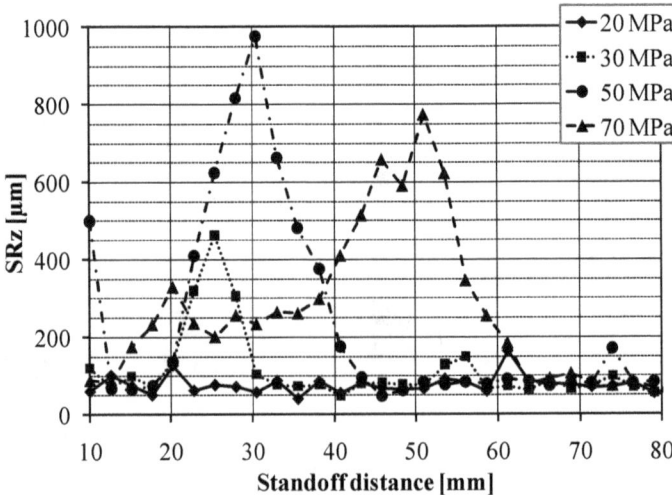

Fig. 18. Influence of a standoff distance on average maximum height roughness (SRz) of the surface created by the action of pulsating fan water jet generated by the fan jet nozzle with equivalent of diameter 1.10 mm and spraying angle of 10°

5. Conclusions

Presented results of the analytical solution and numerical simulation of the transmission of acoustic waves in high-pressure system represent the first step in gaining knowledge regarding processes of generation and propagation of high-frequency pressure pulsations in the liquid under high pressures and their influence on forming and morphology of pulsating liquid jets.

Results obtained from the visualization of pulsating water jets are used in studying of the characteristics of the jets and to verify results obtained from numerical simulation of the process of generating and forming of pulsating water jets. Laboratory and pilot tests of effects of pulsating water jets on various materials showed clearly the potential of pulsating jets to improve the performance of water jetting technology significantly.

It can be concluded that the research presented in the paper contributed to better knowledge of processes occurring in areas of generation and propagation of high-frequency pressure pulsations in the liquid under high pressure, their influence on forming and morphology of pulsating water jets and effects of the jets on materials. However, it is still necessary to further study problems of the efficient transfer of the high-frequency pulsation energy to longer distances in the high-pressure system. This will enable creation of the highly effective pulsating liquid jet with required properties.

6. Acknowledgements

The chapter has been done in connection with project Institute of clean technologies for mining and utilization of raw materials for energy use, reg. no. CZ.1.05/2.1.00/03.0082 supported by Research and Development for Innovations Operational Programme financed by Structural Founds of Europe Union and from the means of state budget of the Czech Republic. Presented work was also supported by the Academy of Sciences of the Czech Republic, project No. AV0Z30860518. Author is thankful for the support.

7. References

Bowden, F. P., & Field, J. E. (1964). The brittle fracture of solids by liquid impact, by solid impact, and by shock. *Proceedings of the Royal Society of London. Series A, Mathematical and Physical Sciences*, Vol. 282, No. 1390, pp. 331-352

de Haller, P. (1933). Untersuchungen über die durch Kavitation hervorgerufenen Korrosionen. *Schweizerische Bauzeitung*, Vol. 101, No. 21& 22, pp. 243-246,260-264

Hancox, N. L., & Brunton, J. H. (1966). The erosion of solids by the repeated impact of liquid drops. *Philosophical Transactions of the Royal Society of London. Series A, Mathematical and Physical Sciences*, Vol. 260, No. 1110, pp. 121-139

Kunaporn, S., Chillman A., Ramulu, M., & Hashish, M. (2009). Effect of waterjet formation on surface preparation and profiling of aluminum alloy. *Wear*, Vol. 265, No. 1-2, pp. 176-185

Pianthong, K., Zakrzewski, S., Behnia, M., & Milton, B. E. (2003). Characteristics of impact driven supersonic liquid jets. *Experimental thermal and fluid science*, Vol. 27, No. 5, pp. 589-598

Raghavan, C. & Ting, E. (1991). Hyper pressure waterjet cutting of thin sheet metal. *Proceedings of 6th American Water Jet Conference*, pp. 493-504, ISBN 1-880342-00-6, Houston, Texas, August, 1991

Thomas, G. P., & Brunton, J. H. (1970). Drop impingement erosion of metals. *Proceedings of the Royal Society of London. Series A, Mathematical and Physical Sciences*, Vol. 314, No. 1519, pp. 549-565

Tsai, S. C., Luu, P., Tam, P., Roski, G., & Tsai, C. S. (1999). Flow visualization of Taylor-mode breakup of a viscous liquid jet. *Physics of fluids*, Vol. 11, No. 6, pp. 1331-1341

Wong, G. S., & Zhu, S. (1995). Speed of sound in seawater as a function of salinity, temperature and pressure. *Journal of the Acoustical Society of America*, Vol. 97, No. 3, pp. 1732-1736

3

Excitation of Periodical Shock Waves in Solid–State Optical Media (Yb:YAG, Glass) at SBS of Focused Low–Coherent Pump Radiation: Structure Changes, Features of Lasing

N.E. Bykovsky and Yu.V. Senatsky
Lebedev Physical Institute, Russian Academy of Sciences, Moscow
Russia

1. Introduction

During several last decades much attention was paid to the processes that occur in solid-state optical media under the interaction with high-power focused laser radiation. A great number of studies were devoted to the phenomena of optical breakdown, structure changes, stimulated scatterings, generation of hypersonic waves in transparent dielectrics under the action of nanosecond (ns) and picosecond (ps) laser pulses (Manenkov & Prokhorov, 1986; Nelson et al., 1982; Ready, 1971; Robinson et al., 1984; Stuart et al., 1995). Recent interest in these studies was stimulated by the appearance of lasers with femtosecond (fs) pulses (Gordienko et al., 2010; Merlin, 1997; Sakakura et al., 2007).

An experimental study of a small region with high pressure and temperature gradients formed in a medium at focusing high-power laser radiation had been performed, as a rule, outside the laser cavity. In our experiment (Basiev et al., 2004), a region with such properties happened to be formed directly in the 2-mirror laser cavity, when Yb:YAG samples were pumped by the focused wide-band (0,89-0,95 µm) radiation from a pulsed LiF: F_2^+ color center laser (ccl). Thus, in contrast to many studies on ytterbium lasers, conditions for generation in Yb-doped samples in this experiment had been distinguished by the very high intensity (over 1 GW/cm²) of the pump, which moreover had a low coherence. Experiments on pumping of Yb-doped and non-doped samples of different optical media (YAG, glass, LiF et al.) by powerful low-coherent radiation from LiF: F_2^+ ccl were continued in subsequent papers (Bykovsky, 2005, 2006; Bykovsky & Senatsky, 2008a,b, 2010). At intensities $I \geq 1$ GW/cm² interaction of ccl pump radiation with the medium in the focal region was essentially nonlinear. The interaction of ccl pulses with samples was accompanied by excitation of stimulated Brillouin scattering (SBS) and stimulated Raman scattering (SRS) of pump radiation. The scattering generated hypersonic waves of high amplitude, which were converted into a periodic sequence of shock waves with sharp pressure jumps on their fronts propagating along the direction of pump. Pressure jumps were so large that they caused a phase transition in an optical medium, which was observed near the sample surface in the form of small domains with spatial modulation of the refractive index caused by the interference of hypersonic waves.

Under ccl pumping due to heat release and generation of intensive hypersonic waves a region with strong temperature, pressure and refractive index gradients and at the same time with a high-level of inversion was formed in the focus of the pump laser in Yb-doped materials. Despite the strong optical inhomogeneity of the medium, Yb lasing in 10-15 ns pulses was observed in Yb:YAG (with 20% concentration of Yb^{3+} ions) and Yb:glass (with 10% Yb^{3+} concentration) during the action of the 20-30 ns pump pulse and after it (due to inversion remaining in the medium). During the SBS of pump radiation the hypersonic wave spatial structure served as a resonator for Yb lasing. Shock waves (with phonon energies up to 1000 cm^{-1}) affected the generation dynamics. The Yb lasing was distinguished by some specific characteristics such as a surprisingly wide spectrum (up to 50 nm) and a high directivity of the emission. In addition to the wide-band generation on the shock-waves grating there was also observed Yb lasing on resonator modes. After the end of the pump pulse another sequence of shock waves diverging outward the focal region affected the build-up of generation between 2 mirrors in the cavity. The line spectra of Yb generation in the resonator contained twisted spectral lines with structures of small-scale spots.

Description of these unusual phenomena observed under the interaction of short intense pump laser pulse of low coherence with optical media and their explanation are presented in this chapter. The optical scheme and the parameters of LiF:F_2^+ ccl are considered. The features of SBS and SRS and the appearance of the periodical shock waves in the optical medium at low-coherent pumping are discussed. The interpretation to observed specific optical damage is given. The mechanism of generation of broadband, high-directional short laser pulses in the spatial structure of thin layers with inversion produced in the region of propagation of intense hypersonic waves in the medium is discussed. Conditions for generation in a 2 mirror resonator containing active medium with a strong refractive index gradient are considered. The interpretation of the observed twisted lines with small-scale structures in generation spectra as well as temporal, spatial-angular characteristics of Yb lasing in the resonator is given.

2. LiF: F_2^+ color center pump laser

Optical pumping of Yb-doped materials can be performed only into the single Yb^{3+} ion absorption band at $^2F_{7/2}$ -$^2F_{5/2}$ transition near 0.9 µm (Figs.1, 2). In this connection, the use of flash lamps as sources of broadband radiation for pumping an Yb-doped medium is ineffective. At the present moment, the most effective and widely used sources of Yb-doped materials pumping are semiconductor laser diodes operating within the spectral range near 0.94 µm. Along with semiconductor diodes solid-state laser pump sources have been used to investigate Yb-doped active media. Cr:LiSAF, Ti:Sa, Nd:YAG (0.94 µm transition) lasers have been used to pump Yb:S-FAP and Yb:YAG (Bykovsky et al., 2000; Kanabe et al., 2000; Marshall et al., 1997, as cited in Bykovsky & Senatsky, 2008b).

In our work a LiF: F_2^+ color center laser (ccl) was used to pump an Yb:YAG crystal (Basiev et al., 2004). The lithium fluorine color center (LiF: F_2^+, LiF: F_2^-) lasers are the sources of radiation in the near IR (0.8-1.3 µm) , and they effectively convert the neodymium and ruby laser radiation into this spectral range (Basiev et al., 1982). Possible room temperature operation, high conversion efficiency (up to 30%), and a large generation tuning range (more than 1000 cm^{-1}) make such lasers very attractive for certain practical applications. The pumping of Yb-doped samples by the ccl had been, undoubtedly, inferior to laser diodes pumping in efficiency. However, at focusing ccl radiation in the Yb-doped medium the pump power densities typical of semiconductor diodes could be easily exceeded. This had been, of course, of interest for research. Therefore, just the ccl had been used later in experiments on study lasing in Yb-doped samples as well as nonlinear interaction of pump radiation with optical materials. The possible generation region of LiF: F_2^+ ccl extends from 0.83 to 1.1 μm (Basiev et al., 1982) and completely covers the absorption band of Yb:YAG. In particular experimental conditions the actual emission band of ccl depends on the selective properties of the resonator, and usually covers only part of the noted wavelength range.

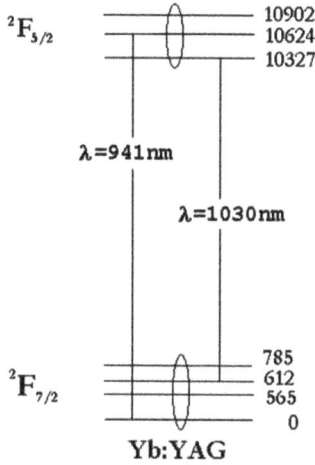

Fig. 1. Energy level diagram of Yb^{3+} ions in a Yb:YAG crystal. Ovals combine the Stark components of levels with rapid ($\approx 10^{-12}$ s) thermal relaxation (Krupke, 2000)

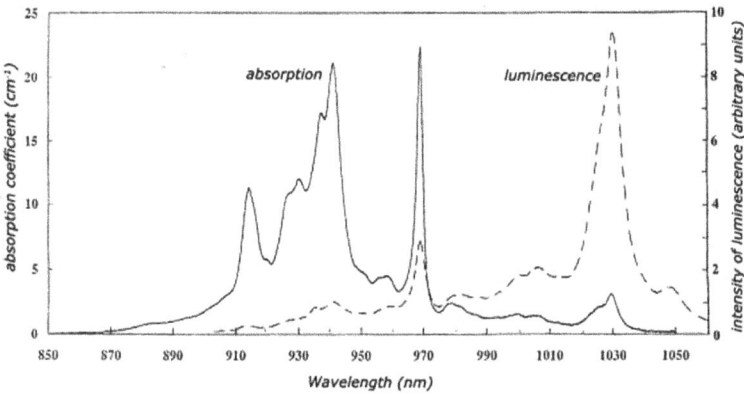

Fig. 2. Absorption and luminescence spectra of Yb^{+3} ions in the YAG crystal (1 mm plate, Yb^{+3} ions concentration 20%)

Figure 3 illustrates the scheme of the ccl with elements of radiation transport and diagnostics. A LiF: F_2^+ crystal (40x20x6 mm) was placed in the resonator formed by a plane mirror M_1 ($\approx$ 100% reflection at 0.9 µm) and a glass plate M_2. The length of the ccl resonator made 30 cm. The LiF: F_2^+ crystal was pumped through glass plate by a ruby laser operating in the single shot regime with pulse duration $\approx$ 30 ns and the energy up to 1 J. A multimode radiation at the wavelength of 0.694 µm was focused into the LiF: F_2^+ crystal by a lens L_1 with the focal length F_1=500 mm. The ccl multimode radiation was, in its turn, focused on the studied samples by a lens L_2 with the focal length F_2=120 mm. The Yb-doped samples were placed into a compact two-mirror resonator.

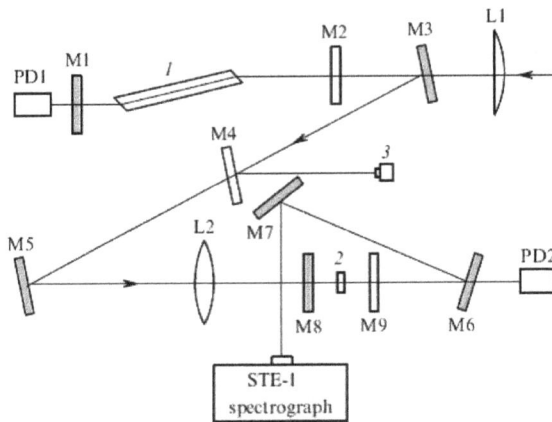

Fig. 3. Scheme of the experimental setup: (1) active element of a ccl; (2) Yb-dopped plate in the resonator with mirrors M8 and M9; (3) calorimeter; (M1, M2) ccl resonator mirrors; (M3 - M7) steering mirrors for ccl and Yb laser radiation; (L1 – L2) lenses; (PD1, PD2) photodiodes

The ruby laser and ccl pulses were registered with the help of photodiodes and a two-channel oscilloscope, and the laser energy was measured by calorimeters. The ccl radiation spectra were analyzed by the STE-1 spectrograph operating in the near IR region. The ccl emitted pulses of 20-30 ns duration and the energy of 100-150 mJ. For the most cases, the ccl pulse shape repeated the shape of the ruby laser pulse, Fig. 4a. Since the ccl resonator round-trip time made $\approx$ 2 ns, then during the ruby laser pumping the radiation made not more than 15 round trips in the ccl cavity. Due to high amplification in the LiF: F_2^+ medium the ccl pulse was formed in several round trips inside the resonator. Though the LiF: F_2^+ crystal was cut at a Brewster angle the ccl radiation was weakly polarized.

Figure 4b presents the densitograms of LiF: F_2^+ laser spectrum. The LiF: F_2^+ laser emission was observed within the range 0.89-0.95 µm. The lines of an argon spectral lamp were used as the wavelengths markers. Large spectrum width of ccl and short time of radiation development in the resonator speak about low coherency of the ccl emission. The ccl multimode radiation divergence was $\approx$ 2x10^{-3} rad. This allowed focusing the ccl pump at the sample into a $\approx$250µm spot. Moving samples along the axis of the focused pump beam one could change the size of the focal region within the limits of 250÷1000 µm, and the power density in the medium within the range of 0.5÷5 GW/cm^2. Such a range of power density variation allowed one to carry out both experiments on Yb lasing (see Section 4-6) and experiments on nonlinear interaction of ccl radiation with optical materials (see Section 3). In some of experiments, the ccl energy density was close to the damage threshold of Yb:YAG, glass and other studied materials. The material being damaged, the further experiments used a fresh part of a sample.

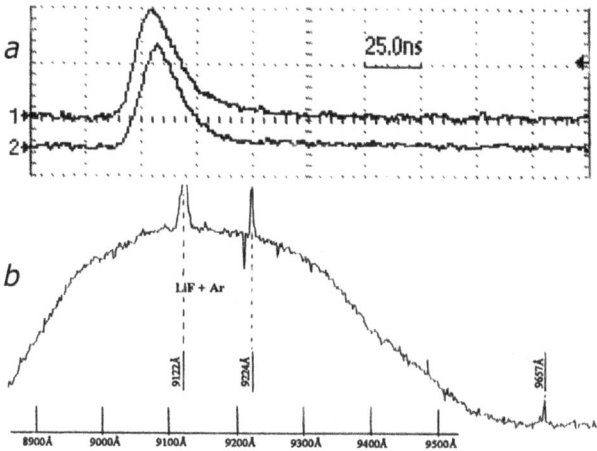

Fig. 4. Oscillograms of the ruby laser (1) and ccl (2) pulses – (a); densitograms of the LiF: F_2^+ laser spectrum with the argon spectral lamp reference lines – (b)

3. Interaction of ccl radiation with optical media

The experiments on nonlinear interaction of ccl radiation with optical media have been mainly performed using non-doped samples out of the cavity, Fig. 5. The ccl radiation was focused at samples by L_2, F=120 mm, and, so, within the sample thickness of 1-3 mm the pump power density changed insignificantly. The lens was tilted at 10^0 to the direction of ccl beam so that reflections from lens surfaces would not come back to the ccl resonator. As the samples there were used plates and slabs with polished surfaces made of the following materials: crystalline quartz (10x10x20mm), Yb:YAG crystal (1-2 mm thick plates), YAG crystal (4,5x30x30 mm), calcite (2 mm plates), LiF crystal (5 mm plate), 2 mm plexiglas plates, glass cube (20x20x30 mm). The ccl radiation was directed onto samples at a normal or at some angle (including the Brewster angle) to the sample's surface. The ccl pulse energy coming to samples varied within the range from 50 to 120 mJ.

In all materials a strong scattering of radiation was registered under the action of a ccl pulse. A diagnostic complex consisting of photodiodes, oscilloscopes, calorimeters and a spectrograph was arranged on the stand (Fig.5) in order to study the scattering of the low coherent wideband (0,89-0,95 μm) radiation of LiF: F_2^+ ccl. Due to very large spectrum width (tens of nm) and complicated space-angular structure of the scattered radiation components, one comes across difficulties in obtaining the spectral data on the scattered radiation, and it was not done. Strong scattering from the ccl focusing region in a wide angular range (tens of degrees) in forward and opposite directions (relative to the ccl beam) was observed for all samples at the ccl power density ≥ 1 GW/cm^2. The scattering intensity grew with the growth of the ccl intensity. The strongest scattering was observed at an angle of 180^0, i.e. back to the ccl aperture. The scattering was also observed at 90^0 to the ccl beam. Using photodiodes FD with ≈ 1 ns time resolution, the shape and duration of the ccl pulse and scattered radiation were registered by two-channel oscilloscopes of 1ns and 4ns resolution. To collect the scattered radiation to photodiodes the spherical mirrors were used, Fig. 5. The registered time and space-angular characteristics of scattering evidenced the rise of stimulated scattering (SBS and SRS) of wideband low-coherent ccl radiation in the samples.

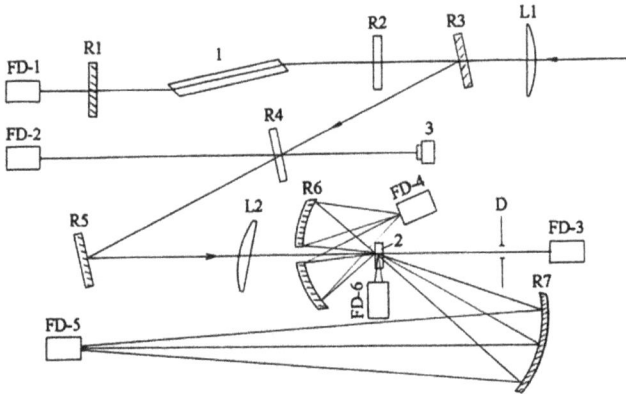

Fig. 5. Scheme of the experiment on the study of nonlinear phenomena in the crystal and glass samples: R1, R2 – LiF: F_2^+ ccl mirrors; R3 – R5 ccl radiation transport mirrors; R6 – R7 – mirrors to collect scattered radiation at diagnostic system; L1, L2 – lenses; FD – photodiodes; 1 – LiF: F_2^+ active element; 2 – sample; 3 – calorimeter; D – diaphragm

Figure 6 shows oscillograms of the ccl and scattered radiation pulses. A series of oscillograms in Fig.6 (left) illustrates the SBS of ccl pump radiation in YAG. Strong back scattering typical of SBS was observed in the YAG sample, placed at different positions to the pump. The duration of backscattered pulses was always smaller than the pump pulse duration ($20\div30$ ns) and constituted $8\div15$ ns. At the oscillogram of the ccl pulse which passed the sample and the diaphragm D, one observes a pump «cut-off» typical for SBS, Fig. 6 (left, I, b). This pulse «cut-off» was connected with the reflection of the ccl beam incident onto the sample under the occurrence of a periodic grating formed by hypersonic waves.

Figure 6 (right) illustrates oscillograms of scattering in the YAG and crystalline quartz samples in forward and backward directions, as well as at 90^0 to the ccl beam. Mirror R6 (Fig.5) focused the backward scattering at angles from 6^0 to 43^0, and mirror R7 - forward scattering at 5^0 to 25^0 in respect to the pump beam. The photodiode FD 6 recorded the 90^0 scattering. The data obtained on the scattering of wide-band low-coherent ccl radiation allow one to consider the pulses observed as different forms of the stimulated scattering. The scattering backward to the ccl aperture proves to be the SBS of pump, as noted above. The forward scattering at large angles may be attributed to the anti-Stokes component of SRS, and the backward scattering at large angle is due to the Stokes component of SRS. Scattering at 90^0, apparently, is due to the SBS and SRS of pump in the transverse direction.

Profiles of the scattered pulses observed in the oscilloscope, Fig.6 (right) illustrate the dynamics of SBS and SRS in samples and the influence of different components of stimulated scattering on each other. It should be noted that the pump intensities at which SBS and SRS were observed in our experiments (0.5 GW/cm²), had been almost an order lower than those typical of scattering at a coherent pumping.

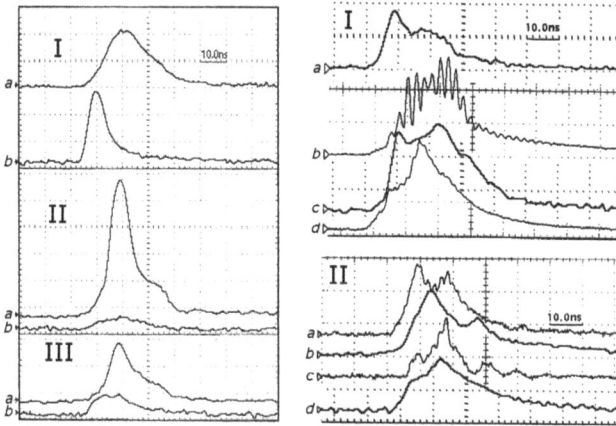

Fig. 6. (Left) SBS of ccl radiation in YAG slab (4,5x30x30mm). Oscillogams of SBS pulses (a) and ccl pulses behind the diaphragm D (b) for different orientations of the slab: I - pumping along the slab long side; II - pump beam along the short side; III - sample at Brewster angle to the pump. (Right) Oscillograms of scattering for YAG (I) and crystalline quartz (II): a–SBS pulse; b–scattering at 90⁰ to the pump; c–anti-Stokes SRS signal, d–Stokes SRS signal

After the action of ccl pulses of 4-5 GW/cm² intensity onto samples multiple marks of such an action in the form of volume and surface damages, and changes of the material structure were observed. Microphotographs of structural changes were collected, data were systematized. The material of transparent dielectrics was found to be damaged in different ways. Cracks in the medium, local regions of structure changes, tracks of self-focusing, and some other types of optical damage were observed near the surface and in the depth of a sample in the direction of the pump beam. Figures 7-9 demonstrate microphotographs of structural changes in YAG and LiF samples. One can observe plane formations of several micron thicknesses with a sharp boundary, Figs.7, 9. Most of such objects were concentrated near the sample surface at depths < 200 μm. Fig.8 demonstrates cracking at the LiF crystal surface. All these figures are illustrative for the energy release in the subsurface layers of samples. The optical damage in transparent dielectrics caused by high-power radiation was observed and studied for a few decades. The structure changes observed in the ccl experiments were compared with literature on the interaction of highly coherent radiation with transparent media. It was found that the low-coherent ccl radiation causes structural changes of some new kinds, which are not described in literature, for example, flat regions with structural changes bordered by a sharp boundary. The effects of nonlinear interaction of ccl pump radiation with transparent optical media (stimulated scattering, structural changes) were interpreted under the assumption that the SBS of low-coherent ccl radiation causes a succession of hypersonic shock waves. The formation of hypersonic shock waves at elastic nonlinearity at SBS was considered theoretically (Polyakova, 1966, 1968). But those results have not been so far supported by SBS experiments using the coherent pumping.

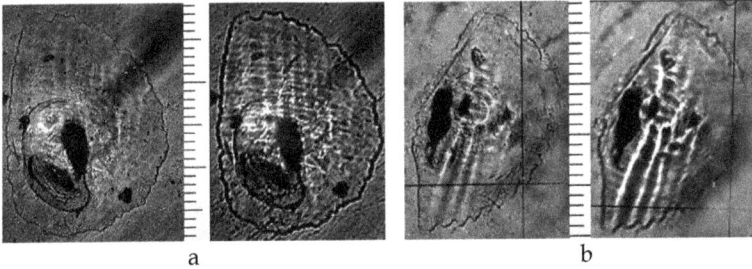

a   b

Fig. 7. Structure changes in YAG plate in 2 different ccl shots (a, b): left – sample surface, right- 7-10 µm under the surface. Wavy lines on the photos are supposed to be caused by the hypersonic waves interference inside the sample, connected with the aberration (coma) of the focusing lens, L$_2$. Scale ~ 11,8 µm/mark

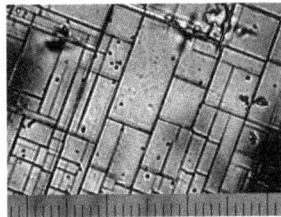

Fig. 8. Cracking at the LiF crystal surface. Scale ~ 11,8 µm/mark

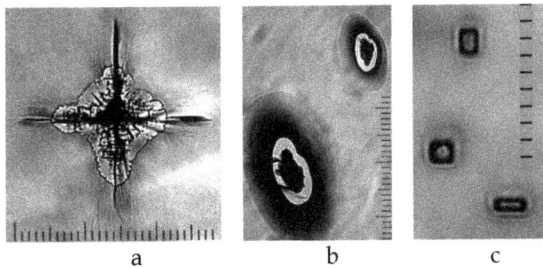

a   b   c

Fig. 9. Cross-like cracking having an additional plane formation with a sharp boundary inside the LiF crystal (a); ellipse-like cracking in the YAG crystal (b); a group of rectangular damages in the LiF crystal (c). Scale ~11,8 µm/mark

Structural changes in samples subjected to ccl pulses allow one to suppose the effect of hypersonic shock waves. Actually, a microphotograph of Fig.7a shows a region of the medium located at the depth of 7-10 µm under a YAG plate surface, where one can clearly see a quasi-periodic structure of stripes "imprinted" into an oval region with a sharp boundary. Figure 7b demonstrates another region, where a similar structure with step 10-15 µm is shown. These structures happened to be visible thanks to the refractive index gradient. However, such structures could not be formed in a medium as a result of the interference of low-coherent ccl radiation. At the same time, these structures could be the result of interference of intensive hypersonic waves propagating at a small angle to each other in the region of SBS of pump radiation. There are some other experimental data, which support a hypothesis of origin of intensive hypersonic waves in the medium. For example, a sharp boundary outlining structural changes in an YAG crystal, Fig.7 is the evidence of matter displacement deep down caused by the motion of dislocations in a surface region probably due to the phase transition in the low-lying layers in the region of hypersonic waves propagation. Note, that no apparent disturbances of the sample surface were observed. It is seen that some elements at the surface inside the oval region are near the same as they are outside: the pre-radiation scratches across the region boundary being continued, Fig.7.

To interpret the phenomena observed (stimulated scattering, structural changes) we consider a physical picture of the ccl radiation interaction with the medium, and a possible mechanism of the hypersonic waves buildup. The pump radiation spectral width of ≈ 600 Å permits one to suppose that a ccl pulse represents a sequence of ultra-short pulses (USP) of fs durations distributed chaotically within the $20 \div 30$ ns envelope of the pump pulse. Energy of a single USP is estimated as $0.1 \div 1$ µJ. Power density of separate ultra-short spikes could exceed 10^{10} W/cm². The interaction of ps and fs pulses with optical media was intensively studied, see for example (Nelson et al.,1982). The interaction of a single fs pulse (10^{13}W/cm² intensity) with transparent dielectrics (quartz, LiF, CaF$_2$, and others) produced generation of coherent optical phonons in the THz region leading to Raman scattering of probe pulses with ps delay (Gordienko et al., 2010; Merlin, 1997). The hypersonic waves (GHz region) were excited in liquids and solid-state media due to SBS on a space grating at the interference of two identical laser beams (10^{10} W/cm² intensity, ps durations) propagating at an angle to each other (Nelson et al., 1982; Robinson et al., 1984). In our conditions samples were irradiated by a succession of the chaotically distributed USP with a spectral range, which may be characterized by the central wavelength λ_p and the wave vector $k_p = 2\pi/\lambda_p$.

The appearance of shock waves at low-coherent pumping and their absence at the coherent pumping may be due to different conditions of SBS of pump radiation for those two cases. As is known, the SBS of coherent pump is developed on density fluctuations, i.e. hypersonic waves of the thermal origin (Ritus, 1982; Starunov & Fabelinsky, 1969). Aside from thermal fluctuations, there exist the stationary inhomogeneities in any solid-state medium, such as micro-inclusions, lattice inhomogeneities, dislocations. The stationary inhomogeneities may be the source of excitation of hypersonic waves and SBS. In fact, the interference of the incident and reflected pump radiation near inhomogeneity gives rise to a standing light wave with a spatial period $\lambda_p/2n_p$ where n_p is the refractive index of the medium at λ_p. The standing wave amplitude is determined by the coefficient of the pump radiation reflection from an inhomogeneity. The standing wave causes modulation of the refractive index, which gives rise to a spatial structure with the period $\Lambda = \lambda_p/2n_p$, which is resonant to the incident pump radiation and can be considered as the "seed" of the hypersonic wave. This structure starts moving due to the absorption of the momentum of light at scattering of pump photons. However, the stationary inhomogeneities do not exert influence on the development of hypersonic waves at the quasi-continuous coherent pumping, because the constant generation of the perturbations near the inhomogeneity is a drawback to the development of hypersonic waves and SBS.

At the pulsed low-coherence pumping, not only fluctuations but stationary inhomogeneities may be the source of SBS and hypersonic waves as well. Let an interaction of a single USP (a spike of pump) with the radiation reflected by an inhomogeneity produces a spatial structure with the period $\Lambda=\lambda_p/2n_p$ in a medium at the distance corresponding to the spike coherence length (for YAG, $n_p \approx 1.82$, $\Lambda \approx 0.25$ μm). For USP with durations ranging 0.01-0.1 ps the coherence length makes 5-50 μm. If the spike duration t_{sp} is less than half period of the hypersonic wave $t_{sp} < T/2$ ($T = \lambda_p/2n_pv_s \approx 50$ ps, $v_s \approx 5\times10^5$ cm/s -sound velocity in YAG) then after the spike end the periodical perturbations should propagate in the direction of pump and backward in the form of two travelling hypersonic waves, whose lifetime in the solid-state medium is $\geq 10^{-9}$ s for the GHz frequencies (Ritus, 1982; Starunov & Fabelinsky, 1969). In this time interval, the hypersonic waves may fall under influence of many other USP pulses, which follow the first spike. The pump radiation affects the hypersonic waves, and as a result it is partially scattered with changing frequency due to the Doppler shift both to the Stokes and anti-Stokes sides. In the anti-Stokes case an atom moving towards pump is decelerated by absorbing momentum from a counter propagating pump photon. That stops eventually the counter-propagating hypersonic wave. At Stokes scattering, however, momentum from a co-propagating pump photon is transferred to an atom in the direction of its motion, supporting the hypersonic wave propagation. Therefore, only one from the two hypersonic waves, i.e. the wave moving along the pump direction is sustained. Herewith the scattered light wave at the Stokes frequency ω_s propagates towards the pump and stimulates emission (at frequency ω_s) of the excited atoms, which absorbed a pump photon (at frequency ω_p). The difference of the photons energy, $\Delta\omega=\omega_p-\omega_s$, determines the amplification of the hypersonic wave, co-propagating the pump. At the same time this wave is extended towards the pump source. In other words, the scattered light wave deflates energy from the pump to the co-propagating hypersonic wave and extends it toward the pump source by half of the coherence length of the USP. The following next pump spikes running over the hypersonic wave should extend it up to the sample input surface.

Because the spatial structure of the co-propagating hypersonic wave, which is resonant with pump, reflects effectively the incident radiation, the maximum photon energy and momentum are transferred to atoms in a "leading" edge of this structure facing the pump source. The energy is accumulated there as a result of a multiple run of USP pulses on a hypersonic wave, which stops near the surface unable to propagate further. Thus, the hypersonic wave amplitude rises. This provides jumps of pressure at the wave fronts, which make the sound velocity higher at the maxima and the leading fronts steeper. The hypersonic wave is transformed into a periodic shock wave due to this growing "anharmonicity". Thus concentration of acoustic energy near the input surface of the sample occurs.

A proposed model of shock waves formation at SBS of low-coherent ccl radiation makes it possible to interpret the experimental data on stimulated scattering and structure changes in the medium. A directed action of high-energy phonons causes structural changes near the sample surface, as shown on microphotographs of Figs.7-9. Release of acoustic energy and the rise of tension in a sub-surface layer may be evidenced by cracking of a LiF crystal surface, Fig.8. Other evidences may be the flat regions of structural changes, Figs.7, 9. The analysis of experimental data on lasing in the Yb-doped samples (see section 4-6) made it possible to estimate pressure at the shock wave fronts (tens of GPa). The pressure above 10 GPa is higher than the pressure of phase transitions in some materials. One may assume that structural changes of YAG crystal with a sharp boundary, Fig.7, 9 are rather due to phase transitions in a medium.

It should be noted that propagation of a hypersonic wave produced at some inhomogeneity deep in the medium may be stopped inside the sample by a "barrier" in the form of a dislocation, a crack, and the like. The wave would be partially stopped by the barrier. Its acoustic energy is released in the sample near the side of the barrier. In the case of acoustic energy exceeding the crystal lattice deformation threshold or the phase transition threshold there occurs local splitting of medium, shift of the barrier's wall. In this way one can explain the observed structural changes in crystals occurring along cleavage planes, Fig.9. Figure 9a demonstrates a typical cross-like cracking of a LiF crystal having an additional plane formation with a sharp boundary. Figure 9b shows an ellipse-like splitting in the YAG crystal. Sharp boundaries in a region of structural changes may be interpreted as phase transition boundaries in the medium. Figure 9c illustrates a group of rectangular damages, which seems to be due to the concentration of acoustic energy on defects (dislocation walls) inside the sample. The formation of tensions around damages, which seen in the form of blurred fringes, Fig.9c, is also the evidence of a considerable release of energy in that region. The comparison of mechanisms for hypersonic wave formation at SBS of the coherent and incoherent pump may be supplemented by the following considerations. The efficiency of the pump energy transfer into the acoustic energy proves to be higher in the case of the low-coherence pumping. At similar pump energy the light–acoustic wave interaction proceeds more effectively when pumping by USP with higher intensities and at small distances (USP coherence lengths) as compared with the case of a high-coherence radiation. Rather large amplitude of light wave scattered on stationary inhomogeneities (as compared to fluctuation scattering) and high intensity of the low-coherence spikes (compared to average level of the pump intensity) should make the SBS threshold lower and stimulate formation of the intensive hypersonic waves. This explains the appearance of intensive hypersonic waves at low-coherence pumping. At the coherent pumping, the hypersonic wave seeded by a small-dimensional fluctuation of pressure, needs large distances for its development, and cannot "grow" to a shock wave. Note, that at the low-coherence pumping, the hypersonic wave may be caused even by a large-scale inhomogeneity, like a plane back surface of a sample. In that case, the produced hypersonic wave has a wide wave front. Even in a thin sample (1-2mm) this may cause structural changes of medium (phase transitions) comparable with dimensions of the pump spot.

Sharpening of the hypersonic wave fronts provides the appearance of high-frequency harmonics up to the optical phonons excitation. Moreover, as steepness of the shock waves increases, the momentum transferred to waves from successive USP pumping increases too, thus exciting oscillations of atoms (coherent phonons) with frequencies inversely proportional to the USP duration. The USP of the pump being scattered on coherent phonons (like probe pulses in works Gordienko et al., 2010; Merlin, 1997) give rise to the stimulated Stokes or anti-Stokes scattering. As optical phonons are concentrated on shock-wave fronts, hence the stimulated scattering is linked to fronts, Fig.10. Due to a large shift of the Stokes wavelength (λ_s), the SRS may be amplified only at the angle α to the pump: $cos\alpha = n_s\lambda_p/n_p\lambda_s$, Fig.10. The anti-Stokes SRS is developed in the opposite direction. The respective directions and angles of scattering are illustrated in Fig.10. It should be noted that high-intensity optical phonons excited at shockwave fronts stir up the molecules at amplitudes and frequencies close to the possible limit in the crystal lattice, and hence make easier structural changing of the medium, i.e. its phase transition to a more dense state.

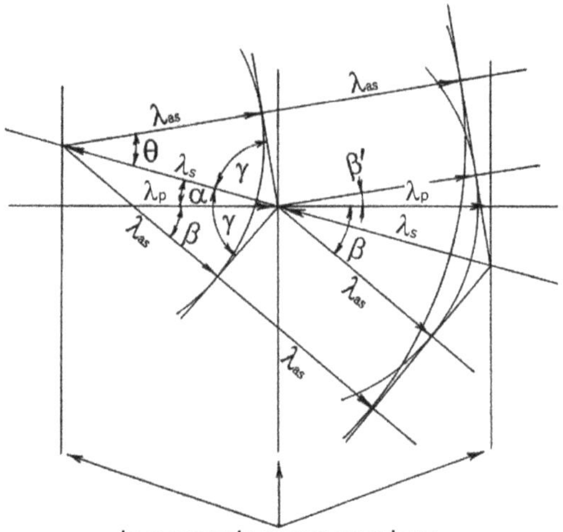

hypersonic wave maxima

Fig. 10. Illustration to Stokes (λ_s) and anti-Stokes (λ_{as}) scattering at hypersonic shock wave fronts: a – angle between Stokes and pump waves; β, β' - angles between anti-Stokes and pump: $n_s\lambda_p/n_p\lambda_s = cosa$; $n_s\lambda_{as}/n_{as}\lambda_s = sin\gamma = cos\theta$; $\theta - a = \beta'$ and $\beta = 90° - (\gamma - a)$, $\beta' = 90° - (\gamma + a)$.

4. Experiments on pumping Yb-doped media by LiF: F_2^+ ccl

Experiments on pumping Yb-doped media were performed using a setup of Fig. 3. Samples: plane – parallel 2-mm-thick Yb :YAG (20% of Yb) or 3-mm Yb-doped phosphate glass (10% of Yb) plates were mounted in a resonator of length L ≈ 20 mm formed by plane mirrors M8 and M9. Samples were pumped by pulses of focused ccl radiation through mirror M8 with the reflectance ≈100% at ≈1 μm transmitting 80%-90% of pump radiation. The pump beam propagated close to the normal to the resonator mirrors. The reflectance of mirror M9 at ≈ 1 μm was 30% (for Yb: YAG) or 70% (for Yb: glass). The energy, spectrum, shape of Yb-laser pulses were registered by calorimeters, photodiodes, and STE-1 spectrograph. Experiments were performed at room temperature in the single-shot regime. In most experiments samples in the resonator were exposed to the ccl radiation of energy 60 - 70 mJ. The focused ccl beam produced the radiation intensity distribution I (r) (r-radius of the excited region) in the subsurface layers of samples with the maximum at the pump beam axis. The minimal size 2r of the focal spot on the sample surface was ≈ 250 μm. The ccl intensity was varied within 0.5 - 5 GW cm^{-2} by moving the lens. Lasing in a Yb :YAG crystal was observed at pump intensities exceeding 0.5 GW cm^{-2}, Yb laser energy did not exceed 0.5 mJ. Figures 11 and 12 show Yb-lasing oscillograms recorded for one shot. The first of a series of laser pulses appeared during the pump pulse, while the next pulses appeared after the end of the pump and were delayed by tens and hundreds of ns, up to 1.2 microsecond (μs). Spectra were recorded by STE-1 on an IR film or on the image converter. The optical scheme of the STE-1 imaged (1:1) the vertical slit of ≈100μm width on the film. At uniform illumination of the slit by the monochromatic radiation of a spectral lamp the film registered separate narrow vertical lines, Fig.4b. In case of Yb-lasing we observed much more complex spectra, Figs 11-14. The lasing spectra were recorded within an angle of ≈10^{-2} rad, which was determined by the slit height and a distance (≈1m) from the sample. Laser radiation going along the resonator axis was directed to the lower part of the slit. No focusing of radiation on the slit was performed. Figs 11, 12 demonstrate lines in the region of Yb transitions in YAG at 1.03 and 1.05 µm. At low pump intensities line lasing appeared first at ≈1.05 µm. As the pump intensity was increased, line lasing was observed simultaneously at 1.03 and 1.05 µm (Fig. 11), and at the maximum pump it was observed at 1.03 µm, Fig.12. Fig.11 demonstrates a noticeable inclination of the spectral line at 1.03 µm from the vertical direction. This means that the lasing wavelength in this line was varied along the slit height. Structure of small-scale (50-200µm) spots was observed in 1.05 µm line, Fig.11. Along with narrowband lasing, broadband unstructured lasing was observed virtually in each pulse. Spectral bands up to 20 nm width were extended to the blue (up to 1.0 µm) and red (up to 1.06 µm) parts of the spectrum far beyond the regions near 1.03 and 1.05 µm in which Yb :YAG lasing is usually observed. As pump was increased, the broadband spectra were shifted to the blue (Fig. 12).

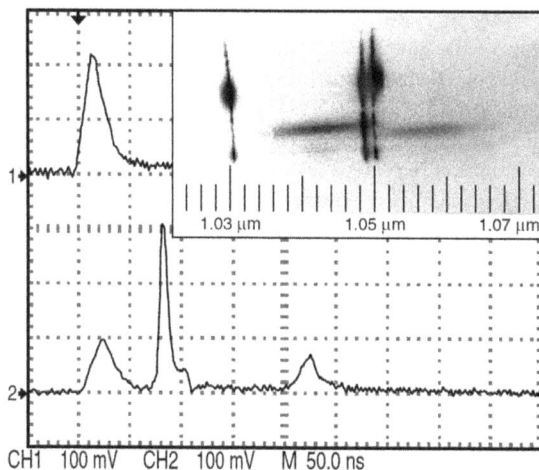

Fig. 11. Oscillograms of pump (1) and Yb:YAG laser (2) pulses, and the lasing spectrum in the 1.03 – 1.06-µm region

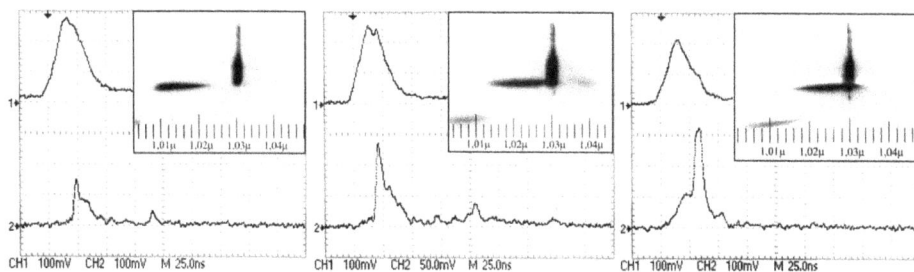

Fig. 12. Oscillograms of pump (1) and Yb:YAG laser (2) pulses and lasing spectra in 1.03 – 1.06-µm region for 3 ccl pulses with energies 100, 125, and 150 mJ (from left to right)

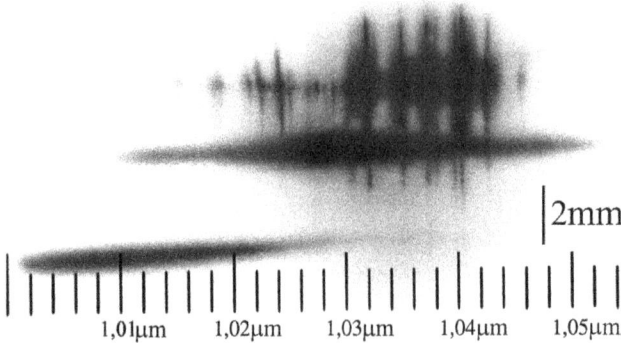

Fig. 13. Lasing spectra of Yb: glass in the 1.00 – 1.05-µm region

Lasing in the Yb: glass (Figs 13,14) appeared at pump intensities 4-5 GW cm^{-2} and was accompanied, as a rule, by the local damage of samples. The lasing energy did not exceed 0.1 mJ. The oscillograms of the glass lasing (similar to the Yb :YAG lasing) exhibit several ns pulses delayed with respect to the pump, Fig.14. The spectrograms demonstrate line (in 1.02-1.05 µm region) and broadband (1.00-1.06 µm) lasing spectra. Structures of small-scale spots are clearly visible in Yb:glass line spectra, Fig.14. The total spectrum of two broad bands of the Yb:glass emission exceeded 50 nm, Fig. 13. The width of stripes of broadband lasing in vertical direction was about 1mm, Fig.13. The line and broadband lasing spectra both in glass and Yb:YAG were recorded from the slit parts located at different heights, Figs 11-14. Note, that spectra of Yb lasing were recorded when the slit was in the far-field diffraction zone with respect to the position of the radiation source. So, the observed narrow stripes and small spots of generation, Figs 11-14, indicated a high directivity of radiation. The angular divergence of radiation, estimated from the size of spots on the slit, did not exceed usually 10^{-3}-10^{-4} rad. At the same time, the angular divergence caused by diffraction from the lasing region with r ≈100 µm should be ≈10^{-2} rad. Thus, the angular directivity of Yb-lasing could exceed the diffraction limit by one- two orders of magnitude, i.e. the generation should occur in highly directed beams of small transverse size.

At pump intensities from 0.5 to 5 GWcm^{-2} used in our experiments, we observed intense pulses of scattered ccl radiation in Yb-doped crystal and glass samples similar to observations of scattering (SBS and SRS) in non-doped samples described in the part 3. Consider the features of the inversion distribution in the Yb-doped medium upon SBS of pump radiation. The dynamic grating of hypersonic waves produced in samples near the surface played the role of an additional external mirror for the ccl and efficiently reflected incident radiation, preventing its propagation inside a crystal or glass sample. The penetration depth of hypersonic waves into the medium is $l \approx 1/\eta$, ($\eta = 10^2$- 10^3 cm^{-1} is the hypersonic waves attenuation coefficient) (Ritus, 1982; Starunov & Fabelinsky, 1969). Thus, the excitation region of SBS (and SRS) in the medium was a surface layer of thickness l < 100 µm. The heat release caused by the dissipation of energy of hypersonic waves and optical phonons excited in the medium upon SRS of ccl radiation occurred in this layer. Inversion was also mainly produced in this thin layer. The maximum energy density stored in Yb :YAG containing 20% of Yb ($N_0 = 2.9$ x10^{21} cm^{-3}) upon excitation of all Yb^{3+} ions to the $^2F_{5/2}$ metastable level was estimated as ≈500 J cm^{-3}. Even for such an extremely large energy in the volume of an Yb :YAG crystal cylinder of diameter 250 µm and length 100 µm, the energy stored in inversion does not exceed, according to estimates, 2.5 mJ, which corresponds to the low level of Yb-lasing energy in experiments. A part of the pump energy was absorbed by Yb ions beyond the region of SBS excitation, and the other part escaped from the sample. So, the population of Yb ions on the metastable level decreased over the sample length along the pump beam. Only a small fraction (estimated as 10%) of the ccl output energy incident on samples was spent to excite ytterbium. A grater part (up to 70%) of the ccl energy was transformed to stimulated scattering due to nonlinear interaction with the medium and was spent to form intense hypersonic waves and optical phonons accompanied by heat release in the medium. The thermal energy required for heating and melting of the above-mentioned Yb :YAG crystal cylinder was estimated as 30 mJ. At the same time, the thermal energy in the medium during the formation of inversion and generation and due to stimulated scattering did not exceed, according to estimates, 10 mJ, i.e. it was insufficient for melting of the material. This suggests that structural changes in the surface layer of samples (Fig.7-9) could be produced by intense hypersonic waves appearing during the SBS of pump radiation. Thus, the Yb-doped medium during the SBS of pump proved to be divided into a layer of thickness ≈100 µm with the very high inversion and the main part of the sample to which pump radiation hardly penetrated.

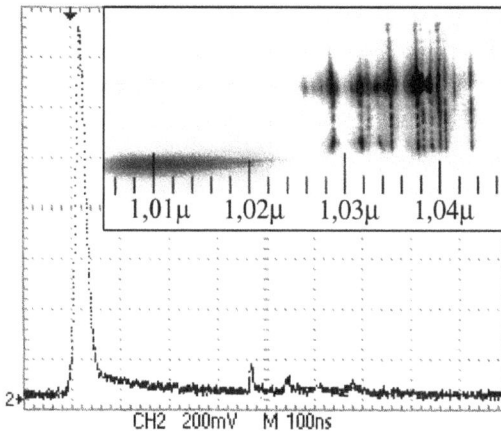

CH2 200mV   M 100ns

Fig. 14. Oscillograms of Yb: glass laser pulses and spectra in the 1.00 – 1.04-µm region

Figures 11-14 demonstrate the integrated (per pulse) temporal pictures and broadband and narrowband lasing spectra of ytterbium. These two types of emission correspond to different Yb-laser pulses appearing at different stages of the development of lasing. We can distinguish two such stages: the first one is the development of Yb-lasing during irradiation of a sample by the pump pulse, and the second one is Yb-lasing after the end of the pump pulse. At the first stage of duration t_1 = 20–30 ns, during the action of a ccl pulse onto the sample the generation could occur without a resonator, just due to the feedback at the dynamic hypersonic wave grating. The broadband unstructured spectra correspond to this stage. At the second stage of duration t_2 up to 1.2 µs there was generation in the Yb laser resonator which was characterized by line spectra, Figs. 11-14.

5. Features of Yb- lasing: broadband spectra, data interpretation

We attribute broadband lasing in the crystal and glass to Yb-laser pulses appearing at the first stage. Consider the conditions of inversion formation and lasing at the first stage by using a model assuming the appearance of shock hypersonic waves in the medium during the SBS of pump radiation. Excitation of Yb ions to the $^2F_{5/2}$ metastable level and formation of hypersonic wave grating (Fig. 15) occurred simultaneously in a thin layer of the medium. The grating period Λ is related to the pump wavelength λ_p and the refractive index of the medium n_p by the expression $\Lambda = \lambda_p/(2n_p)$ (Ritus, 1982; Starunov & Fabelinsky, 1969). By assuming that $\lambda_p \approx 0.92$ µm, we obtain for YAG $\Lambda \approx 0.25$ µm. The period of the hypersonic wave is $T=\lambda_p/(2n_p v_s)$, where v_s is the sound speed. For the glass, we have $T \approx 70$ ps, and for the YAG crystal, $T \approx 50$ ps. Intensive hypersonic waves strongly affected the spatial distribution of inversion. Indeed, conditions were produced for excitation of high-frequency vibrations of the medium (optical phonons) and SRS on the fronts of shock waves. The scattering parameters show that a broad phonon spectrum was excited upon SRS, including phonons with energies ≈ 1000 cm^{-1} near the high-frequency boundary of the phonon spectrum in the medium (Gorelik, 2007). The energy of such phonons is sufficient for the population of the Stark components of the $^2F_{7/2}$ level lying above the ground level of Yb (Fig.1). The redistribution of the population of Stark levels occurs during several ps. Due to the phonon population on the fronts of shock waves, the inversion for transitions from the $^2F_{5/2}$ level to the Stark components of the $^2F_{7/2}$ level was decreased, preventing the development of lasing. At the same time, due to the rapid decrease of pressure and dynamic cooling of the medium, which occurred behind the fronts of shock waves (Fig. 15), the population of the Stark components of the $^2F_{7/2}$ level was rapidly depleted, resulting in the inversion jump. As a result, a structure of thin (<0.3 µm) periodical layers with the high inversion was established in the medium for a short time ($t_{inv} < T$). During the propagation of a hypersonic wave, the inversion regions were displaced. After the time T, the same inversion distribution over layers again reestablished. Thus, upon the SBS and SRS of pump radiation, amplification could appear at transitions between the $^2F_{5/2}$ level and Stark components of the lower $^2F_{7/2}$ level, including transitions in the short-wavelength region (< 1.03 µm), which are not observed usually in Yb- lasers.

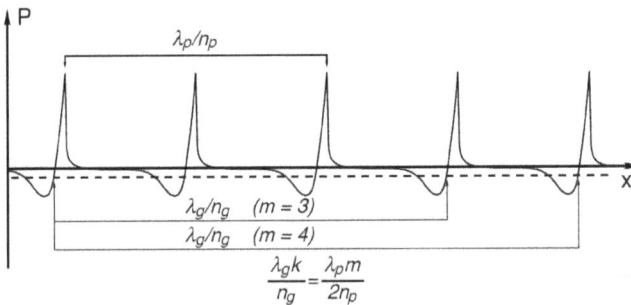

Fig. 15. Illustration to conditions for inversion formation and stimulated emission in the region of shock hypersonic waves in the active medium. The dependence of the pressure profile P in the hypersonic wave on the coordinate x along the pump direction is shown. The dashed straight line indicates the pressure at which inversion is produced. The possible relations between the pump, λ_p and laser, λ_g wavelengths satisfying condition (1) are given

The formation of inversion in the region of propagation of hypersonic waves produces the conditions for generation of short radiation pulses in the medium. Indeed, a distributed feedback (DFB) laser (Kogelnik & Shank, 1971) appeared in fact in our experiments in the hypersonic wave grating region. Such a laser operates without an external resonator and can emit ps pulses (Bor & Muller, 1986; Katarkevich et al, 1996; Kogelnik & Shank, 1971). Unlike dye DFB lasers based on a photo-induced sinusoidal grating in the medium, the high-contrast DFB structure was formed under our conditions due to modulation of the medium parameters (refractive index, inversion) during the propagation of a sequence of intense hypersonic waves. The possibility of creating a DFB laser on a hypersonic wave grating was discussed already in pioneering paper (Kogelnik & Shank, 1971). The length of the DFB structure in the region of SBS is ≈ 100 µm; in this case, the travel time t´ of a photon in the structure is much shorter than the round-trip transit time τ in the Yb- laser resonator (τ = $2L/c$ ≈ 100 ps). The possibility of lasing using the hypersonic wave grating without the resonator was confirmed by experiments. Lasing during the ccl pump pulse was observed both in samples mounted inside the resonator and in samples without the resonator, in particular, in samples mounted at an angle to the pump beam axis. The generation of ps pulses in dye DFB lasers on a stationary grating pumped by ns pulses occurs due to self-Q-switching mechanism: `blowing away' of the gain grating by the structure emission itself (Bor & Muller, 1986; Katarkevich et al, 1996). Unlike this, upon the SBS of ccl pump radiation, a DFB structure moving in the medium appears with period Λ which can depend on the pump intensity (pressure in the medium). Under these conditions, the Q-switching in the DFB structure can be determined, apart from the mechanism considered above, by the rapid movement of the grating in the medium and by variations in its parameters as well. Pulses of ps duration, $t < t_{inv}$ should be generated in our experiments successively in these periodic inversion layers during their movement. Besides, the change in Λ should lead to scanning of the lasing spectral band over the gain profile in the medium. The temporal picture of Yb- lasing at the first stage (during pump) was not investigated in detail due to the lack of recording equipment with a proper time resolution. The shape and duration of Yb-laser pulses were determined by the resolution of the oscilloscope (1- 4 ns).

By considering the conditions for appearing of stimulated emission in the moving layer structure, we can interpret features of broadband lasing spectra of the glass and Yb :YAG crystal (Figs 11-14). The Yb emission appears at wavelengths $\lambda_g > \lambda_p$. This emission is amplified along the normal to the layer structure when the path difference between the layers involved in lasing is equal to the quantity multiple of λ_g. This synchronism condition between hypersonic waves and lasing can be written in the form

$$\frac{\lambda_g k}{n_g} = \frac{\lambda_p m}{2n_p} \tag{1}$$

Here, n_g is the refractive index of the medium for λ_g; and k and m are integers. It is important that the refractive index depends on pressure P in the medium: $n_g, n_p \sim P$. Pressure depends on the intensity of hypersonic waves and the heat release in the medium upon pumping. The maxima of the spatial structure of the laser field are located in regions with a decreased density, behind the fronts of shock waves, Fig. 15. Here n_g can take values lower than the refractive index n_0 at the normal pressure. At the same time, the refractive index n_p on the fronts of shock waves can be larger than at the normal pressure. Under these conditions for $n_g < n_p$, relation (1) is fulfilled for $k = 1$ for small values of m ($m > 2$, $m = 3, 4...$, see, Fig. 15). It follows from the relation $\Lambda = \lambda_p/(2n_p)$ that, as the intensity of hypersonic waves and heat release increase (with increasing P and n_p) at the fixed λ_p, the period Λ of the grating decreases. This in turn leads to the blue shift of λ_g. Correspondingly, as P is decreased, the wavelength λ_g should shift to the red. Experiments confirm these conclusions. The shift of broadband spectra with changing the pump intensity is illustrated in Fig. 12. One can see that the spectrum of Yb:YAG shifts to the blue by 10 nm with increasing the ccl pulse energy from 100 to 150 mJ. Note that the shift of the emission region of DFB lasers with changing of the pressure was observed in dye lasers (Bor & Muller, 1986).

The inhomogeneous distribution of the pump with the maximum at the beam axis gave rise to the transverse gradient of pressure $P(r)$ in the region of propagation of hypersonic waves. Because of this gradient, for regions with different pump intensities (at the beam axis and its periphery) relation (1) is fulfilled for different values of m. These values can correspond to the regions of the broadband spectrum shifted in the wavelength scale and over the slit height, Figs 11-14. In the axial region with the maximum pressure ($r \approx 0$, the lower part of spectrograms), lasing was developed in the short-wavelength part of the spectrum. The spectra in regions with the lower pump ($r > 0$) are shifted to the red (Figs 11-14).

The regions of the broadband spectrum should be located on the wavelength scale in accordance with possible transitions in Yb: glass and Yb:YAG. According to the energy level diagram presented in Fig. 1, the emission of Yb:YAG in the spectral region under study (Figs 11, 12) can be determined by transitions between the two lower components of the metastable $^2F_{5/2}$ level with energies 10327 and 10624 cm^{-1} and three components of the $^2F_{7/2}$ level with energies 565, 612, and 785 cm^{-1}. In each spectral region, lasing in the DFB structure was built up in the active medium within the part of the gain line corresponding to the Yb transition which was resonant for the particular period of hypersonic waves. Thus, a large width of lasing spectra of Yb-doped materials (Figs 11 - 14) could be caused not only by the generation of short pulses in the DFB structure upon emission of thin layers of the active medium but, as explained above, by the shifts of λ_g with changing the pump intensity as well. This is a picture of lasing in Yb-doped media in the region of the propagation of longitudinal hypersonic waves at SBS of pump radiation. Thus, the data on Yb broadband lasing confirm the concept of intense (shock) hypersonic wave formation at powerful low-coherent pumping of the optical media.

The interpretation of data on the "sub-diffraction" angular divergence of Yb:YAG and Yb:glass generation had required the introduction of a new concept for the photon, which means its existence not in the form of a "traveling" wave, but with fixed positions of maxima, minima, and nodes along the photon propagation direction (Bykovsky, 2006; Bykovsky & Senatsky, 2008a,b, 2010). It was assumed that the photon is a combination of the two pinpoint dipoles, which move jointly with light velocity and rotate with frequency ω in the opposite directions. The phase of the photon and its field are connected with the phase of this rotation (Bykovsky, 2006). Such an approach helps to describe the effects observed. The field produced by the photon motion along the coordinate x is described by the function: $E = E_0 \sin(kx - \varphi_0)$, where $k = \dfrac{2\pi}{\lambda}$, and λ is the distance, at which the photon changes its phase by 2π; E_0 –the field amplitude, φ_0 - the initial phase. The new concept does not contradict to the properties of the well-known electromagnetic "traveling" wave: the traveling wave may be "constructed" from separate photons following each other along the same direction and having a continuous relative phase shift from 0 to 2π. If we assume that the excited atoms in a medium irradiate photons along the same direction and with the initial phases $+\pi/2$ or $-\pi/2$ only, and that emitted photons are correlated by the relation $\Delta x = c\Delta t$, where Δx is the distance between the neighbor photons and Δt - the interval between the moments of their emission, then such a combination of photons should form a traveling wave. This assumption follows from an equation describing a field of a neighbor photon irradiated from a point distanced by Δx: $E = E_0 \sin[k(x \pm \Delta x) \pm \varphi_0]$. Taking into account that $k\Delta x = kc\Delta t = \omega\Delta t$ we obtain: $E = E_0 \sin(kx \pm \omega\Delta t \pm \varphi_0)$ - the traveling wave relation. The traveling wave amplitude increases in proportion to the number of equal-phase photons. To provide the traveling wave formation it is necessary that the size of a homogeneous region with inversion, from which a large number of photons is spontaneously irradiated into 4π, exceeded the wavelength λ. Organization of photons in the form of the traveling wave makes it possible to deplete inversion with maximum efficiency, as it is taken off continuously from all points on the path of the wave propagating in the active medium. If the inversion is non-homogeneous and structured with the period $\lambda/2$, such a medium will emit photon in trains, whose maxima should spatially coincide with the inversion maxima. Photons having phases differing by $2\pi n$ form a photon train described by the function: $E_n = E_0 \sin(kx \pm 2\pi n \pm \varphi_0)$. That should not be a traveling wave. `The instant photograph' of the filed distribution in a photon train should represent the contrast periodic structure of antinodes and nodes and can noticeably differ from `the instant photograph' of a traveling sinusoidal wave. However, the summary field produced by many photon trains, which are correlated by the relation $\Delta x = c\Delta t$, becomes "traveling". This "traveling" is like the Christmas-tree garland, whose running lights are due to the time shift of neighboring lamps burning moments.

Because the field produced by the photon is described by a harmonic function, the field superposition of the photon trains propagating at different angles and with different phases gives rise to the wave effects: interference and diffraction. The diffraction is connected with spatial confinement of the light wave front and caused by the interference of photons with different phases contained in the aperture of a beam. The intensity peaks are formed in the directions defined by the beam wave front profile, dimension and shape of a confined region, and the light wavelength. By the angular width of these peaks one may judge about the light beam divergence. The diffraction divergence of the traveling wave arises because the traveling wave restricted by an aperture contains within this aperture at the length of one spatial period the photons with different phases ranging from 0 to 2π. In general the phase dispersion of photons distributed in the interval Δx along the beam direction can be estimated as $\Delta\varphi = 2\pi\Delta x/\lambda$. When the beam is limited by an aperture of diameter d, the direction to the first minimum of the far-field intensity distribution (half the angle α of the diffraction divergence) can be written in the form $\alpha \sim \lambda\Delta\varphi/(2\pi d)$. In the case of a traveling plane monochromatic wave, the phase dispersion $\Delta\varphi$ in the beam can achieve $\pm\pi$. This gives the known estimation $\alpha \sim \lambda/d$ of the divergence angle for a beam limited by an aperture of diameter d. If the phase dispersion of photons is decreased, the diffraction divergence angle α decreases proportionally to $\Delta\varphi$. The group of such phased photons will form a beam with the sharp angular radiation pattern, which was observed in our experiments. Formation of beams with sub diffraction divergence occurs, apparently, in experiments on super-resolution microscopy, when the size of the fluorescent or light scattering object along the direction of beam does not exceed the light wavelength (Galbraith, C.& Galbraith, J., 2011; Tychinskii, V., 2008).

Unlike a usual laser with the extended active medium in the resonator, in our experiments radiation was formed in a structure consisting of thin inversion layers. The regions with reduced pressure behind the front of shock waves, shown in Fig. 15, represent as if the 'instant photograph' of the spatial arrangement of such a structure. The photons with the wavelength resonant with the structure, $\lambda_g/n_g = m\Lambda$, will occupy different spatial positions in the 'photograph' with respect to the layers of thickness $l_{inv} \ll \lambda_g/n_g$. And only a small part of the total number of photons in the 'resonator' of our DFB laser will coincide in position and phase with the inversion maxima in the medium and, therefore, will be amplified efficiently during the lifetime $t_{inv} < T$ of the layers of the given spatial configuration, forming a beam on the basis of photon trains. During the movement of the inversion layers structure (following the movement of the hypersonic wave), new groups of phased photons will be formed, and their frequency will shift with changing the structure parameters. In such a way the model considered above explains the appearance of the sharp angular radiation pattern of the broadband lasing of ytterbium in the crystal and glass. In the case of a 'thick' sinusoidal grating, photons occupying different positions in space (which is equivalent to the large phase dispersion within the photon train) are amplified. This leads to the increase in the radiation divergence. Thus, when a periodic spatial grating of thin inversion layers is produced in the active medium, the high-directional stimulated radiation of an ensemble of excited atoms can be observed in the optical range.

6. Features of ytterbium lasing: line spectra, data interpretation

Line spectra observed in the crystal and glass relate to the second stage of Yb-lasing. The second stage is the Yb-lasing in the resonator after the end of pump due to inversion remained in the medium. Of special interest the details of Yb spectra are. Twisted, inclined and structured lines in spectra were observed in the region 1.03–1.05 µm, Figs.11-14,16. Figures 11, 13, 14 showed an inclination of some lines to the short-wavelength side of the spectrum in respect to the vertical position of the slit. Nearly all line spectra showed quasi-periodic structures of small-scale (50–200 µm) spots (Figs. 11, 14, 16). Twisted lines consisting of separate spots are shown at Fig.16. This picture indicates a wave-like change in the generation wavelength in separate lasing spots along the height of the slit.

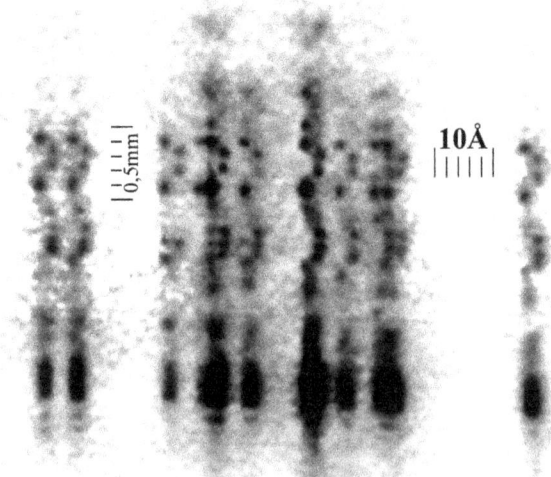

Fig. 16. Twisted lines with multiple spots of Yb:glass generation at several transitions near 1.04 µm; time delay from the pump pulse ≈ 50 ns

Consider the conditions for the development of lasing in a resonator at the end of pump, under disappearance of longitudinal hypersonic waves and relaxation of the excitation region in the active medium. Lasing developed in the resonator in the presence of the region of strong optical inhomogeneity near the active element surface- a thin layer with the inhomogeneous inversion distribution and gradient of the refractive index, n produced by pressure, temperature, and density gradients. To the rest part of the sample the pump penetrated weakly. The refractive index profile was formed during the action of the pump on a medium and it also changed after the end of pump during the relaxation of the excited region. This profile affected the development of Yb-lasing. An experimental study of a non-stationary refractive index profile (so-called "transient lens") arising under short laser pulse interaction with a medium is a rather complicated task. Here we discuss in a qualitative form processes that affected formation of the index profile, $n(r)$ assuming that pumping was symmetrical over the azimuth angle. Consider again a part of the medium in the form of a cylinder of 250 μm diameter and 100 μm length, which fits the size and configuration of the focal region at tight ccl focusing. The heat load occurred in such a cylinder during inversion formation, Yb-lasing, and mainly, due to pump radiation scattering. The heat load was estimated to be <10 mJ, which allows for the average thermal energy density in the cylinder, < 2 × 10^3 J/cm^3. Estimations for the average pressure, P_{av} and temperature, T_{av} just after the end of pump give $P_{av} \approx 1$ GPa and $T_{av} \approx 10^3$ K. In actuality, due to the bell-shaped profile of the pump beam, $P(r)$ and $T(r)$ profiles with maxima exceeding averaged values arose in the medium. The maximum pressure near the axis of the excitation region at the moment near the end of pump pulse may be estimated by using the value P_{pt}, at which a phase transition occurs. Structure modifications similar to phase transitions had been observed in Yb:YAG and Yb:glass samples near the axis, where hypersonic waves of maximum intensity propagated (section 3). We use data for glass $P_{pt} \approx 10$ GPa, $\delta n/\delta P$ =-0.13 × 10^{-5} bar^{-1}, and $\delta n/\delta T$ = –6 × 10^{-6} K^{-1} (Alcock & Emmony, 2002; Koechner, 2006; Mak et al., 1990). At $P(0) \approx P_{pt}$ the drop of $n(r)$ along r from axis to periphery of the cylinder due to the drop in pressure in the medium is $\Delta n(P_{pt}) = \delta n/\delta P \times P_{pt} = 0.13$. Changes of Δn due to temperature gradients are significantly less. For example, at $\Delta T = 10^3$ K, $\Delta n(T) = \delta n/\delta T \times T = -6 \times 10^{-3}$. For YAG $\delta n/\delta T \approx 8 \times 10^{-6}$ K^{-1} (Koechner, 2006; Mak et al., 1990) and $\Delta n(T) \approx 8 \times 10^{-3}$ .Thus, the main contribution to the $n(r)$ profile was made just by pressure. The $n(r)$ profile defines a sign, optical power and aberrations of a "transient" lens in the medium. Optical power of the lens after the end of pump can be estimated at the assumption that the index had been changed by $\Delta n \approx 0.1$ at $\Delta r \approx 150$ μm. That gradient corresponds to appearance in the medium of a positive lens with diameter $2r \approx 300$ μm and $f \approx 1$cm. A simplified scheme for the case of lasing in the resonator with a "transient" lens is given in Fig. 17.

The evolution of profiles $P(r)$ and $n(r)$ after pumping was determined by unloading of the high-pressure region in the focal volume. A problem of unloading of a small cylindrical or spherical region in a solid-state optical medium, where the energy had been stored after ns laser pulse irradiation, had been considered in several studies (Bullough & Gilman, 1966; Conners & Thompson, 1966; Sharma & Rieckhoff, 1970). A typical process of relaxation is the propagation of an elastic dilatational wave across the medium. The time of traveling, t_{tr} of an elastic wave with the sound velocity across the excitation region can be evaluated by t_{tr} =r/v_s. For the glass, at $r_p = 250$ μm and $v_s = 4.5 \times 10^5$ cm/s, $t_{tr} \approx 50$ ns. The propagation of the dilatational wave along radial direction outward the center of the pumped region had led to the pressure profile deformation. With the drop of pressure at the axis, the gradient dn/dr

decreased accordingly. So, in the course of relaxation there occurred smoothing of profiles $P(r)$ and $n(r)$. It should be noted that profile $n(r)$ could include a small-scale modulation as well (imitated at Fig.17). The sources for such modulation might be pressure and density perturbations that had been arising in a medium at propagation of hypersonic waves during and after pumping. Note, that typical time of thermal relaxation of the medium in a cylindrical region with radius r_p is $\tau = r_p^2/4\mu \gg t_{tr}$. Here r_p is the radius of the pumped region and μ is the coefficient of the temperature diffusivity. For Yb:YAG at $r_p = 250$ µm, $\mu = 0.046$ cm²s⁻¹ (Koechner, 2006), $\tau \approx 3$ ms. Hence, the time of thermal relaxation of medium is several orders higher than dynamic unloading. That is, roughly, a picture of formation and evolution of the refractive index profile in the medium in the region of ccl focusing defined by a non-uniform heat load and dynamics of the pressure profile during pumping and under the medium relaxation. So, a sequence of acoustic waves diverging in radial direction outward the focal region should affect the build-up of Yb-lasing in the resonator.

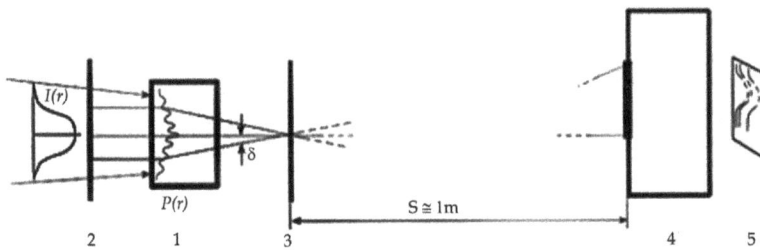

Fig. 17. Scheme of the experiment on excitation of Yb-doped medium in the resonator by focused ccl radiation: 1 - active element; 2, 3 - mirrors; 4 - spectrograph STE-1 located at distance S≈1m from the resonator; 5 - a picture imitating spectra registered; δ –angle between a generation trajectory and resonator axis. The dimensions of the resonator with active element are exaggerated relative to the scale of the scheme. Profiles of pump beam intensity, $I(r)$ and pressure in the medium, $P(r)$ are out of scale as well

The interpretation of data on the Yb-lasing in the resonator has required also the introduction of a new concept for the photon electromagnetic field distribution in space, considered in section 5. This new approach allows the existence of light beams of an aperture $d \approx \lambda$ and with a "sub-diffraction" angle of divergence. Within the framework of this new concept the picture of Yb lasing in the resonator looks as follows. The location of line spectra in the upper part of the slit evidences that Yb radiation emergent from the resonator was deflected from the axis, i.e., off-axis oscillations were built up in the resonator. For the arrangement of mirrors and active element given in our experiment off-axis lasing could be developed over trajectories with reflections at angles δ_1, δ_2 ... at the mirror (3), Fig. 17. Such trajectories located at distances $r_{pt} < r < r_p$ from the axis, should contain curvilinear segments in the area of the optical inhomogeneity. We suppose that beams with aperture of several λ (small generation channels) constituted lasing along these off-axis trajectories. At the high value of dn/dr and the very high level of inversion, in each of the channels the selection of emitted photons (in accordance with their phases) took place in such a way that only a group of photons being approximately in the same phase φ_0 overcame the generation threshold. The photons different by phase from φ_0 were scattered due to diffraction and left the channel aperture. In the given channel with a coordinate r a group of photons of the same phase made several round trips in the resonator during the time t', and a beam of the high directivity with the generation wavelength λ_g corresponding to one of the longitudinal modes was formed. So, generation in different channels over the pumped area developed at their own frequencies, and this radiation, reaching the slit at different points gave us an integrated picture of lasing beams and the spectrum in the form of lines composed of small spots, which dimensions characterized beam divergence in each of the channels.

For Yb-doped plates placed in the middle of the resonator with $L = 20$ mm and for trajectories located in plates at $r < 500$ µm, angles δ should lie inside the spectrograph registration angle $\approx10^{-2}$ rad. Beams of generation left resonator at angles δ and reached the slit. The greater was the distance r between the generation trajectory inside the sample and the resonator axis, the greater was the shift of the corresponding generation beam spot upward in vertical direction on the slit, Fig. 17. So, under the propagation of beams of high-directivity from resonator at angles δ, one, actually, observed at the slit a projection (a magnified "image") of a spatial distribution of generation channels in the active medium over the radial coordinate. Since data on the Yb emission intensity distribution over the azimuth angle were absent, one can speak only about partial mapping on the slit of the spatial distribution of generation channels in the lasing area. The magnification coefficient k of such an image may be estimated from comparison of dimensions of the possible generation area ($r \leq 500$ µm) and the height of line spectra at the slit (≈1 cm), $k \geq 20$. With account of k factor, to observed structures of small-scale spots in line spectra should correspond sources of Yb emission with dimensions of several wavelengths only.

Structures of multiple spots in Yb:YAG and Yb:glass lines (Figs. 11, 14, 16) may find explanation when one takes into consideration the specific spatial configuration of the field of thermo-elastic stresses in the area of ns laser pulse focusing into the medium. The quasi-periodic, alternating in sign, oscillating character of the amplitude of the tangential component of stress in dependence on the radial coordinate inside a small spherical region of a solid-state optical medium at ns laser pulse focusing was ascertained (Conners & Thompson, 1966; Sharma & Rieckhoff, 1970). In spherical ring zones round the centre of the focal region the tangential tensile stresses in the medium are consequently replaced by compressing stresses, and then, again, by the tensile ones, etc. The spatial period of such oscillations calculated for the case of ≈70 mJ laser pulse focusing into the glass constituted 30 µm (Sharma & Rieckhoff, 1970). In conditions of our experiment one can expect the occurrence of oscillating (in space and time) profiles of thermo-elastic stresses in Yb-doped media within the area of ccl radiation focusing as well. Alternating in sign stresses should result in a small-scale modulation at the profile $n(r)$, Fig.17. The oscillation of the $n(r)$ profile should stimulate the Yb generation in ring zones, which may fall into separate generation channels. The structures of bends consisting of generation spots, Figs.16 may be considered as a kind of an "image" of the distribution over the radial coordinate of tangential stress peaks in the focal region projected with a magnification on the slit. It is possible to say that such images were taken by the high-speed photography method. The "illumination" for this high-speed photography came from the Yb laser pulse itself. The "exposure" time of a single frame corresponded to the duration of the generation pulse. Time delay between the pump pulse and the "shooting" moments makes ≈ 50ns (Figs. 16) and over 300 ns (Fig.14). Splitting of a single bend into several generation spots with wavelength shifts (Fig.16) is in agreement with the model of generation channels. The shift of a spot to the long-wavelength side corresponds to increase of pressure (stresses) in the medium, and the shift to the short-wavelength—to decrease of pressure. In the course of medium relaxation, stress amplitudes and, correspondingly, spot wavelength shifts should have been reduced. At ≈50ns delay (twisted lines, Fig.16) spots have noticeable wavelength shifts which correspond to high stress amplitudes. Structures at moments over 300 ns after pump (Fig.14) correspond to a smoothed picture of $n(r)$ profile. Note, that structures recorded with such big delays indicate the continued acoustic "ringing" in the medium. Estimated from Fig. 11, 14, 16 (at $k = 20$) period of stress spatial oscillations in YAG and glass varies from 15 to 40 μm and is in a qualitative agreement with the calculations (Sharma & Rieckhoff, 1970). Acoustic vibrations frequencies corresponding to these values constitute 10^8–10^9 Hz. Attenuation of phonons at these frequencies at a room temperature for YAG is smaller than 0.1 db/μs and for glass— about 10 cm^{-1} (Dutoit, 1974; Zhu et al., 1991). These data confirm that "ringing" of the unloading medium in the focal region in YAG and glass may continue over several μs.

The Yb:YAG line spectra near 1,03 μm were registered usually with 10÷50 ns delays after the end of ccl pulse and even together with the trailing edge of the ccl pulse. Observations of line spectra near 1,03μm, emitted soon after the pump pulse, like Fig.11, reveal noticeable inclination of spectral lines. This bending means that the wavelength of lasing at the same longitudinal mode (the number of nodes is preserved) changes from the centre to periphery of the excited region in the active medium. This corresponds to the development of lasing in some sites of the medium with a pressure (refractive index) gradient from the beam centre to its periphery. It can be easily shown that λ_g shifts to the blue if the refractive index gradient decreases from the beam axis to its periphery, Fig. 11. The pressure drop ΔP in the medium after the end of pump can be estimated from this line spectral shift. The frequency ω_q of the longitudinal mode of the resonator with the number of wavelengths λ_q over the resonator length $2L$ equal to q is described by the expression

$$\omega_q = \frac{\pi c q}{(L - \Delta l)n_0 + \Delta l \cdot n(r)} \tag{2}$$

where Δl is the longitudinal size of the optical inhomogeneity of radius r; n_0 is the averaged refractive index outside the nonlinearity region; and $n(r)$ is the refractive index in the nonlinear region. It follows from (2) that the change in the mode frequency ω_q during the displacement along the radius from r_1 to r_2 is

$$\Delta\omega_q = \frac{\pi c q \Delta l[n(r_1) - n(r_2)]}{[(L - \Delta l)n_0 + \Delta l \cdot n(r_1)][(L - \Delta l)n_0 + \Delta l \cdot n(r_2)]} \tag{3}$$

By substituting the expression for $q = \dfrac{2[(L - \Delta l)n_0 + \Delta l \cdot n(r)]}{\lambda_q(r)}$ into (3) and assuming that

$\Delta l n(r) < L n_0$, we obtain the dependence of the change in the refractive index, Δn on r

$$\Delta n = \frac{\Delta\lambda_q(r)L n_0}{\lambda_q \Delta l} \tag{4}$$

Here, $\Delta\lambda_q(r)$ is the wavelength shift along the radius. For $\Delta\lambda_q(r) \approx 1.4 \times 10^{-7}$ cm, $\lambda_q \approx 10^{-4}$ cm, L ≈ 2 cm, $n_0 \approx 1$, and $\Delta l \approx 10^{-2}$ cm, the change in the index is $\Delta n \approx 0.28$. By assuming that the change in the index is produced only by the change in pressure along the radius and using the value of dn/dP for glass (Alcock & Emmony, 2002), we obtain $\Delta P \approx 30$ GPa. So, this estimates the pressure, which arises at the axis of the focal region after the end of pump. Similar estimations are possible for the pressure jumps, which arise during the propagation of dilatational waves outward the center of the focal region. Knowing a value of the wavelength shift in a bend at Fig.16, $\Delta\lambda_q$ (r), one can estimate changes of the index, Δn and pressure, ΔP in the medium, using the expression (4). As seen from Fig.16, for glass at $\Delta\lambda_q$ (r) $\approx 0.3 \times 10^{-7}$ cm, $\lambda_q \approx 10^{-4}$ cm, $L \approx 2$ cm, $n_0 \approx 1$, $\Delta l \approx 10^{-2}$ cm, $\Delta n \approx 0.05$, and $\Delta P \approx 5$ GPa. The estimated ΔP considerably exceeds the glass fracture strength < 0.1 GPa (Sharma & Rieckhoff, 1970), which was measured usually for applied static load. The role of tensile stresses in laser damage of transparent dielectrics was discussed in many publications (Koldunov et al., 2002; Sharma & Rieckhoff, 1970; Strekalov, 2000). It was considered that due to oscillating character of the stress amplitude in the focal area there must be observed laser damage of the medium in the form of periodically spaced spherical rings (Sharma & Rieckhoff, 1970; Strekalov, 2000). It is known that material strength sharply grows under pulsed load as compared to the static load. Under the high-speed deformation (in the ns range), strength of material becomes comparable to the theoretical limit-tens GPa (Kanel et al., 2007). So a fast periodic change of the stress sign in the medium should decelerate the development of material destruction in the form of rings. This is the reason why the laser damage in the form of multiple rings usually was not observed in many experiments when ns pulses of laser radiation were focused into the volume of transparent dielectrics. The damage in the form of rings was not observed in our experiments as well. One of the few observations of the multiple ring damage (Martinelli, 1966) relates to the case when glass samples were exposed to focused free-running laser radiation. Anyway, according to calculations (Sharma & Rieckhoff, 1970) oscillating profile of thermo-elastic stresses should occur in the region of ns laser pulse focusing. The presented material provides experimental data which confirm in the quality form the calculated (Sharma & Rieckhoff, 1970) picture of thermo-elastic stresses distribution and elastic wave propagation across the medium.

7. Conclusion

The study of the interaction of powerful ns pulses of low-coherence radiation of the LiF: F_2^+ color center laser (ccl) with optical materials (Yb:YAG, glass, et al.) was carried out. Efficient SBS of low-coherence pump, accompanied by SRS and formation of hypersonic waves reaching the intensity of shock waves were found. A physical model of excitation of SBS and hypersonic waves at scattering of ultrashort pulses of low-coherence pump at stationary inhomogeneities in optical materials is presented. It is shown that ns laser pulse, whose duration is much higher than its inverse spectral width, causes SBS much more efficient than a pulse of high coherence with the same duration and energy. Unlike SBS of a coherent radiation caused by a pressure fluctuation, scattering of low-coherence pump may be caused by any stationary inhomogeneities in a medium: cracks, dislocations, micro-inclusions, or just by a plane back surface of a sample. An effective energy contribution of light pulses into hypersonic waves on a small coherence length near the input surface of a sample leads to their transformation into a periodic succession of high-pressure shock waves, which results in structure changes of a crystal lattice (phase transition) in that region. The appearance of structural changes in optical materials that are specific to the interaction of powerful pulses of low coherence radiation with matter was found. The mechanisms of structural changes based on the action of intense hypersonic waves were considered.

Nanosecond pulses of Yb lasing in the region 1.00-1.06 µm with the spectral width up to 20 nm in Yb :YAG and 50 nm in the Yb: glass samples were observed. The divergence of the broadband laser radiation (10^{-3} -10^{-4} rad) was one or two orders of magnitude smaller than the diffraction limit respectively to the source of Yb radiation in a sample. The mechanism of generation of broadband laser pulses of short duration and high directivity in the spatial structure of thin layers with inversion produced in the region of the propagation of intense hypersonic waves in the medium is discussed. The interpretation of experimental data is based on a new concept of the spatial distribution of the electromagnetic field of a photon not in the form of a "traveling" wave but with the field structures located in fixed positions along the photon propagation direction. The new approach allows the existence of light beams of an aperture $d \approx \lambda$ and with a "sub-diffraction" angle of divergence. Such beams must consist of groups of phase-synchronized photons with a small phase difference distanced by an interval $\approx \lambda$. This synchronized group is no longer a "traveling" wave, its angle of divergence is defined by the phase difference of photons in a group.

Spectral lines in 1.03–1.05 µm region structured by 50–200 µm spots as well as lines with inclinations were found at Yb lasing in a resonator. Structures of multiple spots in spectra reflect the specific spatial configuration of the field of thermo-elastic stresses in the unloading region of ccl pulse focusing after the end of the pump. Inclinations of spectral lines reflect the pressure gradient from the center to the periphery of the region of ccl focusing. Basing on inclinations of spectral lines the pressure in the region of shock hypersonic wave propagation was estimated. Estimations show that for some of the studied media the pressure values may exceed the phase transition threshold.

8. Acknowledgment

Authors thank O. Yaremchuk for the help in preparing this article for publication.

9. References

Alcock, R. & Emmony, D. (2002). Sensitivity of reflection transducers *J. Appl. Phys.*, Vol. 92, No.3 (August 2002), pp. 1630-1643, ISSN 0021-8979.

Basiev, T. et al. (1982). Solid-state tunable lasers based on color centers in ionic crystals *Bulletin of the Academy of Sciences of the USSR, ser. phys.*, Vol. 46, No.8, pp.145-154.

Basiev, T.T, Bykovsky, N.E, Konyushkin, V.A & Senatsky, Yu.V. (2004). Use of a LiF colour centre laser for pumping an Yb:YAG active medium. *Quantum. Electron.* Vol.34, No.12, pp. 1138-1142, ISSN 0368-7147.

Bor, Z. & Muller, A. (1986). Picosecond distributed feedback dye lasers *IEEE J. Quantum Electron.*, 22 (8), 1524-1533. ISSN: 0018-9197.

Bullough, R. & Gilman, J.J. (1966). Elastic Explosions in Solids Caused by Radiation. *J. Appl. Phys.* Vol.37, pp.2283-2288 , ISSN 0021-8979.

Bykovsky, N. & Senatsky, Yu. (2008a). Broadband collimated generation in YAG:Yb crystal and ytterbium glass under LiF:F2+color center laser pumping. *Laser Phys. Lett.* Vol.5, Iss. 9. pp. 664–670, ISSN 1612-2011.

Bykovsky, N.& Senatsky, Yu. (2008b). Spectra, temporal structure, and angular directivity of laser radiation of a Yb:YAG crystal and ytterbium glass pumped by low-coherence radiation from a F2+:LiF colour centre laser. *Quantum Electron.* Vol.38, No.9, pp. 813-822, ISSN 0368-7147.

Bykovsky, N.E. & Senatsky, Yu.V. (2010). Twisted lines and small-scale structures in generation spectra of Yb-doped media excited by focused radiation of LiF:F 2+ color center laser. *Laser Physics* Vol.20, No.2, pp. 478-486, ISSN 1054-660X.

Bykovsky, N.E. (2005). *Preprint FIAN* No. 16. (http://ellphi.lebedev.ru/12/pdf16/pdf).

Bykovsky, N.E. (2006). *Preprint FIAN* No. 36. (http://ellphi.lebedev.ru/17/pdf36/pdf).

Conners, G. & Thompson, R. (1966). A Continuum Mechanical Model for Laser-Induced Fracture in Transparent Media. *J. Appl. Phys.* Vol.37, pp.3434-3441, ISSN 0021-8979.

Dutoit, M. (1974). Microwave phonon attenuation in yttrium aluminum garnet and gadolinium gallium garnet. *J. Appl. Phys.* Vol.45, pp.2836-2841, ISSN 0021-8979.

Galbraith, C. & Galbraith, J. (2011). Super-resolution microscopy at a glance *Journal of Cell Science* Vol.124, №10, pp.1607-1611 ***ISSN***: 0021-9533

Gordienko,V., Mikheev P. & Potemkin F. (2010). Generation of coherent terahertz phonons by sharp focusing of a femtosecond laser beam in the bulk of crystalline insulators in a regime of plasma formation.*JETP Letters*, Vol.92, No.8, pp.502-506, ISSN 0021-3640.

Gorelik, V. (2007). Optics of globular photonic crystals. *QuantumElectron*, Vol.37(5), pp. 409-432, ISSN 0368-7147.

Kanel, G., Fortov, V. & Razorenov, S. (2007). Shock waves in condensed-state physics. *Phys. Usp.* Vol.50, pp.771-791, ISSN: 0038-5670.

Katarkevich, V., Kurstak V., Rubinov, A., & Efendiev T. (1996). Kinetics of the operation of a distributed-feedback dye laser with nanosecond excitation. *Quantum Electron.*,Vol. 26, pp.1061- 1064, ISSN 0368-7147.

Koechner, W. (2006). *Solid State Laser Engineering*, 6th ed., Springer, Berlin.

Kogelnik, H., & Shank, C. (1971). Stimulated emission in a periodic structure. *Appl. Phys. Lett.*, Vol.18 (4), pp.152-155, ISSN 0003-6951.

Koldunov, M., Manenkov, A. & Pokotilo, I. (2002). Mechanical damage in transparent solids caused by laser pulses of different durations. *Quantum Electron.*, Vol.32. No.4, pp.335–340, ISSN 0368-7147.

Krupke W.F. (2000). Ytterbium solid-state lasers. The first decade *IEEE J. Sel. Top. Quantum Electron.*, Vol.6 (6), pp.1287 – 1296, ISSN1077-260X.

Mak, A & Soms, L. (1990). *Nd:Glass Lasers*, Nauka, Moscow.

Manenkov, A. & Prokhorov, A. (1986). Laser-induced damage in solids. *Sov. Phys. Usp.*, Vol.148, pp.104–122. ISSN: 0038-5670.

Martinelli, J. (1996). Laser-Induced Damage Thresholds for Various Glasses. *J. Appl. Phys.*, Vol.37, pp.1939-1941, ISSN 0021-8979.

Merlin R. (1997). Generating coherent THz phonons with light pulses. *Solid State Commun.*, Vol.102, pp.207-220, ISSN 0038-1098.

Nelson K.A., Miller R.J.D. & Fayer M.D., (1982). Optical generation of tunable ultrasonic waves. *J. Appl. Phys.*, Vol.53, pp.1144-1149, ISSN 0021-8979.

Polyakova A.L. (1968). Elastic Nonlinearity in Stimulated Mandel'shtam-Brillouin Scattering. *JETP Letters*, Vol. 7, Iss. 2, pp. 57-59, ISSN: 0021-3640.

Polyakova, A. (1966). Nonlinear Effects in a Hypersonic Wave. *JETP Letters*, Vol.4, Iss. 4, p.90-92, ISSN: 0021-3640.

Ready, J. F. (1971). *Effects of High Power Laser Radiation*, Academic, New York, London.

Ritus, A. (1982). Study of Mandel'shtam-Brillouin light scattering in crystals and glasses respecting to problems of quantum electronics and fiber optics. *Trudy FIAN* 137, pp. 3-80.

Robinson M.M., Yan Y.-X., Gamble E.B., Williams Jr., L.R., Meth J.S, & Nelson K.A. (1984). Picosecond impulsive stimulated Brillouin scattering: Optical excitation of coherent transverse acoustic waves and application to time-domain investigations of structural phase transitions. *Chem.Phys. Lett.*, Vol.112, pp.491-496, ISSN 0009-2614.

Sakakura, M., Terazima, M., Shimotsuma, Y., Miura, K. & Hirao, K, (2007) Observation of pressure wave generated by focusing a femtosecond laser pulse inside a glass *Opt. Exp.* 15, 5674-5686, ISSN 1094-4087.

Sharma B.S. & Rieckhoff, K.E. (1970). Laser-induced dielectric breakdown and mechanical damage in silicate glasses. *Can. J. Phys.* Vol.48(10), 1178–1191, ISSN 0008-4204.

Starunov V.S. & Fabelinskii I.L. (1970) Stimulated Mandel'shtam-Brillouin scattering and stimulated entropy (temperature) scattering of light. *Sov. Phys. Usp.*, Vol.13, pp.428–428. ISSN: 0038-5670.

Strekalov, V. (2000).Mechanical damage of transparent dielectrics by focused laser radiation *J. Tech. Phys. Lett.* Vol.26, No. 24, pp.19-22. ISSN: 0320 - 0116 .

Stuart, B, Feit, M, Rubenchik, A., Shore, B. & Perry, M. (1995). Laser-Induced Damage in Dielectrics with Nanosecond to Subpicosecond Pulses. *Phys. Rev. Lett.* Vol.74, pp.2248–2251, ISSN 0031-9007.

Suguira, H., Ikeda, R., Kondo, K. & T. Yamadaya, T. (1997). Densified silica glass after shock compression. *J. Appl. Phys.*, Vol.81, pp.1651-1656, ISSN 0021-8979.

Tychinskii, V. (2008). Super-resolution and singularities in phase images *Phys.Usp.* Vol.178, №11, pp. 1205-1214. *ISSN*: 0038-5670.

Zhu, T., Maris, H., & Tauc, J. (1991). Attenuation of longitudinal-acoustic phonons in amorphous SiO_2 at frequencies up to 440 GHz. *Phys. Rev. B* Vol.44, pp.4281–4289, ISSN 1098-0121.Dictionary -

4

An Optimal Distribution of Actuators in Active Beam Vibration – Some Aspects, Theoretical Considerations

Adam Brański
Rzeszow University of Technology
Poland

1. Introduction

The reduction of the effects of mechanical vibration fall into the of vibration isolation, design for vibration or vibration control (de Silva, 2000). The vibration control is subdivided into two group: passive control and active one. The core of the vibration control is to detect the level of vibration in a system and to counteract the effects of the vibration, so it needs two devices.

Hence, the passive devices do not require external power for their operation. Hence, passive control is relatively simple, reliable and economical. But it has limitations namely, the control force depends entirely on the natural dynamics and it may not be adjust on line. Furthermore, in a passive device, there is no supply of power from an external source. It leads to the incomplete control, particularly in complex and high-order systems.

The shortcomings of passive control can be overcome using an active one. In this case, the system response is directly sensed on line and on that basis, the specific control actions are applied to any locations of the system. But the active control needs external power, namely to apply control forces to vibrating system through actuators and to measure vibration response using sensors.

Two different types of actuators can be applied (Shimon et al., 2005). The first, inertial actuators, make up a piezoelectric material to vibrate large masses. Their vibrations are used to counteract the vibrations of the structure (Jiang et al., 2000). The advantages and disadvantages are enumerated in above reference.

The second type of actuators is a layer of smart or intelligent materials. The sensors also belong to these materials; together they are well–known as piezoelectric elements (Tylikowski & Przybyłowicz, 2004). It was shown that these elements can offer excellent potential for an active vibration reduction of the structure vibrating with low frequencies (Croker, 2007; Fuller at al, 1997; Hansen & Snyder, 1997; Kozień, 2006; Przybyłowicz, 2002; Wiciak, 2008). As a general, piezoelectric elements are glued to the host structure. It makes the advantage, namely their incorporating into the structure is that the actuating mechanism becomes part of the structure. Both sensors and actuators are relatively light, compared to the structure, and can be made in arbitrary shape. The disadvantage is that they once bonded and they cannnot be used again. In recent years the measure of the vibration with the sensors are replaced by touch less measures. For this reason, hereafter in research the sensors are omitted and only second type actuators will be considered. Nowadays actuators are used to very original structures for example to the satellite boom (Moshrefi-Torbati et al., 2006) or to sun plate (Qiu et al., 2007).

To make the reduction more effective, many problems should be solved.

- dynamic effects (mass loading and stiffness) of the actuators on the structure vibration (Charette et al., 1998; Gosiewski & Koszewnik, 2007; Hernandes et al., 2000; Q. Wang & C. Wang, 2001)
- dynamic effects of the glue (between actuators and structure) on the structure vibration (Pietrzakowski, 2004; Sheu et al., 2008).
- actuators' geometric-technical features (Frecker, 2003; Hong et al., 2007; Wang, 2007),
- orientation of the actuators on the structure (Bruant et al., 2010; Ip & Tse, 2001; Qiu et al., 2007),
- appropriate actuators distribution on the structure (Bruant et al., 2010),
- others, but they play a minor part.

Reviewing the literature, it appears that the actuators distribution play a major part. Now, a question arises about an optimal distribution of actuators. In the recent year, a great number of papers has been published on this subject. It is obvious that there are a lot of optimization techniques; an excellent survey is given in (Bruant et al., 2010). Two main approaches are distinguished to this problem.

First of them is the coupling of the optimization of actuators/sensors locations and controller parameters. In this case the following criterions are taken into account for the optimization:

- quadratic cost function of the measure error and the control energy (Bruant et al., 2001),
- maximization of dissipation energy during the control (Yang, 2005),
- spatial H_2 norm of the closed-loop transfer matrix from the disturbance to the distributed controlled output (Liu et al., 2006),
- simultaneous simple H_∞ controller (Guney & Eskinat, 2007).

As can be seen, the optimization criterions are dependent on the choice of controllers. Therefore, the optimal location obtained using one controller may not be a suitable choice for another one.

At the latter approach, the optimal location is obtained independently of the controller definition. In this case, the following criterions are used:

- maximization controllability/observability criterion using the gramian matrices (Bruant & Proslier, 2005; Jha & Inman, 2003),
- modal controllability index based on singular value analysis of the control vector (Dhuri, & Seshu, 2006),
- maximization of the control forces transmitted by the actuators to the structure (Q. Wang & C. Wang, 2001),
- using the H_2 norm (Halim & Reza Moheimani, 2003; Qiu et al., 2007).

In the quoted references, it was not provided the actuators distribution in explicite; only the general rules (criterions) were formulated. However, this problem was partially solved; it was proved in (Brański & Szela, 2007; Brański & Szela, 2008; Szela, 2009; Brański & Szela 2010; Brański & Lipiński, 2011) that the most effective actuators distribution was on the structure sub-domains with the largest curvatures; such distribution was called quasi-optimal one. As the research object, a right-angled triangle plate with clamped-free-free boundary conditions was taken into account. The quasi-optimal distribution was deduced based on the heuristic reasons and the conclusions were confirmed only numerically. Furthermore, the problem was solved merely for the separate modes.

Basing on the quasi-optimal distribution of the actuators, the protection beam vibration is achieved (Brański et al., 2010; Brański & Lipiński, 2011). In this case always the separate modes were considered. The problem was solved based on heuristic reasons and was confirmed analytically. In the latest own research, the results presented in (Brański et al., 2010) were substantiated analytically (Brański & Lipiński, 2011).

In this chapter, the above attitude to the optimal actuators distribution is continued and extended. First at all, the optimal problem is formulated. For this purpose, the optimization criterion is defined. It is assumed that a measure of the vibration reduction is a reduction coefficient (Szela, 2009; Brański & Szela 2010) and here it becomes the objective function. This attitude is quite similar to the maximization of the control forces transmitted by the actuators to the structure (Q. Wang & C. Wang, 2001).

Dynamics effects of the glue and actuators are also considered. Furthermore, the solution of active vibrations reduction is derived for general solution, not only for separate modes. Since analytical solution was attained with separation of variables method, first of all the modes of the problem are derived. Next, the orthogonality condition of the modes is derived too.

The simple supported beam is chosen as the research object. The study of beams is very important in a variety of practical cases, noteworthy, the vibration analysis of structures like bridges, tall buildings, and so on. Loosing a bit on generality, it is considerably easier to realize the aim of the paper. It is assumed that the beam is excited with evenly spread and harmonic force. The material inner damping coefficients of all elements of the research system are taken into account. It seems that all main factors having the influence on the beam vibration were considered.

To solve the problem analytically, a few simplifications are made. Namely, the energy provided to the system is in the form of voltage applied to the surface of the actuators. Assuming that the charge is homogenously distributed, as a result of piezoelectric effect, the actuators interact with the beam with moments for couple of forces homogenously distributed along the actuators' edges. Next, these moments are replaced with the couple of forces and finally, they are counteracted the vibrations.

All problems were considered only theoretically; no calculations are run. It seems that presented considerations will be the base to many numerical simulations and experiments. To the author's knowledge, the theoretical description of the optimal actuators distribution on even simple structure like the beam, up to now have not been brought up.

2. Active beam vibration reduction with additional elements

In this problem, the additional elements make the concentrated masses and actuators and all constitute the mechanical set beam-actuators-masses. Adding actuators (and the glue at the same time) is the technical necessity but they introduce to the mechanical set the additional dynamics effects namely, local stiffness and concentrated masses. As far as concentrated masses are concerned, adding them is substantiated as follows. The proposed optimal distribution of the actuators needs asymmetrical beam vibrations and these ones may be ensured by at least one concentrated mass.

2.1 Uniform beam vibration with damping

There are four theories (models) for the transversely vibrating uniform beam (Han et al., 1999): Euler-Bernoulli, Rayleigh, shear and Timoshenko. The first of them, called the classical beam theory, is applied here. It is simple and provides reasonable results for formulated problem.

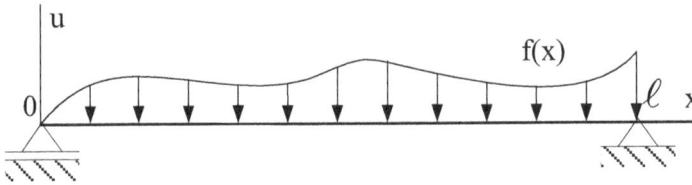

Fig. 1. The geometry of the simple supported beam

Let be the beam as depicted in Fig. 1. The Bernoulli-Euler equation governs transverse vibration (or bending or lateral vibration) of the beam has a following standard form (Kaliski, 1986; Pietrzakowski, 2004),

$$EJD^4u + \mu EJD^4(D_tu) + \rho SD_t^2u = -f \tag{1}$$

where $u = u(x,t)$ – beam deflection at the point x and the time t, $f = f(x,t)$ – load force, $D^4(.) = \partial^4(.)/\partial x^4$, $D_t(.) = \partial(.)/\partial t$; hereafter the rest symbols are jointly explained.
To solve Eq. (1) explicitly, four boundary conditions, at the ends of the beam, are needed. In general, boundary conditions represent displacement, slope, moment and shear respectively. Here, it is assumed that the beam is simple supported, then both displacement and the bending moment equal zero

$$u(0,t) = 0, \qquad D^2u(0,t) = 0 \tag{2}$$

$$u(\ell,t) = 0, \qquad D^2u(\ell,t) = 0 \tag{3}$$

To solve over determined problem, one needs to know initial conditions. But here, the harmonic steady state plays a major part, so that the initial conditions are omitted.

2.2 Beam vibration with concentrated masses

To solve the intended problem, Eq. (1) must be rounded out. First of all, to obtain asymmetric modes and consistently asymmetric general vibration, a few concentrated masses are added to the beam (Low & Naguleswaran, 1998; Majkut, 2010; Naguleswaran, 1999). They are marked by $\{m_r\}$, and their distribution is described with set of coordinates $\{x_r\}$, see Fig. 2, hence

$$\sum_r m_r \delta(x - x_r) = m_1 \delta(x - x_1) + m_2 \delta(x - x_2) + ... + m_r \delta(x - x_r) + ... \tag{4}$$

where $r = 1,2,...,n_r$, $\delta(.)$ – Dirac's delta function.

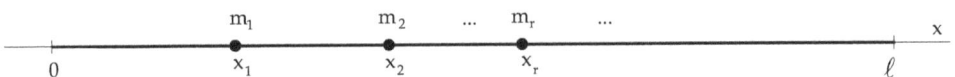

Fig. 2. Distribution of the concentrated masses

Furthermore, the dynamic effects of the actuators and glue on the beam vibration are introduced. The location and length of separate actuators, and the glue layers simultaneously, are denoted commonly with coordinates $\{x_s\}$ and $\{\ell_s\}$ respectively and they are arranged as depicted in Fig. 3.

Fig. 3. Distribution of actuators and glue layers

For simplicity, let $P = \{E, J, h, \rho, S, \mu\}$ means the physical and geometrical parameters of the beam, actuators and glue, i.e. {Young's modulus, surface moment of inertia, thickness, mass density, surface of the rectangular cross-section, inner damping factor} respectively. Furthermore all parameters are supplemented with following index $\vartheta = \{b, a, g\} = \{[b]eam, [a]ctuator, [g]lue\}$, for example $S_\vartheta = bh_\vartheta$ means the surface of the rectangular cross-section, b – beam / glue layer width. Moments of inertia are calculated relatively of y-axis, see Fig. 4, where the neutral axis displacement d is neglected , hence $J_b = (bh_b^3)/12$, $J_g = (bh_g^3)/12 + S_g(h_b/2 + h_g/2)^2$, $J_a = (bh_a^3)/12 + S_a(h_b/2 + h_g + h_a/2)^2$.

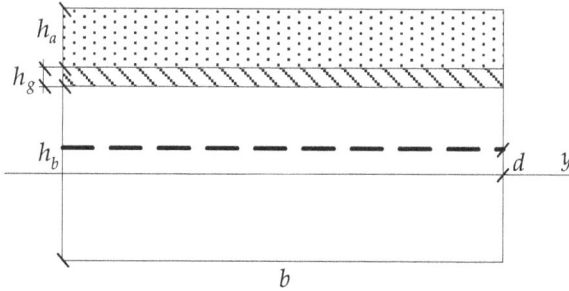

Fig. 4. Cross-sections of the set beam–actuator–glue

The parameters of the set beam-actuators-glue may be written as

$$P = P_b + \sum_s P_s H(x_{1s} - x_{2s}) = P_b + \sum_s P_s \langle H \rangle^0 \tag{5}$$

where $s = 1, 2, \dots, n_s$, $P_s = P_a + P_g$, $\langle H \rangle^0 = H(x_{1s} - x_{2s}) = H(x - x_{1s}) - H(x - x_{2s})$, $H(x - x_{1s})$ – Heaviside step function in point x_{1s} and so on, $\{x_{1s}, x_{2s}\} = \{x_s - \ell_s/2, x_s + \ell_s/2\}$. For n_s aktuators (n_s glue layers) and n_r concentrated masses, Eq. (1) takes the form

$$\left(E_b J_b + \sum_s E_s J_s \langle H \rangle^0\right) D^4 u + \left(\mu_b E_b J_b + \sum_s \mu_s E_s J_s \langle H \rangle^0\right) D^4(D_t u) + \left(\rho_b S_b + \sum_s \rho_s S_s \langle H \rangle^0\right) D_t^2 u +$$
$$+ \sum_r m_r \delta(x - x_r) D_t^2 u = -f \tag{6}$$

The Eq. (6) may be written down quite similar like Eq. (1), namely

$$EJ D^4 u + \mu EJ D^4 (D_t u) + (\rho S + \alpha_r) D_t^2 u = -f \tag{7}$$

where hereafter

$$EJ = E_b J_b + \sum_s E_s J_s \langle H \rangle^0, \qquad \mu EJ = \mu_b E_b J_b + \sum_s \mu_s E_s J_s \langle H \rangle^0,$$

$$\rho S = \rho_b S_b + \sum_s \rho_s S_s \langle H \rangle^0, \qquad \alpha_r = \sum_r m_r \delta(x - x_r) \tag{8}$$

On the ground of the EJ, ρS and α_r form, Eq. (7) can not be understood in a classical manner. To solve it, some methods may be applied. One of them is presented in (Ercoli & Laura, 1987; Kasprzyk & Wiciak, 2007; Majkut, 2010); another attitude may be found in (C.N. Bapat & C. Bapat, 1987) and it is applied here.

Fig. 5. Geometry of the set beam–one actuator–one mass

At the latter attitude, the beam is divided into some uniform elements. The division may not be coincidental. To clearly explain this problem, for simplicity consider a set beam-one actuator (and glue)-one concentrated mass, Fig. 5. The division is imposed out of the change of physical properties namely, properties of the actuators (and glue) and concentrated masses. So, the beam is divided into $j = 1, 2, ..., n_j = 4$ elements. All elements may be considered separately and the solution to Eq. (7) can be expressed as

$$u(x,t) = \sum_j u_j(x,t) \tag{9}$$

where $u_j(x,t)$ is the solution on j-element and it is fulfilled the following equation

$$E_j J_j D^4 u_j + \mu_j E_j J_j D^4 (D_t u_j) + (\rho_j S_j + \alpha_r) D_t^2 u_j = -f_j \tag{10}$$

To find $u_j(x,t)$ with the separation of variables method, the eigenvalues and eigenfunctions for each element are needed.

2.3 Eigenvalues and eigenfunctions problem
In this problem it is assumed that $f_j(x,t) = 0$ and $\mu_j = 0$, hence based on Eq. (10) one obtains

$$E_j J_j D^4 u_j + \rho_j S_j D_t^2 u_j = 0 \tag{11}$$

where $E_j J_j$ and $\rho_j S_j$ may be different on the separate elements, but here, as depicted in Fig. 5, is

$$E_1 J_1 = E_3 J_3 = E_4 J_4 = E_b J_b, \qquad E_2 J_2 = E_b J_b + E_a J_a + E_g J_g \tag{12}$$

$$\rho_1 S_1 = \rho_3 S_3 = \rho_4 S_4 = \rho_b S_b, \qquad \rho_2 S_2 = \rho_b S_b + \rho_a S_a + \rho_g S_g \tag{13}$$

The boundary conditions for the j-element consist of boundary conditions of the problem and coupling conditions between neighboring elements. The concentrated mass m_r is considered in coupling conditions between third and forth elements and therefore it is omitted in Eq. (11).

Let the solution be represented by a product of spatial and temporal functions

$$u_j(x,t) = X_j(x) \, T(t) \tag{14}$$

Substituting (14) into (11) gives

$$E_j J_j \, D^4 X_j T + \rho_j S_j X_j \, D_t^2 T = 0 \tag{15}$$

or

$$\frac{E_j J_j}{\rho_j S_j} \frac{D^4 X_j}{X_j} = -\frac{D_t^2 T}{T} = \omega^2 \tag{16}$$

hence

$$D^4 X_j - \lambda_j^4 X_j = 0 \tag{17}$$

$$D_t^2 T + \omega^2 \, T = 0 \tag{18}$$

where the dispersion relationship is given by

$$\lambda_j^4 = \omega^2 \frac{\rho_j S_j}{E_j J_j} = \frac{\omega^2}{\gamma_j} \tag{19}$$

The Eq. (17) is very important and the solution to it is

$$X_j(x) = A_j K_1(\lambda_j x) + B_j K_2(\lambda_j x) + C_j K_3(\lambda_j x) + D_j K_4(\lambda_j x) \tag{20}$$

where Krylov functions are defined as, (Kaliski, 1986),

$$K_1(z) = \big(ch(z) + \cos(z)\big)/2, \qquad K_2(z) = \big(sh(z) - \sin(z)\big)/2,$$
$$K_3(z) = \big(ch(z) - \cos(z)\big)/2, \qquad K_4(z) = \big(sh(z) + \sin(z)\big)/2 \tag{21}$$

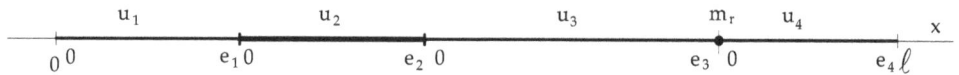

Fig. 6. Geometry of the set beam-one actuator-one mass in local coordinates

The boundary conditions in local coordinates, $x \in [0, e_j]$, to the separate j-element have the form, Fig. 6,

- boundary conditions at the left end of the 1st–element

$$X_1(0) = 0, \qquad D^2 X_1(0) = 0 \tag{22}$$

- coupling conditions between 1st and 2nd–elements

$$X_1(\lambda_1 e_1) = X_2(\lambda_2 0), \qquad DX_1(\lambda_1 e_1) = DX_2(\lambda_2 0),$$

$$E_1 J_1 D^2 X_1(\lambda_1 e_1) = E_2 J_2 D^2 X_2(\lambda_2 0), \qquad E_1 J_1 D^3 X_1(\lambda_1 e_1) = E_2 J_2 D^3 X_2(\lambda_2 0) \qquad (23)$$

- coupling conditions between 2nd and 3rd–elements

$$X_2(\lambda_2 e_2) = X_3(\lambda_3 0), \qquad DX_2(\lambda_2 e_2) = DX_3(\lambda_3 0),$$

$$E_2 J_2 D^2 X_2(\lambda_2 e_2) = E_3 J_3 D^2 X_3(\lambda_3 0), \qquad E_2 J_2 D^3 X_2(\lambda_2 e_2) = E_3 J_3 D^3 X_3(\lambda_3 0) \qquad (24)$$

- coupling conditions between 3rd and 4th–elements

$$X_3(\lambda_3 e_3) = X_4(\lambda_4 0), \qquad DX_3(\lambda_3 e_3) = DX_4(\lambda_4 0), \qquad E_3 J_3 D^2 X_3(\lambda_3 e_3) = E_4 J_4 D^2 X_4(\lambda_4 0)$$

and

$$E_3 J_3 D^3 X_3(\lambda_3 e_3) + m_r \omega^2 X_3(\lambda_3 e_3) = E_4 J_4 D^3 X_4(\lambda_4 0)$$

or

$$E_3 J_3 D^3 X_3(\lambda_3 e_3) = m_r \omega^2 X_4(\lambda_3 0) + E_4 J_4 D^3 X_4(\lambda_4 0) \qquad (25)$$

- boundary conditions at the right end of the 4th–element

$$X_4(\lambda_4 e_4) = 0, \qquad D^2 X_4(\lambda_4 e_4) = 0 \qquad (26)$$

Since $\lambda_1 \neq \lambda_2 \neq \lambda_3 \neq \lambda_4$ then, to calculate them, the Eq. (19) must be used. It is convenient to express $\{\lambda_2, \lambda_3, \lambda_4\}$ as a function λ_1, hence

$$\lambda_1^4 \gamma_1 = \lambda_2^4 \gamma_2 = \lambda_3^4 \gamma_3 = \lambda_4^4 \gamma_4 = \omega^2 \qquad (27)$$

or

$$\lambda_2^4 = \lambda_1^4 (\gamma_1/\gamma_2), \quad \lambda_3^4 = \lambda_1^4 (\gamma_1/\gamma_3), \quad \lambda_4^4 = \lambda_1^4 (\gamma_1/\gamma_4) \qquad (28)$$

Substituting Eq. (20) into boundary conditions (22) it appears that $A_1 = 0$, $C_1 = 0$. In the same way, the rest of conditions given by Eqs. (23) – (26) lead to the set of algebraic equations and it may be written in the matrix form

$$\mathbf{A}\,\mathbf{x} = \mathbf{0} \qquad (29)$$

The matrix $\mathbf{A}$ is too large, to presented it in explicit form. Hence, its elements fall into blocks so that the matrix $\mathbf{A}$ can be written as

$$\mathbf{A} = \begin{bmatrix} A_1' & A_2'' & 0 & 0 \\ 0 & B_1'' & B_2''' & 0 \\ 0 & 0 & C_1''' & C_2''' \\ 0 & 0 & 0 & D_1''' \end{bmatrix} \qquad (30)$$

In the current boundary problem, the separate blocks take the form

$$
\begin{bmatrix} A_1' & A_2'' \\ 0 & B_1'' \end{bmatrix} =
\begin{bmatrix}
K_2' & K_4' & -1 & 0 & 0 & 0 \\
\lambda_1 K_3' & \lambda_1 K_1' & 0 & 0 & 0 & -\lambda_2 \\
\lambda_1^2 E_1 J_1 K_4' & \lambda_1^2 E_1 J_1 K_2' & 0 & 0 & -\lambda_2^2 E_2 J_2 & 0 \\
\lambda_1^3 E_1 J_1 K_1' & \lambda_1^3 E_1 J_1 K_3' & 0 & -\lambda_2^3 E_2 J_2 & 0 & 0 \\
& & K_1'' & K_2'' & K_3'' & K_4'' \\
& 0 & \lambda_2 K_2'' & \lambda_2 K_3'' & \lambda_2 K_4'' & \lambda_2 K_1'' \\
& & \lambda_2^2 E_2 J_2 K_3'' & \lambda_2^2 E_2 J_2 K_4'' & \lambda_2^2 E_2 J_2 K_1'' & \lambda_2^2 E_2 J_2 K_2'' \\
& & \lambda_2^3 E_2 J_2 K_4'' & \lambda_2^3 E_2 J_2 K_1'' & \lambda_2^3 E_2 J_2 K_2'' & \lambda_2^3 E_2 J_2 K_3''
\end{bmatrix}
\tag{31}
$$

$$
\begin{bmatrix} B_1'' & B_2''' \\ 0 & C_1''' \end{bmatrix} =
\begin{bmatrix}
& & -1 & 0 & 0 & 0 \\
B_1'' & & 0 & 0 & 0 & -\lambda_3 \\
& & 0 & 0 & -\lambda_3^2 E_3 J_3 & 0 \\
& & 0 & -\lambda_3^3 E_3 J_3 & 0 & 0 \\
& & K_1''' & K_2''' & K_3''' & K_4''' \\
& 0 & \lambda_3 K_2''' & \lambda_3 K_3''' & \lambda_3 K_4''' & \lambda_3 K_1''' \\
& & \lambda_3^2 E_3 J_3 K_3''' & \lambda_3^2 E_3 J_3 K_4''' & \lambda_3^2 E_3 J_3 K_1''' & \lambda_3^2 E_3 J_3 K_2''' \\
& & \lambda_3^3 E_3 J_3 K_4''' & \lambda_3^3 E_3 J_3 K_1''' & \lambda_3^3 E_3 J_3 K_2''' & \lambda_3^3 E_3 J_3 K_3'''
\end{bmatrix}
\tag{32}
$$

$$
\begin{bmatrix} C_1'' & C_2''' \\ 0 & D_1''' \end{bmatrix} =
\begin{bmatrix}
& & -1 & 0 & 0 & 0 \\
C_1''' & & 0 & 0 & 0 & -\lambda_4 \\
& & 0 & 0 & -\lambda_4^2 E_4 J_4 & 0 \\
& & -m_r \omega_v^2 & -\lambda_4^3 E_4 J_4 & 0 & 0 \\
& 0 & K_1'''' & K_2'''' & K_3'''' & K_4'''' \\
& & K_3'''' & K_4'''' & K_1'''' & K_2''''
\end{bmatrix}
\tag{33}
$$

where the symbols in matrices are given by

$$\{K_v\} = \{K_1, K_2, K_3, K_4\},$$

$$K_v' = K_v(\lambda_1 e_1), \qquad K_v'' = K_v(\lambda_2 e_2), \qquad K_v''' = K_v(\lambda_3 e_3), \qquad K_v'''' = K_v(\lambda_4 e_4) \tag{34}$$

The unknowns are collected in column matrix

$$\mathbf{x} = [B_1, D_1, A_2, B_2, C_2, D_2, A_3, B_3, C_3, D_3, A_4, B_4, C_4, D_4]^{\mathrm{T}} \tag{35}$$

To solve of the homogeneous matrix equation (29), one assumes that $\det \mathbf{A}(\lambda_1) = 0$ and it gives the set $\{\lambda_{1v}\}$, $v = 1, 2, ..., n$. Based on Eq. (28) one can calculate $\{\lambda_{2v}, \lambda_{3v}, \lambda_{4v}\}$ and finally, based on Eq. (19), the frequency $\{\omega_v\}$ of the system beam-actuator-mass.

Now, the unknowns put down in column matrix, Eq. (35), should be determined. Let the main matrix elements **A** be written as two suffix quantities $A_{\alpha\beta}$, where α and β label the rows and columns respectively. Let $M_{\alpha\beta}$ be the minor of the $A_{\alpha\beta}$ element. The general solution to Eq. (29) is

$$B_1 : D_1 : A_2 : ... = (-1)^{\alpha+1}M_{\alpha 1} : (-1)^{\alpha+2}M_{\alpha 2} : (-1)^{\alpha+3}M_{\alpha 3} : ... \tag{36}$$

Substituting $\{\lambda_{j\nu}\}$ and unknowns **x** to Eq. (20), the ν –eigenfunctions (ν –modes) assigned to the j –element are obtained. The solution to Eq. (10) is given by

$$X(x) = \sum_j X_j(x) = \sum_{j\nu} X_j(\lambda_{j\nu}x) = \sum_{j\nu} X_{j\nu}(x) = \sum_\nu X_\nu(x) \tag{37}$$

where $\sum_{j\nu}(...) = \sum_j \sum_\nu (...)$ and the separate modes are equal

$$X_{j\nu}(x) = A_j K_1(\lambda_{j\nu}x) + B_j K_2(\lambda_{j\nu}x) + C_j K_3(\lambda_{j\nu}x) + D_j K_4(\lambda_{j\nu}x) \tag{38}$$

2.4 Orthogonality of modes

Orthogonality condition of the uniform beam modes may be found in (Kaliski, 1986; de Silva, 2000). First of all, based on twice integration by parts, one has

$$\int_0^\ell X_\nu(x)D^4X_\mu(x)dx = \left(X_\nu(x)D^3X_\mu(x) - DX_\nu(x)D^2X_\mu(x)\right)\Big|_0^\ell + \int_0^\ell D^2X_\nu(x)D^2X_\mu(x)dx \tag{39}$$

The separate modes $X_\mu(x)$, $X_\nu(x)$, fulfill the following modal equations

$$EJD^4X_\mu(x) = \omega_\mu^2 \rho S X_\mu(x) \tag{40}$$

$$EJD^4X_\nu(x) = \omega_\nu^2 \rho S X_\nu(x) \tag{41}$$

Multiplying above equations by $X_\nu(x)$ and $X_\mu(x)$ respectively, integrate both in range o integration $x \in [0, \ell]$, use Eq. (39), subtract the second result from the first one, one obtains (for simplicity an argument (x) is omitted)

$$(\omega_\nu^2 - \omega_\mu^2)\rho S \int_0^\ell X_\nu X_\mu dx = EJ\left[\left(X_\mu D^3X_\nu - DX_\mu D^2X_\nu\right) - \left(X_\nu D^3X_\mu - DX_\nu D^2X_\mu\right)\right]\Big|_0^\ell \tag{42}$$

For standard boundary conditions, the right-hand-side equals zero.
The procedure outlined above can be used to the problem presented in Fig. 6, but Eq. (39) must be applied to the separate j –element, namely

$$\int_0^{e_j} X_{j\nu}(x)D^4X_{j\mu}(x)dx = \left(X_{j\nu}(x)D^3X_{j\mu}(x) - DX_{j\nu}(x)D^2X_{j\mu}(x)\right)\Big|_0^{e_j} +$$
$$+ \int_0^{e_j} D^2X_{j\nu}(x)D^2X_{j\mu}(x)dx \tag{43}$$

Considering both boundary conditions of the problem and coupling conditions between neighboring elements, Eqs. (22)–(26), instead of Eq. (42) one has

$$\left(\omega_v^2 - \omega_\mu^2\right)\left(\rho_1 S_1 \int_0^{e_1} X_{1v} X_{1\mu}\, dx + \rho_2 S_2 \int_0^{e_2} X_{2v} X_{2\mu}\, dx + \rho_3 S_3 \int_0^{e_3} X_{3v} X_{3\mu}\, dx + \rho_4 S_4 \int_0^{e_4} X_{4v} X_{4\mu}\, dx + \right.$$

$$\left. + m_r X_{4v}(0) X_{4\mu}(0)\right) = E_4 J_4 \cdot$$

$$\cdot \left[\left(X_{4\mu}(e_4) D^3 X_{4v}(e_4) - DX_{4\mu}(e_4) D^2 X_{4v}(e_4)\right) - \left(X_{4v}(e_4) D^3 X_{4\mu}(e_4) - DX_{4v}(e_4) D^2 X_{4\mu}(e_4)\right)\right] \quad (44)$$

Because of Eq. (26), the right-hand-side is zero, hence

$$\left(\omega_v^2 - \omega_\mu^2\right)\left(\rho_1 S_1 \int_0^{e_1} X_{1v} X_{1\mu}\, dx + \rho_2 S_2 \int_0^{e_2} X_{2v} X_{2\mu}\, dx + \rho_3 S_3 \int_0^{e_3} X_{3v} X_{3\mu}\, dx + \rho_4 S_4 \int_0^{e_4} X_{4v} X_{4\mu}\, dx + \right.$$

$$\left. + m_r X_{4v}(0) X_{4\mu}(0)\right) = 0 \quad (45)$$

Fig. 7. Geometry of set with n_j –elements

The orthogonality condition, Eq. (45), may be generalized in a simple way. Let the system beam-actuators-masses be divided into n_j –elements as depicted in Fig. 7. In this case one has

$$\left(\omega_v^2 - \omega_\mu^2\right)\left[\sum_j\left(\rho_j S_j \int_j X_{jv} X_{j\mu}\, dx + m_j X_{jv}(0) X_{j\mu}(0)\right) + m_{n_j+1} X_{n_j v}(e_{n_j}) X_{n_j \mu}(e_{n_j})\right] = 0 \quad (46)$$

Since the term $\omega_v^2 - \omega_\mu^2$ is canceled for $\mu = v$, the general orthogonality condition is given by

$$\sum_j\left(\rho_j S_j \int_j X_{jv} X_{j\mu}\, dx + m_j X_{jv}(0) X_{j\mu}(0)\right) + m_{n_j+1} X_{n_j v}(e_{n_j}) X_{n_j \mu}(e_{n_j}) = \begin{cases} 0, & v \neq \mu \\ \beta_v^2, & v = \mu \end{cases} \quad (47)$$

The Eq. (47) in particular case is used in deriving the solution to the forced vibration problem.

2.5 Forced vibrations with damping

A point departure for further consideration is Eq. (7); for j –element one has

$$E_j J_j D^4 u_j + \mu_j E_j J_j D^4\left(D_t u_j\right) + \rho_j S_j D_t^2 u_j = -f_j \quad (48)$$

The solution to Eq. (48) is forced vibrations with damping. Let be the load force in the form

$$f_j(x,t) = f_j(x) \exp(i\omega_f t) \quad (49)$$

where $i = (-1)^{1/2}$, ω_f - excited frequency.
Applying separation of variables method, the solution to Eq. (48) is assumed as

$$u_j(x,t) = X_{jf}(x) \exp(i\omega_f t) \quad (50)$$

Substituting Eqs. (49) and (50) to Eq. (48) one obtains

$$\gamma_j(1+i\mu_j\omega_f)D^4X_{jf}(x) - \omega_f^2 X_{jf}(x) = -\frac{1}{\rho_j S_j}f_j(x) \tag{51}$$

The solution of the above equation is given by

$$X_{jf}(x) = \sum_v C_{jv} X_{jv}(x) \tag{52}$$

where C_{jv} – constants, $X_{jv}(x)$ – Eq. (38).

After some calculation, the constants C_{jv} are expressed by

$$C_{jv} = \frac{1}{\rho_j S_j}\frac{1}{\alpha_{jv}^2}\frac{1}{\beta_v^2}I_{jv} = C_{jv}^* I_{jv} \tag{53}$$

where

$$C_{jv}^* = \frac{1}{\rho_j S_j}\frac{1}{\alpha_{jv}^2}\frac{1}{\beta_v^2}, \quad \frac{1}{\alpha_{jv}^2} = \frac{1}{(1+i\mu_j\omega_f)\omega_v^2 - \omega_f^2}, \quad I_{jv} = -\int_j f_j(x)X_{jv}(x)dx \tag{54}$$

In the end, the problem of the forced j –element beam vibration with damping, excited with the force $f_j(x)$ is solved; in the harmonic steady state it is given by

$$X_f(x) = \sum_j X_{jf}(x) = \sum_{jv}C_{jv} X_{jv}(x) \tag{55}$$

In current problem, two form of the forces have the practical meaning namely, the force with constant amplitude $f_{j0}(x) = f_0$ and the force acting at discrete point $f_{ja}(x_i)$. The former may be interpreted as the spread excitation forced, for example with plane acoustic wave, but the latter is the control force due to actuators, henceforth

$$f_j(x) = f_0 + f_{ja}(x_i) \tag{56}$$

2.6 Interaction between beam and actuators

It is assumed that the actuator is perfectly bonded to the beam surface. Exciting actuator, the interaction between actuator and the beam is appeared. The interaction process is explained in (Hansen & Snyder, 1997; Fuller at al, 1997) in detail and references cited therein. Assuming the spatially uniform actuator, it provides boundary induction solely in terms of the external line moment distributed along its edges (Burke & Hubbard, 1991; Sullivan et al., 1996). So, the bending moment in y –direction is given by the formula (Hansen & Snyder 1997), Fig. 8,

$$M_x = -M_0\left(<x-x_1>^{-2} - <x-x_2>^{-2}\right) \tag{57}$$

where $<.>^{-1} = \delta(.)$ and $<.>^{-2} = D\delta(.)$ – doublet function, M_0 – line moment amplitude

$$M_0 = C_a\frac{d_{31}}{h_a}V \tag{58}$$

where C_a – constant depending on geometry and mechanical properties of the actuator and plate, d_{31} – piezoelectric material strain constants, V – voltage in the direction of polarization.

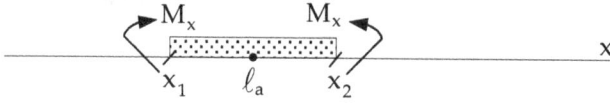

Fig. 8. External line moments of the actuator

The problem is to determine of the C_a, because it depends on the analysis method of the mutual interaction between beam-actuator (Hansen & Snyder 1997; Pietrzakowski, 2004). Let the static force coupling model is taken into account. If relatively thin actuator compared with beam thickness is assumed (so uniform normal stress distribution is accepted) and furthermore by ignoring the neutral axis displacement d , see Fig. 4, the constant C_a is come down to the form

$$C_a = \frac{E_b h_b E_a h_a (h_b + h_a)}{2(E_b h_b + E_a h_a)} \tag{59}$$

Since the beam vibration equation is the forces equation then to consider the action of actuator with the beam, moments M_x are replaced with two couples of forces, Fig. 9,

$$M_x = f_a \ell_a / 2 \tag{60}$$

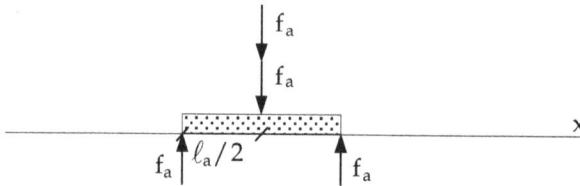

Fig. 9. External pair of forces of the actuator

Next, the separate forces are considered in Eq. (56).

2.7 Beam vibration reduction through actuators

For the problem presented in Fig. 5, the total load of the beam, described by Eq. (56), is given by

$$f_j(x) = -f_0 + \left(f_{js} \delta(x - x_{1s}) - 2 f_{js} \delta(x_s) + f_{js} \delta(x + x_{2s}) \right) \tag{61}$$

where the symbol f_{ja} is replace by f_{js} in order to express, in the future, the interaction sum of actuators and the glue on the beam.

An expression in brackets is the sum of interacting forces actuator-beam. Hence, the integral I_{jv} , Eq. (54), for $f_j(x)$ expressed by above equation is given by

$$I_{j\nu} = -\int_j f_j(x)\,X_{j\nu}(x)\,dx = -f_0\int_j X_{j\nu}(x)\,dx + f_{js}\int_j \left(\delta(x-x_{1s}) - 2\delta(x_s) + \delta(x+x_{2s})\right)X_{j\nu}\,dx =$$
$$= -f_0\int_j X_{j\nu}(x)\,dx + f_{js}\left[X_{j\nu}(x_{1s}) - 2X_{j\nu}(x_s) + X_{j\nu}(x_{2s})\right] \tag{62}$$

The expression in square bracket constitutes the second-order central finite difference. Since the distance between nodes ℓ_s is constant, then the difference can be transformed into

$$\frac{1}{\ell_s^2}\left[X_{j\nu}(x_{1s}) - 2X_{j\nu}(x_s) + X_{j\nu}(x_{2s})\right] = D^2 X_{j\nu}(x_s) \tag{63}$$

where

$$D^2 X_{j\nu}(x_s) = \pm\kappa_{j\nu}(x_s) \tag{64}$$

The $\kappa_{j\nu}(x_s)$ is the curvature of the mode $X_{j\nu}(x)$ at the point $x = x_s$ (Brański & Szela, 2007; Brański & Szela, 2008). The sign of the $\kappa_{j\nu}(x_s)$ is contractual namely, if the bending of the beam is directed upwards, the sign is positive and vice versa. Substituting Eq. (64) into Eq. (62), one obtains

$$I_{j\nu} = -f_0\int_j X_{j\nu}(x)\,dx + f_{js}\ell_s^2\,\kappa_{j\nu}(x_s) = I_{j\nu 0} + I_{j\nu s} \tag{65}$$

Next, substituting Eq. (65) into Eq. (52) through Eq. (53), the reduction vibration is obtained

$$X_{jf}(x) = \sum_\nu C_{j\nu}^* I_{j\nu} X_{j\nu}(x) = \sum_\nu C_{j\nu}^*\left(I_{j\nu 0} + I_{j\nu s}\right)X_{j\nu}(x) = \sum_\nu A_{jf\nu} X_{j\nu}(x) \tag{66}$$

Note, that the amplitude $A_{jf\nu}$ is the direct quantity which is liable to the reduction, in explicit form is

$$A_{jf\nu} = C_{j\nu}^* I_{j\nu} = C_{j\nu}^*\left(I_{j\nu 0} + I_{j\nu s}\right) \tag{67}$$

At the same time, together with the vibration reduction amplitude $A_{jf\nu}$, the curvature is subjected to the reduction and based on Eq. (66) is

$$\kappa_{jf} = \pm D^2 X_{jf} = \sum_\nu \kappa_{jf\nu} = \pm\sum_\nu A_{jf\nu}\,\kappa_{j\nu} \tag{68}$$

Furthermore, the reduction of the $A_{jf\nu}$ leads to the reduction of the shear force $Q_{jf}(x)$ and bending moment $M_{jf}(x)$ (Brański et al., 2010; Kaliski, 1986; Kozień, 2006),

$$Q_{jf}(x) = \pm E_j J_j\,D\kappa_{jf}(x) = \pm E_j J_j\sum_\nu D\kappa_{jf\nu}(x) = \pm E_j J_j\sum_\nu A_{jf\nu}\,D\kappa_{j\nu}(x) \tag{69}$$

$$M_{jf}(x) = \pm E_j J_j\,\kappa_{jf}(x) = \pm E_j J_j\sum_\nu \kappa_{jf\nu}(x) = \pm E_j J_j\sum_\nu A_{jf\nu}\,\kappa_{j\nu}(x) \tag{70}$$

As can be seen, the vibration reduction undergo on the following amplitudes: of the beam vibration $X_{jf}(x)$ – Eq. (66), of the shear force $Q_{jf}(x)$ – Eq. (69) and of the bending moment $M_{jf}(x)$ – Eq. (70). Hereafter, the notion e.g. "shear force reduction" is used instead of "the reduction of the amplitude of the shear force", and so on.

The Eqs (66), (69) and (70) and may be written commonly

$$\Psi_{jf}(x) = \sum_{\nu} \Psi_{jf\nu}(x) = C_j \sum_{\nu} A_{jf\nu} \, \Phi_{j\nu}(x) \tag{71}$$

where

$$\Psi_{jf} = \{X_{jf}, Q_{jf}, M_{jf}\}, \quad \Phi_{j\nu} = \{X_{j\nu}, D\kappa_{j\nu}, \kappa_{j\nu}\}, \quad C_j = \{1, \pm E_j J_j, \pm E_j J_j\} \tag{72}$$

Omitting for simplicity the index "f", for entire system beam-actuators one has

$$\Psi(x) = \sum_{j} \Psi_j(x) = \sum_{j\nu} \Psi_{j\nu}(x) \tag{73}$$

or

$$\Psi(x) = \sum_{j\nu} C_j A_{j\nu} \, \Phi_{j\nu}(x) = C A \Phi(x) \tag{74}$$

where

$$A = C^* I = C^* \left(I_0 + \sum_s f_s \ell_s^2 \kappa(x_s) \right) = C^* \left(I_0 + I_\Sigma \right) \tag{75}$$

$$\Phi(x) = \{X(x), D\kappa(x), \kappa(x)\} \tag{76}$$

It is appeared from Eq. (75) that the active vibration reduction depends on the following parameters:

- ω_f – excited frequency, it is contained in C_ν^*,
- x_s – distribution of the actuators on the beam,
- $\kappa(x_s)$ – value of the beam curvature at the point of the actuators distribution,
- f_s – interacting forces between beam-actuators or more generally – mechanical properties of the actuators,
- ℓ_s – actuators lengths or more generally – geometrical properties of the actuators,
- n_s – number of actuators.

As mentioned above, the optimal actuators distribution described with $\{x_s\}$ has an important meaning and finding of the $\{x_s\}$ is the aim of the chapter.

3. Optimal actuators distribution problem

Before the optimization problem will be formulated, any coefficients of the vibration reduction should be defined.

3.1 Reduction and effectiveness coefficients

Let be the difference between any quantities of the beam vibration

$$\Delta\Psi(x) = \Psi(x) - \Psi_R(x) \tag{77}$$

where $\Psi(x)$, $\Psi_R(x)$ – quantities calculated without and with actuators respectively; $\Psi(x)$, $\Psi_R(x)$ are given together by Eq. (74), where

$$A = C^* I_0 \tag{78}$$

and

$$A_R = C^* \left(I_0 + I_\Sigma \right) \tag{79}$$

The difference $\Delta \Psi(x)$ is interpreted as the quantity of the vibration reduction and it is the first measure of this reduction namely, the quantity reduction coefficient.
The second measure of the vibration reduction is defined as

$$R_\psi(x) = \frac{\Delta \Psi(x)}{\Psi(x)} = \frac{\Psi(x) - \Psi_R(x)}{\Psi(x)} \tag{80}$$

It is called as the reduction coefficient and it may be expressed in per cent. Note, that if the reduction coefficient equals one, the vibration reduction is total, $\Psi_R(x) = 0$.
An effectiveness of the vibration reduction is defined as a quotient of some vibration reduction measure by an amount of the energy W provided to the system in order to excite actuators. Hence, thirst measure of the vibration reduction may be defined by so called the effectiveness coefficient

$$E_\psi(x) = R_\psi(x) / W \tag{81}$$

The energy W provided to the system is translated into couples of forces, Fig. 9. Therefore, the energy W may be replaced by forces $f_R = \sum_s 4 f_s$, hence

$$E_\psi(x) = R_\psi(x) / f_R \tag{82}$$

The Eqs. (77) – (82) define the appropriate factors of the vibration reduction at the point x. In many cases, it is convenient to calculate mean values of these coefficients at whole beam domain or at the beam sub-domains. First of them is the mean quantity reduction coefficient and it is defined by the formula

$$\Delta \Psi_m = \frac{1}{n_i} \sum_i \left(\Psi(x_i) - \Psi_R(x_i) \right) \qquad i = 1, 2, ..., n_i \tag{83}$$

Consequently, the mean reduction coefficient and the mean effectiveness coefficient are defined respectively

$$R_{\Psi m} = \Delta \Psi_m / \Psi_m , \qquad \Psi_m = \sum_i \Psi(x_i) / n_i \tag{84}$$

$$E_{\Psi m} = R_{\Psi m} / f_R \tag{85}$$

The coefficients defined above may constitute the base to formulate the optimization problem; hereafter the $R_\psi(x)$ is chosen.

3.2 Formulation of the optimization problem
In this chapter, one formulates the following problem: find the optimal actuators distribution $\{ x_s \}$ which maximize of the reduction coefficient $R_\psi(x)$; hence $R_\psi(x)$ is assumed as an objective function. In this case the maximal value of $R_\psi(x)$ equals one and it

means p-reduction; such instance is considered in (Brański et al., 2010; Brański & Lipiński, 2011) and it seems that it is possible only in for separate mode.

Let the energy provided to the actuators be constant and hence, the f_s is always constant. Now, for clarity of the disquisition, rewrite the effectiveness coefficient in explicit form

$$R_\Psi(x) = \frac{\Psi(x) - \Psi_R(x)}{\Psi(x)} = \frac{CA\Phi(x) - CA_R\Phi(x)}{CA\Phi(x)} \tag{86}$$

Working out on above assumption, the $R_\Psi(x)$ will be maximal, if $\Psi_R(x)$ is minimal. Hence, the optimal condition $R_\Psi(x) = R_{\Psi;max}(x)$ leads to the next condition $\Psi_R(x) \equiv \Psi_{R;min}(x)$. Note, that the $\Psi_R(x)$ depends on the reduction amplitude A_R. So, the $\Psi_R(x) \equiv \Psi_{R;min}(x)$, if the amplitude A_R is minimal and instead of the above condition, it leads to

$$A_R = A_{R;min} \tag{87}$$

3.3 Heuristic analysis of the optimization problem

Note, that the amplitude A_R comprises the factor $C^* \neq 0$, but it is constant and this is the factor $I_0 + I_\Sigma = I_R$ which is changed. In practice, instead of the condition (87), the following condition of the reduction must be fulfilled

$$I_R = I_0 + I_\Sigma = I_0 + \sum_s f_s \ell_s^2 \kappa(x_s) = I_{min} \tag{88}$$

For future considerations the sign of I_R is very important. The vibrations are reduced with actuators, if the I_Σ is positive, but must be fulfilled the following condition: $I_R = I_0 + I_\Sigma \geq 0$; $I_R = 0$ assures the total reduction. If this condition is not fulfilled, the actuators excite vibrations and thereby they are not accomplished owns role. Note, that the sign of I_0 is always negative, see Eq. (65). Then, the sign of I_Σ must be positive and it depends on the signs both forces f_s and curvatures $\kappa(x_s)$.

From physical point of view, the sign of $\kappa(x)$ is changed and as established above; it is positive, if the bending of the beam is directed upwards. Then, for many actuators one has

$$I_\Sigma = f_1 \ell_1^2(+)\kappa(x_1) + f_2 \ell_2^2(-)\kappa(x_2) + \dots = \sum_s f_s \ell_s^2(-1)^{s+1}\kappa(x_s) > 0 \tag{89}$$

To obtain the positive sign of I_Σ, the signs of f_s should alternates; this problem clearly expressed in the following way

$$I_\Sigma = (+)f_1 \ell_1^2(+)\kappa(x_1) + (-)f_2 \ell_2^2(-)\kappa(x_2) + \dots = \sum_s (-1)^{s+1}f_s \ell_s^2(-1)^{s+1}\kappa(x_s) > 0 \tag{90}$$

First at all, it is possible if the signs of $\kappa(x_s)$ and f_s are the same, namely positive or negative, and they take their extremes. To fulfill this requirement, the actuators should be specially distributed on the beam. An idea of description of the sign of $\kappa(x_s)$ is advance determined. The value of $\kappa(x_s)$ is determined by means of the distribution of the actuators; they are bended at $\{x_s\}$. Hence, the distribution has a great significance; this problem was solved in (Brański & Szela, 2007; Brański & Szela, 2008; Brański et al., 2010; Brański & Szela 2010; Szela, 2009). Interpreting Eq. (88) through Eq. (90) it is appear that the actuators ought to be bonded on the beam sub-domains in which the curvatures reach their extremum and consequently the highest and lowest values respectively, see Fig. 10. This is so called quasi-optimal actuators distribution and it is described with $x_Q \equiv x_s$ points, their number is $n_Q \equiv n_s$.

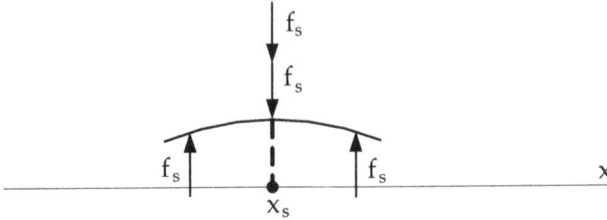

Fig. 10. Optimal distribution of the actuators

As far as signs and values of $\{f_s\}$ are concerned, it was assumed that the added energy exciting actuators is constant. So, the values $\{f_s\}$ of the separate actuators are known and always are constant, while the sign f_s springs from the physical interpretation of the interaction beam-actuator. As can be seen in Fig. 10, the forces $2f_s$ are placed at the point of local extreme, namely at the x_s, with the opposite direction to the bending of the beam $X(x)$. At the same time, the forces f_s on the actuator edges are in the direction of the beam bending and let assume that this sign of f_s is positive. Another way, the vectors f_s and the beam bending $X(x)$ are in the same direction. In such sign convection, both f_s and $\kappa(x_s)$ in Eq. (90) have the same signs and all terms are positive. Furthermore, the actuators distribution described with $x_Q \equiv x_s$ ensures the maximum of the reduction coefficient.

The heuristic analysis described above was substantiated numerically for the separate beam and triangular modes and the details may be found in own papers.

3.4 Analytical analysis of the optimization problem

The aim of this section is to work out of the analytical method, which will describe such distribution of the actuators in order to assure the maximum of the reduction coefficient. It is expected that the analytical method will confirm the quasi-optimal distribution which has been found above with heuristic method. Therefore the assumptions are the same like in heuristic method, namely n_s, f_s and ℓ_s are settled.

Let the distribution of actuators be marked with the set of unknown coordinates $\{x_s\}$ for the moment; that are exactly these coordinates $\{x_s\} \equiv \{x_Q\}$ of which are looked for. One starts from Eq. (88), hence

$$I_R = I_0 + I_\Sigma = I_0 + f_1 \, \ell_1^2 \, \kappa(x_1) + f_1 \, \ell_2^2 \, \kappa(x_2) + \dots \tag{91}$$

Since the $\kappa(x)$ is the function which changes the sign, it is appropriate to search the points x_s which assure the extreme $I_R(x)$, not minimum only. The function $I_R(x)$ can have the extreme only at points x_s, at which $DI_R(x)$ is equal to zero or $f_1 \, \ell_1^2 \, D\kappa(x_1) + f_1 \, \ell_2^2 \, D\kappa(x_2) + \dots = 0$ does not exist (Fichtenholtz, 1999). Because I_0 is constant than a necessary condition for existing extreme value is

$$DI_R(x_s) = 0 \tag{92}$$

where $DI_R(x_s) = DI_R(x = x_s)$ and hence

$$f_1 \ell_1^2 D\kappa(x_1) + f_1 \ell_2^2 D\kappa(x_2) + \dots = 0 \tag{93}$$

Because of $f_s \ell_s^2 \neq 0$ then instead of Eq. (93) one has

$$D\kappa(x_s) = 0 \tag{94}$$

From the condition (94), a set of stationary points $\{x_s\}$ is obtained. The sufficient condition for existing extreme demands, in order to the function be determined on either side of the point x_s and $D\kappa(x)$ must change sign at this point (turning point); it is sufficient condition formulated in the first form. This condition is expressed in the other form

$$D^2\kappa(x_s) \neq 0 \tag{95}$$

If this condition is not fulfilling, then this point should be omitted.

One still needs to consider the biggest and the lowest values of $\kappa(x)$; they are in hypothetical points $\{x_{max}, x_{min}\}$. In order to find them, the values of $\kappa(x)$ at the stationary points $\{x_s\}$ are calculated and they are compared to the values calculated at the end points of the appropriate interval. In the future consideration, the n_s points among stationary $\{x_s\}$ and $\{x_{max}, x_{min}\}$ ones, at which $\kappa(x)$ takes in turn its absolute values, are taken into account. The problem of the signs of the $\kappa(x_s)$ and f_s is quite the same as in heuristic analysis.

Analytical analysis was applied for p-reduction and for the separate beam modes (Brański & Lipiński, 2011). As pointed out there, the analytical solution to the optimal actuators distribution problem confirms the results obtained with heuristic solution.

4. Conclusion

Deriving the shape of $\kappa(x)$, the influence both masses and stiffness of the actuators and glue on the shape of $X(x)$, and consistently on the shape of $\kappa(x)$, were omitted; if not, an adaptation method must be applied. But after determining shape of $\kappa(x)$, all these parameters were considered.

As can be note, the actuators optimal distribution is attained assuming that the added energy to excite actuators is constant. It is translated into constant f_s. Having the optimal distribution, the reduction coefficient may be improved by adding more energy or in order words, by increasing f_s. This way, presented optimal method corresponds to that one presented in (Q. Wang & C. Wang, 2001), namely "maximization of the control forces transmitted by the actuators to the structure".

Based on theoretical considerations, and numerical ones presented in own papers, the following conclusion may be formulated.

1.   The optimization problem of the actuators distribution assuring the maximal active vibration reduction of the beam, measured with reduction coefficient, may be solved both heuristically and analytically. In analyzed problem, it turned out that both methods give the same results.
2.   The following algorithm of analytical method may be worked out:

- to search of stationary points $\{x_s\}$ of the beam curvature,
- to search of $\{x_{max}, x_{min}\}$ points of the beam curvature,
- n_Q points among stationary $\{x_s\}$ and $\{x_{max}, x_{min}\}$ ones, at which $\kappa(x)$ takes in turn its maximum absolute values, are selected, they are denoted by $\{x_Q\}$,
- to bond the actuators at the $\{x_Q\}$ points,
- to determine the value of the reduction coefficient,
- to increase the value f_s, through the energy increase which excites actuators, until the reduction coefficient will attain its maximum.

It seems that proposed optimization method is very simple and may be useful in many technical problems of active vibration reduction. This work is a starting point for many computer simulations and experiments.

5. References

Bapat, C.N. & Bapat, C. (1987). Natural frequencies of a beam with non-classical boundary conditions and concentrated masses, *Journal of Sound and Vibrations*, Vol.112, No.1, pp. 177–182.

Brański, A. & Szela, S. (2007). On the quasi optimal distribution of PZTs in active reduction of the triangular plate vibration. *Archives of Control Sciences*, Vol.17, No.4, pp. 427–437.

Brański, A. & Szela, S. (2008). Improvement of effectiveness in active triangular plate vibration reduction. *Archives of Acoustics*, Vol.33, No.4, pp.521-530.

Brański, A.; Borkowski, M. & Szela, S. (2010). The idea of the selection of PZT-beam interaction forces in active vibration protection problem. *Acta Physica Polonica*, Vol.118, pp.17-22.

Brański, A. & Szela S. (2010). Quasi-optimal PZT distribution in active vibration reduction of the triangular plate with P-F-F boundary conditions. *Archives of Control Sciences*, Vol.20, No.2, pp.209–226.

Brański, A. & Lipiński, G. (2011). Analytical determination of the PZT's distribution in active beam vibration protection problem. (in press in *Acta Physica Polonica*).

Brański, A. & Szela, S. (2011). Evaluation of the active plate vibration reduction via the parameter of the acoustic field. (in press in *Acta Physica Polonica*).

Bruant, I.; Coffignal, G.; Lene, F. & Verge, M. (2001). A methodology for determination of piezoelectric actuator and sensor location on beam structure. *Journal of Sound and Vibrations*, Vol.243, No.5, pp.861–882.

Bruant, I. & Proslier, L. (2005). Optimal location of actuators and sensors in active vibration control, *Journal Inteligent Material System Structures*, Vol.16, pp.197–206.

Bruant, I.; Gallimard, L. & Nikoukar, S. (2010). Optimal piezoelectric actuator and sensor location for active vibration control, using genetic algorithm. *Journal of Sound and Vibrations*, Vol.329, pp.1615-1635.

Burke, S.E. & Hubbard, J.E. (1991). Distributed transducer vibration control of thin plate, *J.A.S.A.*, Vol.90, No.2, pp.937–944.

Charette, F.; Berry, A. & Guigou, C. (1998). Dynamic effect of piezoelectric actuators on the vibration response of a plate. *Journal Inteligent Material System Structures*, Vol.8, pp.513-524.

Croker, M.J. (2007). *Handbook of noise and vibration control*, John Wiley & Sons.

Dhuri, K.D. & Seshu, P. (2006). Piezo actuator placement and sizing for good control effectiveness and minimal change in original system dynamics. *Smart Material Structure*, Vol.15, pp.1661–1672.

Ercoli, L. & Laura, P.A.A. (1987). Analytical and experimental investigation on continuous beam carying elastically mounted masses. *Journal of Sound and Vibrations*, Vol.114, No.3, pp.519–533.

Fichtenholtz, G.M. (1999). Differential and integral calculus, PWN, Warsaw.

Frecker, M. (2003). Recent advances in optimization of smart structures and actuators. *Journal Intelligent Material System Structures*, Vol.14, pp.207–215.

Fuller, C.R.; Elliot, S.J. & Nielsen, P.A. (1997). *Active control of vibration*, Academic Press, London.

Gosiewski, Z. & Koszewnik, A. (2007). The influence of the piezoelements placement on the active vibration damping system, *Proceedings Active Noise and Vibration Control Method*, pp.69–79, Krakow,

Guney, M. & Eskinat, E. (2007). Optimal actuator and sensor placement in flexible structures using closed-loop criteria, *Journal of Sound and Vibrations*, Vol.312, pp.210–233.

Halim, D. & Reza Moheimani, S.O. (2003). An optimization approach to optimal placement of collocated piezoelectric actuators and sensors on a thin plate. *Mechatronics*, Vol.13, pp.27–47.

Han, S.M., Benaroya, H. & Wei, T. (1999). Dynamics of transversely vibrating beams using four engineering theories, *Journal of Sound and Vibrations*, Vol.225, No.5, pp.935–988.

Hansen, C.H. & Snyder, S.D. (1997). *Active control of noise and vibration*, E&FN SPON, London.

Hernandes, J.A.; Almeida, S.F.M. & Nabarrete, A. (2000). Stiffening effects on the free vibration behavior of composite plates with PZT actuators, *Composite Structures*, Vol.49, pp.55–63.

Hong, C.; Gardonio, P. & Elliott, S.J. (2007). Active control of resiliently mounted beams using triangular actuators, *Journal of Sound and Vibrations*, Vol.301, pp.297–318.

Ip, K.H. & Tse, P.C. (2001). Optimal configuration of a piezoelectric path for vibration control of isotropic rectangular plate, *Smart Material Structure*, Vol.10, pp.395–403.

Jha, A.K. & Inman, D.J. (2003). Optimal sizes and placement of piezoelectric actuators and sensors for an inflated torus. *Juornal Inteligent Material System Structures*, Vol.14, pp.563–576.

Jiang, T.Y.; Ng, T.Y. & Lam, K.Y. (2000). Optimization of a piezoelectric ceramic actuator. *Sensors and Actuators*, Vol.84, pp.81–94.

Kaliski, S. (1986). *Vibrations and Waves*, PWN, Warsaw.

Kasprzyk, S. & Wiciak, M. (2007). Differential equation of transverse vibrations of a beam with a local stroke change of stiffness, *Opuscula Mathematica*, Vol.27, No.2, 245–252.

Kozień, M. (2006). *Acoustic radiation of plates and shallow shells*, PK, Monograph 331, ISS 0860-097X, Krakow.

Liu, W.; Hou, Z. & Demetriou, M.A. (2006). A computational scheme for the optimal sensor/actuator placement of flexible structures using spatial H2 measures, *Mechanical Systems and Signal Processing*, Vol.20, pp.881–895.

Low, K.H. & Naguleswaran, S. (1998). On the eigenfrequencies for mass loaded beams under classical boundary conditions, *Journal of Sound and Vibrations*, Vol.215, No.2, pp.381–389.

Majkut, L. (2010). Eigenvalue based inverse model of beam for structural modification and diagnostics. Part I: Theoretical formulation, *Latin American Journal of Solids and Structures*, Vol.7, pp.423436.

Moshrefi-Torbati, M.; Keane, A.J.; Elliott, S.J.; Brennan, M.J.; Anthony, D.K. & Rogers, E. (2006). Active vibration control (AVC) of a satelite boom structure using optimally positioned stacked piezoelectric actuators, *Journal of Sound and Vibrations*, Vol.292, pp.203-220.

Naguleswaran, S. (1999). Lateral vibration of a uniform Euler-Bernoulli beam carrying a particle at an intermediate point, *Journal of Sound and Vibrations*, Vol.227, No.1, pp.205–214.

Pietrzakowski, M. (2004). *Active damping of transverse vibration using distributed piezoelectric elements*, Monograph 204, PW, ISSN 0137-2335, Warsaw.

Przybyłowicz, P.M. (2002). *Piezoelectric vibration control of rotating structures*, Monograph 197, PW, ISSN 0137-2335. Warsaw.

Qiu, Z.; Zhang, X.; Wu, H. & Zhang, H. (2007). Optimal placement and active vibration control for piezoelectric smart flexible cantilever plate, *Journal of Sound and Vibrations*, Vol.301, pp.521-543.

Sheu, W.J.; Huang, R.T. & Wang, C.C. (2008). Influence of bonding glues on the vibration of piezoelectric fans, *Sensors and Actuators A*, Vo.148, pp.115–121.

Shimon, P.; Richer, E. & Hurmuzlu, Y. (2005). Theoretical and experimental study of efficient control of vibration in a clamped square plate, *Journal of Sound and Vibrations*, Vol.282, pp.453-473.

de Silva, C.W. (2000). *Vibration, Fundamentals and practice*, CRC Press.

Sullivan, J.M.; Hubbard, J.E. & Burke, S.E. (1996). Modeling approach for two-dimensional distributed transducers of arbitrary spatial distribution, *J.A.S.A.*, Vol.99, No.5, pp.2965–2974.

Szela, S. (2009). *Distribution method of the actuators in an active vibration reduction of the triangular plate*, AGH, Krakow, Ph.D. thesis, in polish.

Tylikowski, A. & Przybyłowicz, P.M. (2004). *Non-classical piezoelectric materials in vibrations stability and dumping*, PW, Warsaw.

Wang, F. (2007). *Shape optimization for piezoceramics*, Paderborn, Ph.D. thesis.

Wang, Q. & Wang, C. (2001). A controllability index for optimal design of piezoelectric actuators in vibration control of beam structures, *Journal of Sound and Vibrations*, Vol.242, No.3, pp.507–518.

Wiciak, J. (2008). *Vibration and Structural Acoustic Control-Selected Aspects*, AGH, Monografh 175, ISSN 0867-6631, Krakow.

Yang, Y.; Jin, Z. & Soh, C.K. (2005). Integrated optimal design of vibration control system for smart beam using genetic algorithms, *Journal of Sound and Vibrations*, Vol.282, pp.1293–1307.

Part 2

Acoustic Wave Based Microdevices

5

Multilayered Structure as a Novel Material for Surface Acoustic Wave Devices: Physical Insight

Natalya Naumenko
Moscow Steel and Alloys Institute (Technological University)
Russia

1. Introduction

Since 70-ies, when the first delay lines and filters employing surface acoustic waves (SAW) were designed and fabricated, the use of SAW devices in special and commercial applications has expanded rapidly and the range of their working parameters was extended significantly (Hashimoto, 2000; Ruppel, 2001, 2002). In the last decade, their wide application in communication systems, cellular phones and base stations, wireless temperature and gas sensors has placed new requirements to SAW devices, such as very high operating frequencies (up to 10 GHz), low insertion loss, about 1 dB, high power durability, stable parameters at high temperatures etc.

The main element of a SAW device is a piezoelectric substrate with an interdigital transducer (IDT) used for generation and detection of SAW in the substrate. The number of single crystals utilized as substrates in SAW devices did not increase substantially since 70-ies because a new material must satisfy the list of strict requirements to be applied in commercial SAW devices: sufficiently strong piezoelectric effect, low or moderate variation of SAW velocity with temperature, low cost of as-grown large size crystals for fabrication of 4-inch wafers, long-term power durability, well developed and non-expensive fabrication process for SAW devices etc. Today only few single crystals are utilized as substrates in SAW devices: lithium niobate, LiNbO$_3$ (LN), lithium tantalate, LiTaO$_3$ (LT), quartz, SiO$_2$, lithium tetraborate, Li$_2$B$_4$O$_7$ (LBO), langasite, La$_3$Ga$_5$SiO$_{14}$ (LGS) and some crystals of LGS group (LGT, LGN etc.) with similar properties.

The SAW velocities in these single crystals do not exceed 4000 m/s, which limit the highest operating frequencies of SAW devices by 2.5-3 GHz because of limitations imposed by the line-resolution technology of IDT fabrication. The minimum achievable insertion loss and maximum bandwidth of SAW devices depend on the electromechanical coupling coefficient, which can be evaluated for SAW as $k^2 \approx 2\Delta V/V$, where ΔV is the difference between SAW velocities on free and electrically shorted surfaces. The largest values of k^2 can be obtained in some orientations of LN and LT. Ferroelectric properties of these materials are responsible for a strong piezoelectric effect. As a result, k^2 reaches 5.7% in LN and 1.2% in LT, for SAW. For leaky SAW (LSAW) propagating in rotated Y-cuts of both crystals, the coupling is higher and can exceed 20% for LN and 5% for LT. However, LSAW attenuates because of its leakage into the bulk waves when it propagates along the crystal surface. As a result, insertion loss of a SAW device increases. Attenuation coefficient depends on a crystal cut and IDT geometry. For example, in 36° to 48° rotated YX cuts of LT and in 41° to 76° YX rotated YX cuts of LN, high electromechanical coupling of LSAW can be combined with low attenuation coefficient via simultaneous optimization of orientation and electrode structure (Naumenko & Abbott, US patents, 2003, 2004). When these substrates are utilized in radiofrequency (RF) SAW filters with resonator-type structures, low insertion loss of 1dB or even less can be obtained. Today such low loss filters are widely used in mobile communication and navigation systems. The main drawback of these devices is high sensitivity of the characteristics to variations of temperature because the typical values of temperature coefficient of frequency (TCF) vary between -30 ppm/°C and -40 ppm/°C for LT and between -60 ppm/°C and -75 ppm/°C for LN.

Contrary to LN and LT, quartz is characterized by excellent temperature stability of SAW characteristics but low electromechanical coupling coefficient, $k^2 < 0.15\%$. Hence, even in resonator-type SAW filters with very narrow bandwidths, about 0.05%, where the loss of radiated energy is minimized due to the energy storage in a resonator, the best insertion loss achieved in a SAW device with matching circuits is only 2.5-4 dB.

In some orientations of LBO, LGS and other crystals of LGS group, zero TCF is combined with a moderate electromechanical coupling coefficient. However, these crystals have limited applications in commercial SAW devices because low SAW velocities restrict high-frequency applications on LGS and LBO dissolves in water and acid solutions, which prohibits application of conventional wafer fabrication processes to this material and finally results in an increased cost of SAW devices.

Hence, none of available single crystalline materials provides a combination of large piezoelectric coupling, zero TCF and high propagation velocity. A strong need in such material exists today, especially for application in SAW duplexers and multi-standard cellular phones, where the temperature compensation is the key issue because of necessity to divide a limited frequency bandwidth into few channels with no overlapping allowed in a wide range of operating temperatures. As an alternative to conventional SAW substrates, layered or multilayered (stratified) materials were studied extensively since 80-ies but only in the last decade some of these structures found commercial applications in SAW devices, due to the recent successes of thin film deposition technologies and development of robust simulation tools for design of SAW devices on layered structures.

2. Multilayered structures as materials for SAW devices

As described above, the increasing requirements to the substrate materials, on one side, and rapid development of thin film deposition technologies, on the other side, gave rise to the novel class of materials for SAW devices – layered or multilayered structures. One or more films of different materials deposited on a regular substrate can improve its characteristics significantly. A proper combination of a substrate and a thin film helps to overcome the limitations of the conventional SAW substrates. In this section, a brief overview of the layered structures will be given. The examples presented here are currently investigated by different research groups as promising compound SAW substrate materials or found already some applications in SAW devices. The focus is made on the recent achievements in material research mostly motivated by challenges of the rapidly developing market of communication devices. Based on this overview, a generalized multilayered structure, which includes all described examples, will be derived and a method of investigating this structure will be presented and then applied to few different structures, for which more detailed discussion of SAW propagation characteristics will be given.

The first example of a layered structure is a dielectric isotropic silicon dioxide (SiO$_2$) film deposited on one of rotated YX-cuts of LN or LT characterized by a high electromechanical coupling. Today these structures attract attention of researchers and SAW designers as the most promising candidates for application in SAW duplexers required in most of popular mobile phone systems (Kovacs et al., 2004; Kadota, 2007; Nakai et al., 2008; Nakanishi et al., 2008]. These RF devices must separate the transmitted and received signals in a narrow frequency interval and in a wide range of operating temperatures, e.g. between -30°C and +85°C. Therefore, substrate materials combining high propagation velocity, high electromechanical coupling and low TCF are strongly required. Due to the opposite signs of TCF in SiO$_2$ and LT or LN, a layered SiO$_2$/LT or SiO$_2$/LN structure allows to obtain the desired combination of characteristics. When it is utilized in a resonator-type filter, the electrodes of IDT and metal gratings are commonly built at the interface between SiO$_2$ and LN or LT. Besides, a heavy metal, such as copper, is utilized as an electrode material (Nakai et al., 2008). A layered structure with such electrode configuration is schematically presented as *Type 1* in Fig.1. The location of electrodes at the interface helps to keep a high electromechanical coupling and combine it with a large reflection coefficient in SAW resonators. A large metallization thickness effectively reduces resistive losses and results in a high Q factor of a SAW resonator. Another advantage of using heavy electrodes is a reduced propagation loss, which is achieved due to the transformation of LSAW into SAW. SAW devices with low insertion loss of 1-2 dB and low TCF of -7ppm/°C, have been successfully realized on such substrates (Kadota, 2007).

A thickness of SiO$_2$ film in SiO$_2$/LT and SiO$_2$/LN structures can vary within a wide range, from a few percent of a SAW wavelength up to a few wavelengths, to provide the required combination of electromechanical coupling, TCF and propagation loss. Moreover, SiO$_2$ film helps to isolate the working surface of a SAW device from environmental influences and facilitates packaging of SAW chip.

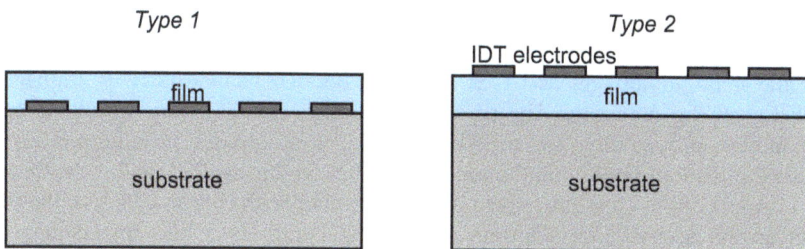

Fig. 1. Two typical structures with one thin film

Another electrode configuration in a structure with one thin film is schematically presented as *Type 2* in Fig. 1, with IDT located on the top surface. It is typical for a piezoelectric film on a non-piezoelectric substrate or on a substrate with low electromechanical coupling coefficient. As a piezoelectric film, zinc oxide ZnO is widely used (Kadota & Minakata, 1998; Nakahata et al., 2000; Emanetoglu et al., 2000; Brizoual et al., 2008). ZnO films are cheap and provide sufficiently high values of electromechanical coupling. The film deposition technique (e.g. magnetron sputtering) has been well developed for this material. Another piezoelectric film, which is extensively studied as a promising material for high-frequency SAW devices is aluminum nitride (AlN) (Benetti et al., 2005, 2008; Fujii et al., 2008; Omori et al., 2008). It is characterized by chemical stability, mechanical strength, high acoustic velocity and good dielectric quality. Some other piezoelectric films, like CdS or GaN, were investigated previously but did not receive as much attention as ZnO or AlN.

A piezoelectric film is usually combined with silica glass, silicon, sapphire or diamond substrate. Silica glass is cheap, the use of silicon as a wafer enables simple integration of IF and RF components in one chip, sapphire is characterized by high SAW velocities, up to 6000 m/s, and diamond provides the highest SAW velocities among all materials, up to 11000 m/s, and is being used for high frequency SAW devices in the GHz range. For example, SAW resonator with center frequency about 4.5 GHz was built on AlN/diamond structure characterized by SAW velocity about 10000 m/s and $k^2 \approx 1\%$ (Omori et al., 2008). A combination of SAW velocity about 5500 m/s and electromechanical coupling about 0.25% can be obtained in AlN/sapphire structure (Ballandras et al., 2004). To reduce TCF of a SAW device, ZnO film can be combined with quartz or LGS. For example, nearly zero TCF and k^2 about 1.8% was achieved for SAW in ZnO/quartz structure, via optimization of quartz orientation (Kadota et al., 2008).

One more structure, which can be referred to the *Type 1*, recently found application in SAW devices. It is a thin plate of a piezoelectric crystal, such as LN or LT with thickness 10-15 wavelengths, which is directly bonded to a dielectric or semiconductor wafer. The bonding technology (Eda et al., 2000) provides excellent contact between the two materials and allows fabrication of SAW devices with reproducible characteristics on a thin LN or LT plate bonded to a thick silicon or glass wafer. In these structures, high values of electromechanical coupling coefficients typical for LN and LT are combined with improved TCFs, due to low thermal expansion coefficients (TCE) determined by massive silicon, glass or sapphire wafer (Tsutsumi et al., 2004). An example of bonded wafer will be numerically investigated in section 4.

The quality of a contact between LN or LT plate and a silicon wafer can be improved if a thin SiO_2 film is deposited between these materials (Abbott et al., 2005). Such two-layered structure with IDT on the top surface is schematically shown as *Type 3* in Fig. 2. With silicon as a substrate, SiO_2 as the first film and LN or LT as the second film (plate), this structure can give the same advantages as LT/Si or LN/Si bonded wafers but with higher quality contact between the materials. The presence of additional SiO_2 film results in spurious acoustic modes propagating in a SAW device. These modes deteriorate the device performance and should be simulated properly to achieve the desired device characteristics.

Another example of the *Type 3* structure is a silicon wafer with isotropic SiO_2 as the first film and ZnO as the second film. Optimization of SiO_2 and ZnO film thicknesses enables obtaining of a structure with TCF=0 (Emanetoglu et al., 2000). A high frequency SAW device can be built if SiO_2 and ZnO films are deposited atop of a diamond or a sapphire substrate. With ZnO as the first film and isotropic SiO_2 as the second film, the preferential location of IDT electrodes is at the substrate-film interface (*Type 4*). Alternatively, IDT can be built on ZnO surface and then buried in SiO_2 overlay (*Type 5*). For example, Nakahata (Nakahata et al., 2000) reported on a SAW resonator using $SiO_2/ZnO/diamond$ structure with two different electrode configurations (*Type 4* and *Type 5*). Zero TCF, high velocity about 10000 m/s and $k^2 \approx 1.2\%$ were obtained for shear horizontally (SH) polarized SAW mode. A resonator with center frequency about 2.5 GHz, temperature compensated characteristics and low insertion loss was fabricated on this structure.

Nakahata (Nakahata et al., 1995) reported one more example of the two-layered structure, which can be referred to *Type 4*. It is a ZnO film on a silicon wafer with thin isotropic diamond layer between them. The following SAW characteristics have been obtained: velocity $V \approx 8050$ m/s, $k^2 \approx 1.42$ %, TCF≈ 0.

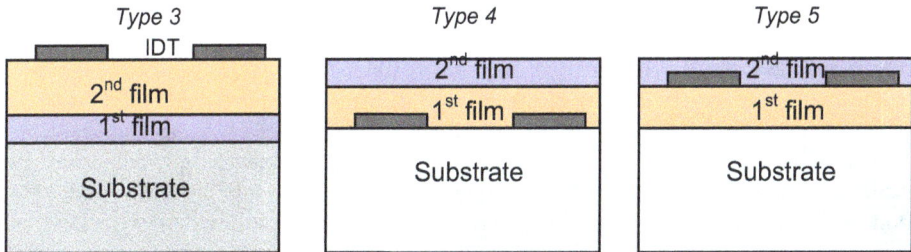

Fig. 2. Three typical structures with two thin films

The examples described above are not aimed at comprehensive survey of layered structures potentially applicable in SAW devices but demonstrate that a variety of layered structures can be referred to a few basic types. A unified approach to analysis of acoustic modes in different layered structures would be beneficial for optimization of SAW devices, because such approach allows comparing characteristics of the same SAW design built on different combinations of film and substrate materials.

The simulation of SAW characteristics is an important part of the SAW device design procedure. In a specified structure, such simulation must take into account orientation of each material if it is anisotropic, film thicknesses, a thickness and shape of IDT electrodes, electrode width to pitch ratio etc. Besides, the accurate analysis of all modes propagating in the investigated structure is required, including the main SAW or LSAW mode and all spurious modes generated by IDT in the specified frequency interval. A number of spurious modes grows with a number of layers and increasing of their thicknesses, which makes the simulation procedure more complicated. Moreover, with increasing film thickness SAW changes its nature and eventually transforms into a new type of acoustic wave. However, the characteristics of any acoustic mode change continuously with this transformation.

The variation of film thicknesses within wide range helps to obtain a variety of *novel* materials with different combinations of characteristics demanded for SAW devices of different applications. After a proper combination of materials is selected, the geometrical parameters of a multilayered structure must be optimized to satisfy the desired electrical specification, including frequency bandwidth, insertion loss, out-of-band rejection, shape factor of frequency response or Q factor of a SAW resonator, temperature deviation of frequency etc. It is a common practice to optimize film and electrode thicknesses and other geometrical parameters of IDTs simultaneously with orientations of anisotropic materials included in the layered structure, to achieve the best SAW device performance.

The challenges described above require a robust, fast and universal numerical technique, which could be applied to different types of multilayered structures, with film thicknesses varying within wide range and allowing transformation of SAW into boundary waves, plate modes or other types of acoustic waves. Such technique is described in the next section.

3. An advanced numerical technique for analysis of acoustic waves in multilayered structures

The most popular numerical technique used for simulation of SAW characteristic in multi-layered structures is *Transfer Matrix Method* (TMM) (Adler, 1990). It is based on the matrix formalism suggested by Stroh (Stroh, 1965) for solution of a SAW problem in anisotropic media. For each material of a multilayered structure, TTM assumes building of a *fundamental acoustic tensor* dependent on the material constants and the analyzed orientation. Then the characteristics of the partial acoustic modes are found as the eigen vectors and eigen values of the matrix associated with this tensor. Finally, the *transfer matrix* is calculated, which characterizes the change of acoustic fields within the analyzed layer. The method is fast and convenient and does not impose any limitations on the number of layers. However, it is known to work unstable when the film thickness exceeds 3-5 wavelength because of bad conditioned matrices built of elements, some of which exponentially decay and others exponentially grow with film thickness. These elements are associated with *incident* and *reflected* modes. As suggested by Tan (Tan, 2002) and Reinhardt (Reinhardt et al., 2003), a separate treatment of these two groups of partial modes helps to avoid the instability and extends the range of the analyzed thicknesses from zero to infinite value.

Another limitation of the previously reported numerical techniques developed for analysis of SAW in multilayered structures is their focusing on a certain type of acoustic waves, which is related to a fixed type of a multilayered structure, with analytically defined boundary conditions at the interfaces and external boundaries. For example, acoustic waves propagating in a substrate with a thin film of finite thickness and *stress-free* boundary conditions at the top surface are different from acoustic waves propagating in the same combination of materials when the film thickness tends to infinite value. In practice, the results obtained with different versions of software using fixed boundary conditions often diverge and do not allow seeing how the wave characteristics change continuously with variation of film thickness within wide range. For example, the software *FEMSDA* (Endoh et al., 1995; Hashimoto et al., 2007, 2008), which is very popular among SAW device designers, includes separate versions for analysis of SAW in a substrate with a thin film and for investigation of boundary waves. In the first version, a film thickness providing robust calculations does not exceed a half-wavelength.

To overcome the described limitation, an advanced numerical technique was developed (Naumenko, 2009, 2010). It can be applied to a variety of multilayered structures and types of acoustic waves. The universal character of the software is achieved due to characterization of the air as a dielectric medium with a very small density and elastic stiffness constants and treatment of this medium as an example of a dielectric. The same numerical methods are applied to this and other materials, which compose a layered structure. With such approach, it is not necessary to fix the stress-free boundary conditions at the top surface of a thin film or at the boundaries of a finite-thickness plate to find acoustic waves propagating in these structures. The stress-free boundary conditions are automatically simulated with high accuracy when the adjacent medium is specified as the air.

The developed technique refers to a multilayered structure schematically shown in Fig. 3, in which a metal film or IDT is located between N upper and M lower layers, where M and N can vary between one and ten or more if necessary.

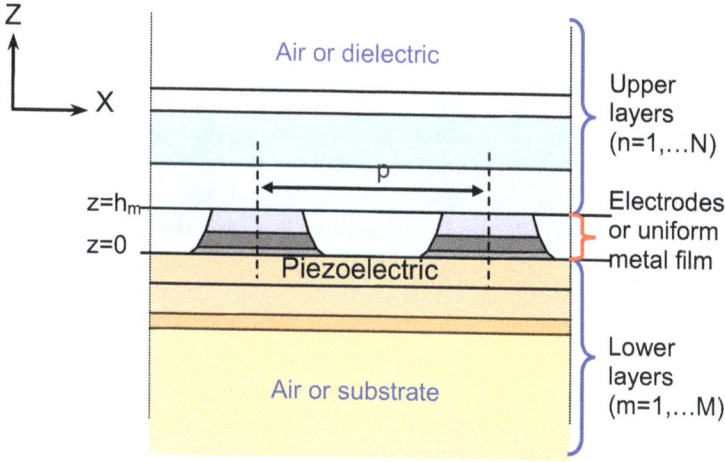

Fig. 3. Schematic drawing of analyzed multilayered structure

Analysis starts from the uppermost or lowermost half-infinite material, in which the wave structure is calculated. It can be a dielectric, a piezoelectric material or the air. In each adjacent finite-thickness layer, the transformation of the wave structure is deduced via separate treatment of incident and reflected partial modes. It means that the reflection and transmission matrix coefficients replace the transfer matrix to escape numerical noise at film thicknesses exceeding 3-5 wavelengths. For the structures with few dielectric (isotropic) films, the variation of the dielectric permittivity within each film characterized by the finite thickness h and dielectric permittivity ε_{film} is taken into account via the well known recursive equation (Ingebrigtsen, 1969):

$$\varepsilon(z+h) = \varepsilon_{film} \frac{\left[\varepsilon(z) + \varepsilon_{film} \cdot th(kh)\right]}{\left[\varepsilon_{film} + \varepsilon(z) \cdot th(kh)\right]} , \tag{1}$$

where k is the wave number. Analysis of the lower and upper multilayered half-spaces is considered completed when the wave structure has been determined at $z=0$ and $z=h_m$, where h_m is a metal film thickness, and the surface impedance matrices $\hat{Z}_{UP}(k)$ and $\hat{Z}_{LOW}(k)$ have been calculated at the upper and lower boundaries of the metal film. Each of these matrices characterizes the ratio between the vectors of displacements $\mathbf{u}$ and normal stresses $\mathbf{T}$ at the analyzed interface, $\hat{Z} = \mathbf{u}\mathbf{T}^{-1}$. A piezoelectric material is characterized by the generalized 4-dimensional displacement and stress vectors, with added electrostatic potential ϕ and normal electrical displacement D, respectively. The matrices $\hat{Z}_{UP}(k)$ and $\hat{Z}_{LOW}(k)$ comprise the information about the layers located above and below the metal film and enable simple formulation of electrical boundary conditions at $z=0$ and $z=h_m$. If the mass load of metal film is included in the analysis of the upper N layers, then the function of *effective dielectric permittivity* (EDP) $\varepsilon_s(k)$ can be calculated at $z=0$ (Ingebrigtsen, 1969; Milsom et al., 1977). This function relates the electric charge σ at the surface to the electrostatic potential φ,

$$\sigma = |k| \cdot \varepsilon_s(k) \cdot \varphi \tag{2}$$

and can be used for extraction of the velocities and electromechanical coupling coefficients of SAW and other acoustic modes propagating in a multilayered structure. EDP function was originally introduced for semi-infinite piezoelectric medium, but it is also an efficient tool for analysis of acoustic waves in layered structures.

The numerical method described above was extended to a periodic metal grating sandwiched between two multilayered structures. In this case, the spectral domain analysis (SDA) of the upper and lower multi-layered half-spaces is combined with the finite-element method (FEM) applied to simulation of the electrode region. To some extent, the developed *SDA-FEM-SDA* technique (Naumenko, 2010) can be considered as an advanced FEMSDA method. In this case, the function of Harmonic Admittance $Y(f,s)$ (Blotekjear et al, 1973; Zang et al, 1993) is calculated. Similar to EDP function, Harmonic Admittance relates the electric charge on the electrodes of the infinite periodic grating Q to the applied harmonically varying voltage V_e,

$$Q = (j\omega)^{-1} Y(f,s) \cdot V_e \tag{3}$$

and depends on frequency f and the normalized wave number, $s = p/\lambda$, where p is a pitch of the grating. This function can be used for simulation of a SAW resonator and calculation of its main parameters: resonant and anti-resonant frequencies, reflection coefficient etc. Also it comprises the information about SAW and other acoustic modes, which can be generated in the analyzed layered structure, and their characteristics can be extracted from $Y(f,s)$.

It should be mentioned that a numerical procedure of finding eigen modes of the fundamental acoustic tensor can be successfully applied to the air as an example of isotropic medium and the calculated SAW characteristic do not differ noticeably from that obtained with stress-free conditions set analytically at the film/air interface. The method and software SDA-FEM-SDA enable analysis of electrodes composed of few different metal layers and having a complicated profile, with different edge angles in metal layers. The gaps between electrodes may be empty or filled with a dielectric material. Due to these options, some important physical effects can be simulated, such as the effect of a sublayer (e.g. titanium) often used for better adherence of electrode metal to the substrate or the effect of nonrectangular electrode profile on a SAW device performance.

The developed numerical technique can be applied to different types of multilayered structures and different acoustic waves can be investigated, for example,

a.   SAW and LSAW in a piezoelectric substrate;
b.   SAW and LSAW in a substrate with one or few thin films (e.g. Love modes);
c.   plate modes generated by IDT and propagating in a thin plate (e.g. Lamb waves);
d.   boundary waves propagating along the interface between two half-infinite media;
e.   acoustic waves propagating in a thin piezoelectric plate bonded to a thick wafer.

In addition, a continuous transformation between different types of acoustic waves can be observed. It gives a physical insight into the mechanisms of wave transformation with increasing film thickness. An example of wave transformation will be considered in Section 5.

4. Multilayered structures: Examples of analysis

In this section, few examples of application of the developed numerical technique to the structures of practical importance are presented.

4.1 SiO$_2$/42°YX LT with Al film at the interface

The first example is a dielectric film on a piezoelectric substrate, which can be referred to the *Type 2* structure shown in Fig.1. The calculated characteristics of LSAWs propagating in SiO$_2$/42°YX LT with uniform Al film atop of the structure are presented in Fig. 4.

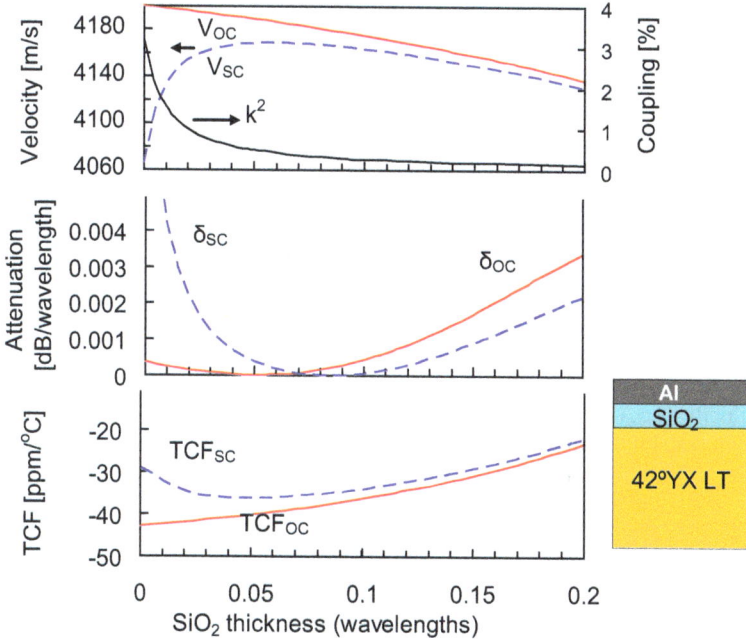

Fig. 4. Characteristics of leaky SAW propagating in SiO$_2$/42°YX LT with uniform Al film (h_{Al}=5%λ) atop of SiO$_2$ film, as functions of normalized SiO$_2$ film thickness. OC or SC electrical conditions are analyzed

SiO$_2$ is an isotropic dielectric film. LSAW velocities V, attenuation coefficients δ (in dB/λ, where λ is LSAW wavelength) and TCF are presented as functions of the normalized SiO$_2$ film thickness, h/λ. These characteristics have been calculated for the open-circuited (OC) and short-circuited (SC) electrical conditions in Al film. The finite thickness of a metal film (h_{Al}=5%λ) was taken into account. The difference between the OC and SC velocities determines the electromechanical coupling coefficient k^2, which decreases rapidly with increasing dielectric film thickness. The behavior of attenuation coefficients depends on the electrical condition. The functions $\delta_{OC}(h_{SiO2})$ and $\delta_{SC}(h_{SiO2})$ reach nearly zero values at h_{SiO2}=5%λ and h_{SiO2}=8%λ, respectively. Therefore, the variation of SiO$_2$ film thickness can be used for minimization of propagation losses in a SAW device. Due to the opposite signs of TCF in SiO$_2$ film and LT substrate, in the layered structure the absolute value of TCF reduces with increasing film thickness but does not reach zero in the investigated interval of film thickness. However, larger SiO_2 thicknesses are not considered because the electromechanical coupling coefficient becomes too low for practical applications. Further improvement of TCF is possible if IDT is located at the SiO_2/LT interface (*Type 1* structure). Such example will be considered in Section 5.

4.2 ZnO/sapphire: Existence of high-velocity SAW

The next example, ZnO film on a sapphire substrate, is a layered structure with a piezoelectric film on a non-piezoelectric substrate, which is potentially useful for high-frequency SAW device applications. Also this example demonstrates that a deposition of a thin film on a substrate can result in the existence of a high-velocity SAW, which cannot exist in a crystal without a thin film.

The typical SAW velocities in layered structures using a sapphire substrate are about 5500 m/s. Leaky SAWs, which have higher velocities, propagate with certain attenuation dependent on orientation of a substrate. Two types of LSAW can exist in crystals and layered structures: common or low-velocity LSAW, with velocities confined in the interval between that of the slow quasi-shear and fast quasi-shear *limiting* bulk acoustic waves (LBAWs), and high-velocity LSAW with velocities between that of the fast quasi-shear and quasi-longitudinal LBAWs. The *limiting* BAW is a bulk wave, which propagates in the sagittal plane (i.e. the plane, which is normal to the substrate surface and parallel to the propagation direction of SAW or LSAW) and is characterized by the group velocity parallel to the substrate surface. Usually a high velocity LSAW is not suitable for SAW device applications because of its fast attenuation. In some crystals with strong acoustic anisotropy, a high-velocity LSAW degenerates into the quasi-longitudinal LBAW in selected orientations, and low-attenuated LSAW can propagate around such orientations (Naumenko, 1996). For example, such waves exist in some orientations of quartz and LBO.

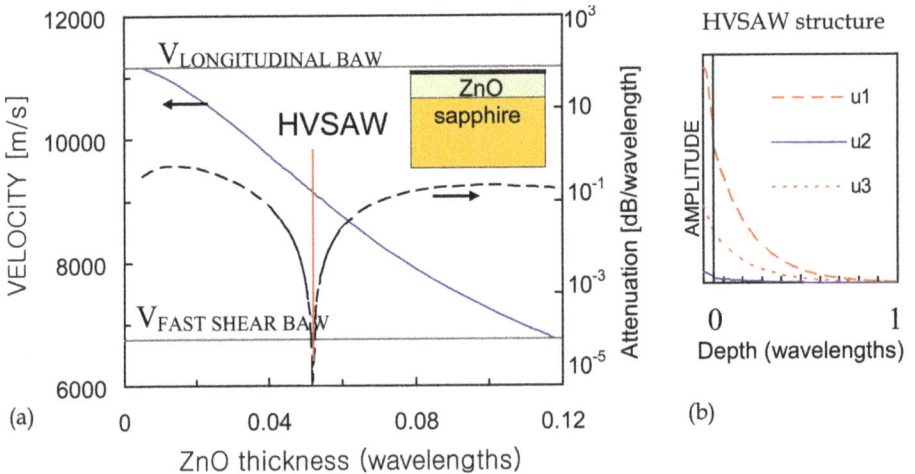

Fig. 5. (a) Velocity and attenuation coefficients of high-velocity leaky waves propagating in sapphire, Euler angles (0°, -20.3°, 0°), with ZnO film, as functions of normalized ZnO thickness, and (b) Displacements as function of depth, for HVSAW existing at h_{ZnO}=5.16%λ

Fig. 5, a shows the calculated velocities and attenuation coefficients of high-velocity LSAWs propagating in ZnO/sapphire structure, when the sapphire orientation is defined by the Euler angles (0°, -20.3°, 0°). The thickness of a metal film deposited atop of ZnO is not taken into account. The velocities of the fast quasi-shear and quasi-longitudinal LBAWs in sapphire are shown as $V_{FAST\ SHEAR\ BAW}$ and $V_{LONGITUDINAL\ BAW}$, respectively. The non-attenuated SAW solution, which occurs on a high-velocity LSAW branch at $h_{ZnO}\approx5.16\%\lambda$ (Fig.5, a), is not a quasi-bulk wave described above. Fig.5,b shows the mechanical displacements, which follow the propagation of this wave. The analyzed solution is a sagittally polarized surface wave, which attenuates exponentially into the depth of sapphire, similar to Rayleigh SAW. Such *high velocity SAW* (HVSAW) can not exist without perturbation of a free crystal surface, e.g. by deposition of a thin film or a metal grating. The existence of this type of waves was revealed via numerical analysis of experimental data on SAW modes in ZnO/SiC structure (Didenko et al., 2000) and confirmed by other examples of layered structures, which support propagation of these waves, such as ZnO/diamond, Zno/sapphire etc. (Naumenko & Didenko, 1999).

The HVSAW found in ZnO/sapphire may be attractive for applications in high-frequency SAW devices because it combines a high propagation velocity exceeding 9000 m/s with electromechanical coupling about 0.3%. With deposition of a metal film or a periodic metal grating the wave with attractive properties does not disappear but a combination of cut angle and ZnO thickness should be optimized properly to provide low LSAW attenuation.

4.3 Al grating on 46°YX LT/Si bonded wafer

In this example, SAW modes are investigated in LT/Si structure with a periodic Al grating atop of LT, which can be obtained experimentally by bonding LT plate to a silicon wafer. In such structure, the TCF may be dramatically reduced compared to regular LT wafer.

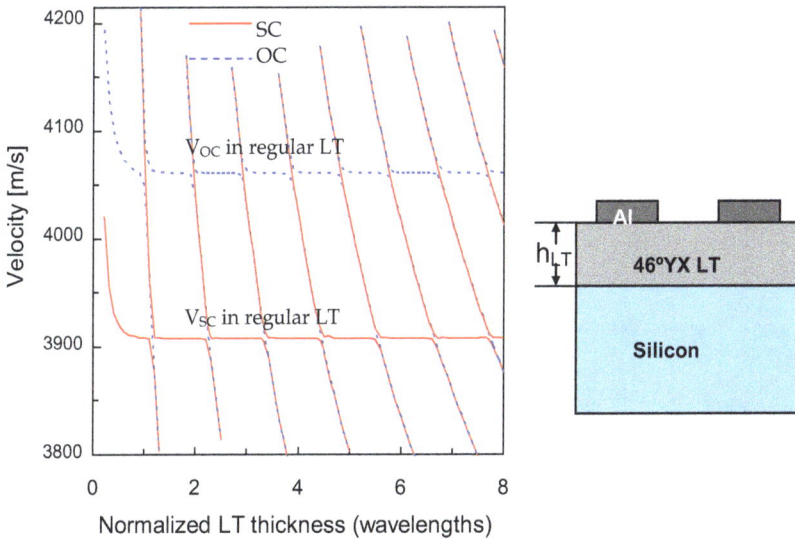

Fig. 6. Velocities of acoustic modes propagating in 46°YX LT plate bonded to silicon wafer, with OC and SC Al grating, as functions of normalized LT thickness, $h_{Al}=9\%\lambda$

The results shown in Fig.6 were obtained with the software SDA-FEM-SDA because the effect of a periodic metal grating is different from that of a uniform metal film. When LT thickness is about 1-2 wavelengths, the velocity of the SAW mode propagating in the layered structure is nearly the same as in a regular LT substrate with electrode thickness $h_{Al}=9\%\lambda$ but the wave characteristics are perturbed by interactions with multiple plate modes, the number of which grows with increasing LT thickness. It should be noted that in a regular 46°YX LT substrate the acoustic wave propagating with nearly the same velocity has a leaky wave nature. The bonding of LT plate to a silicon wafer results in the transformation of LSAW into SAW. The leakage of the wave becomes impossible because in silicon the shear BAW propagates faster than the analyzed SAW mode.

4.4 Al grating on 46°YX LT/SiO₂/Si bonded wafer
The next example differs from the previous one by the additional SiO₂ film between LT and Si wafer. A silicon dioxide layer is required to be deposited on the LT wafer to enable a stronger bond to silicon (Abbott et al., 2005). The presence of additional SiO₂ film impacts the acoustic and electrical properties of a bonded wafer and SAW resonators built on its surface. The spectrum of acoustic modes propagating in LT/SiO₂/Si bonded wafer looks more complicated than in LT/Si structure and depends on SiO₂ and LT thicknesses. In Fig. 7, the velocities of acoustic modes are shown as functions of the normalized SiO₂ film thickness when LT thickness is fixed, $h_{LT}=6\lambda$.

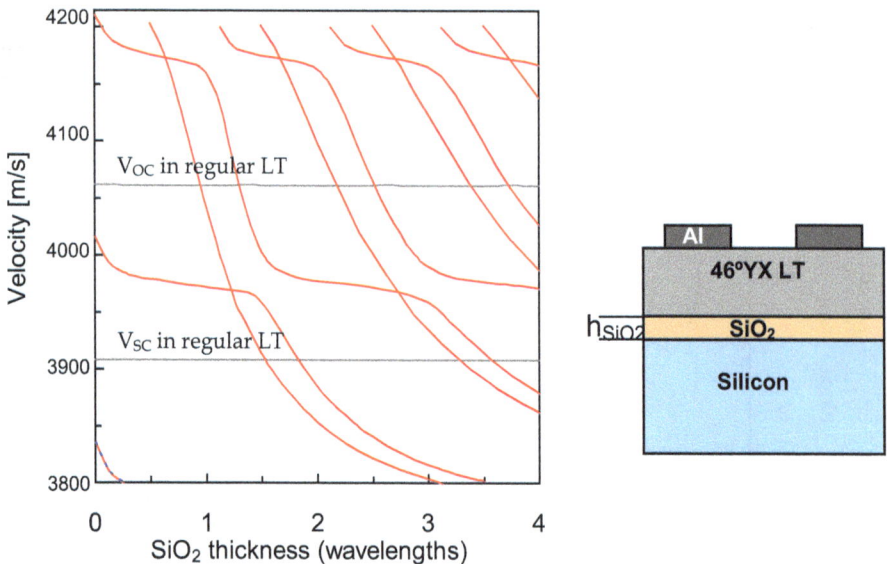

Fig. 7. Velocities of acoustic modes propagating in 46°YX LT/SiO₂/Si structure with Al grating, as functions of normalized thickness of SiO₂ film, when $h_{LT}=6\lambda$ and $h_{Al}=9\%\lambda$

4.5 Cu grating buried in SiO₂, on 46°YX LT/SiO₂/Si bonded wafer
This numerical example refers to the same bonded structure as described above but demonstrates the effect of additional SiO₂ overlay deposited over periodic metal grating

electrodes. Such layer is aimed at further improvement of the temperature characteristics of SAW devices.

In this case, Cu electrodes are investigated because this metal provides higher reflection coefficients in SAW resonators and hence lower insertion loss in RF SAW filters than Al, when electrodes are buried in SiO_2 overlay. The velocities of acoustic modes are shown in Fig.8 as functions of the normalized intermediate SiO_2 film thickness, while the thicknesses of SiO_2 overlay, LT plate and Cu electrodes are fixed, $h_{OVL}=0.25\%\lambda$, $h_{LT}=4\lambda$ and $h_{Cu}=2.5\%\lambda$, respectively. The dispersion of SAW velocities in $46°LT/SiO_2/Si$ (Fig. 8,a) and $SiO_2/46°LT/SiO_2/Si$ (Fig. 8,b) structures demonstrates that in practice very accurate simulation is required to account for all spurious modes, because these modes may affect the admittance of a SAW resonator.

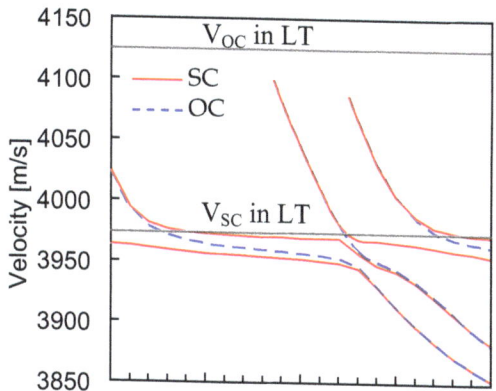
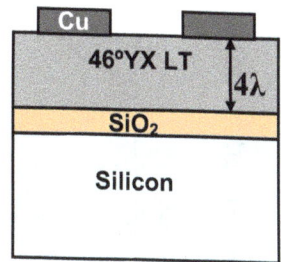

(a)

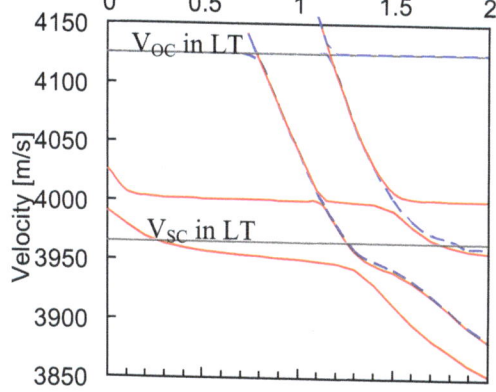
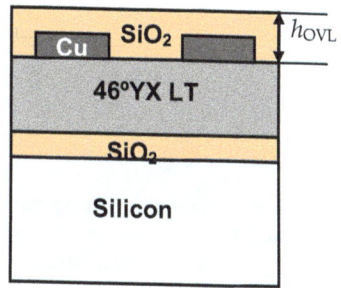

(b)

Fig. 8. Velocities of acoustic modes as functions of normalized thickness of intermediate SiO_2 film, when $h_{LT}=4\lambda$ and $h_{Cu}=2.5\%\lambda$, (a) in $46°LT/SiO_2/Si$ structure with Cu grating, and (b) in $SiO_2/46°LT/SiO_2/Si$ structure with Cu grating atop of $46°LT$ plate and overlay thickness $h_{OVL}=0.25\%\lambda$

Fig. 9 demonstrates an example of calculated admittance of a periodic Cu grating used with different multilayered structures described above. In addition to the main SAW mode, which exhibits resonance and anti-resonance, the multiple spurious modes propagate in the analyzed layered structures and can disturb a SAW resonator performance. The frequencies of the spurious responses are very sensitive to the thicknesses of the layers in a multilayered structure. However, the accurate simulation of these modes enables optimization of the layered structure to minimize the effect of the spurious modes on a resonator performance.

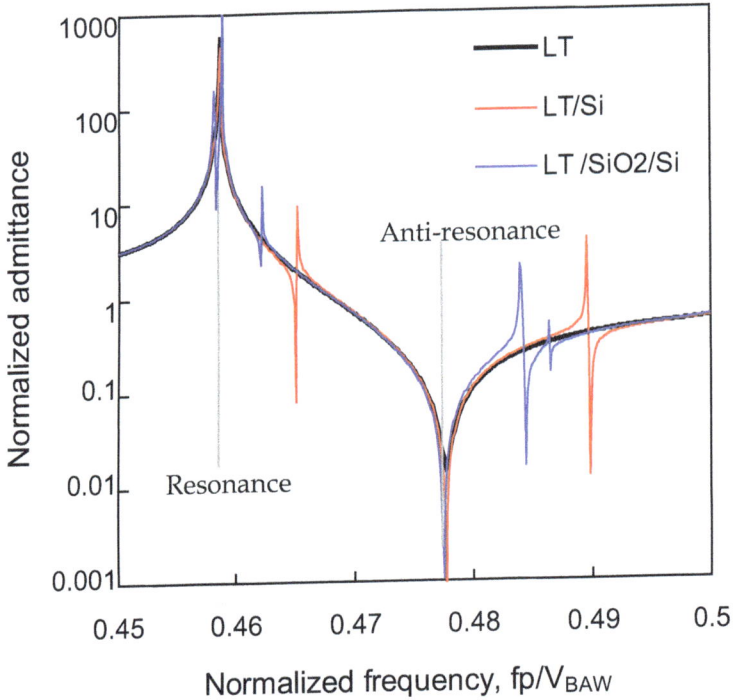

Fig. 9. Admittance of infinite periodic grating with Al electrodes (h_{Al}=10%λ), as function of normalized frequency, when the grating is built on regular 42°YX LT substrate, on LT plate (h_{LT}=5.5λ) bonded to Si substrate and atop of LT/SiO$_2$/Si structure, with h_{SiO2}=2λ. V$_{BAW}$=4214.636 m/s

The effect of spurious modes on a resonator performance can be minimized by the variation of rotation angle of LT plate. If one of rotated YX cuts of LN is used as a piezoelectric plate, the insertion loss of a resonator SAW device can be reduced in a wider bandwidth.

The examples presented in this section illustrate possible applications of the developed numerical technique and demonstrate that being a part of design tools for SAW device simulation it can be also an efficient tool for analysis of acoustic modes in multilayered structures.

5. Transformation of acoustic waves in anisotropic layered structures

In isotropic or highly symmetric materials, acoustic waves are characterized by mechanical displacements either belonging to the sagittal plane, $u=(u_1, 0, u_3)$, or normal to this plane, $u=(0, u_2, 0)$, i.e. such waves are Rayleigh-type or SH polarized modes. Similar solutions occur in some symmetric orientations of materials belonging to the lower symmetry classes. Such solutions have been extensively studied analytically. To the best author's knowledge, the most comprehensive overview of different types of acoustic waves existing in substrates with thin films, in thin plates and at the boundary between two half-infinite media was made by Viktorov (Viktorov, 1967, 1981). Some statements, which refer to acoustic waves propagating in isotropic structures, are listed below.

a. In isotropic substrate with isotropic thin film, SH-polarized Love waves can propagate if the shear BAW propagates faster in a substrate than in a film.

b. Along the boundary between two rigidly connected isotropic half-infinite media, the sagittally polarized Stonely waves can propagate. These waves are usually trapped near the interface between two media, with penetration depth about one wavelength.

c. SH-polarized boundary waves can exist if additional thin film with lower shear BAW velocity is added between two half-infinite media.

d. In isotropic thin plates, two types of waves can exist: sagittaly polarized Lamb waves (symmetric and anti-symmetric modes) and SH-polarized plate modes. With increasing plate thickness, higher-order modes appear and their number increases.

e. With plate thickness decreasing to zero, the first-order symmetric Lamb mode degenerates into the longitudinal BAW. With increasing plate thickness, this mode finally degenerates into two Rayleigh SAWs propagating along the boundaries of the plate.

f. Higher-order plate modes arise from the shear and longitudinal BAWs at certain cut-off thicknesses and have a structure of standing waves propagating between two interfaces.

The layered structures used in SAW devices must include at least one anisotropic material to provide a piezoelectric coupling of SAW with IDT. Anisotropy results in mixed polarizations of acoustic modes propagating in thin films, plates and along the interface between two media. The transformation of each mode with increasing film thickness is unique and requires separate investigation. Whereas analytical study of such waves is possible only in some symmetric orientations, the numerical technique presented in this chapter enables calculation of the wave characteristics and analysis of displacements associated with different acoustic modes. Its application to multilayered structures can reveal the mechanisms of wave transformation. The understanding of these mechanisms helps to select properly the thicknesses of metal and dielectric or piezoelectric layers to ensure the propagation of a required acoustic wave.

An example of such investigation is presented here. It refers to 42°YX LT with SiO$_2$ film. Similar structure was considered in Section 4.1, but in the present example a periodic grating is analyzed instead of a thin metal film and this grating is located at the interface between LT substrate and SiO$_2$ film. As a metal of the grating, copper is considered. Such structure is of great practical importance as potential material for RF SAW devices with improved temperature characteristics.

The calculated velocities of acoustic modes propagating in LT substrate with copper grating are shown in Fig.10,a as functions of the Cu electrode thickness. Fig.10,b shows the velocities of acoustic waves propagating in SiO_2/Cu grating/LT structure, as functions of the normalized SiO_2 thickness, with Cu thickness fixed, $h_{Cu}=0.2\lambda$. The analyzed Cu thicknesses look too high for application in SAW resonators but provide better insight into the wave transformation mechanisms, which is a purpose of this investigation.

When the metal thickness is small, two acoustic waves exist in LT, SAW1 and LSAW. With increasing Cu thickness, the velocities of both modes decrease rapidly and at $h_{Cu}=0.075\lambda$ LSAW transforms into the second SAW mode, SAW2. With further increasing of electrode thickness, two SAW modes interact with each other. To avoid discontinuities in the characteristics of two SAW modes, these modes are distinguished by their velocities: $V_{SAW2}>V_{SAW1}$.

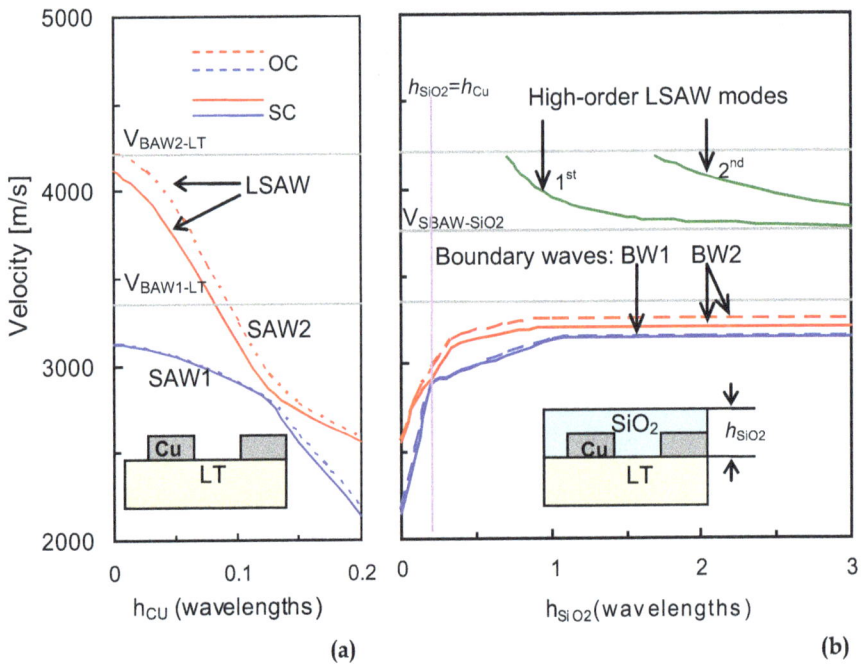

Fig. 10. SAW and leaky SAW velocities in 42°YX LT, (a) with Cu film, as functions of film thickness; (b) with Cu and SiO_2 films, as functions of normalized SiO_2 film thickness

If metal thickness is fixed ($h_{Cu}=0.2\lambda$) and the gaps between electrodes are filled with SiO_2, the velocities of two SAW modes grow rapidly (Fig.10,b). Another interaction between SAW1 and SAW2 occurs at $h_{SiO2}\approx h_{Cu}$, i.e. when the top surface of the whole structure becomes flat. With increasing SiO_2 film thickness two SAW modes finally transform into the boundary waves, BW1 and BW2. The boundary waves propagate with velocities lower than that of the shear BAW in SiO_2. The wave BW2 shows electromechanical coupling sufficient for application in resonator SAW devices, $k^2=3.49\%$. For this mode, TCF grows from -31 ppm/°C in LT substrate up to 10 ppm/°C in a layered structure with SC grating and from -43 ppm/°C up to 6 ppm/°C with OC grating. At $h_{SiO2}>0.7\lambda$, higher-order plate modes arise from the fast shear limiting BAW in LT.

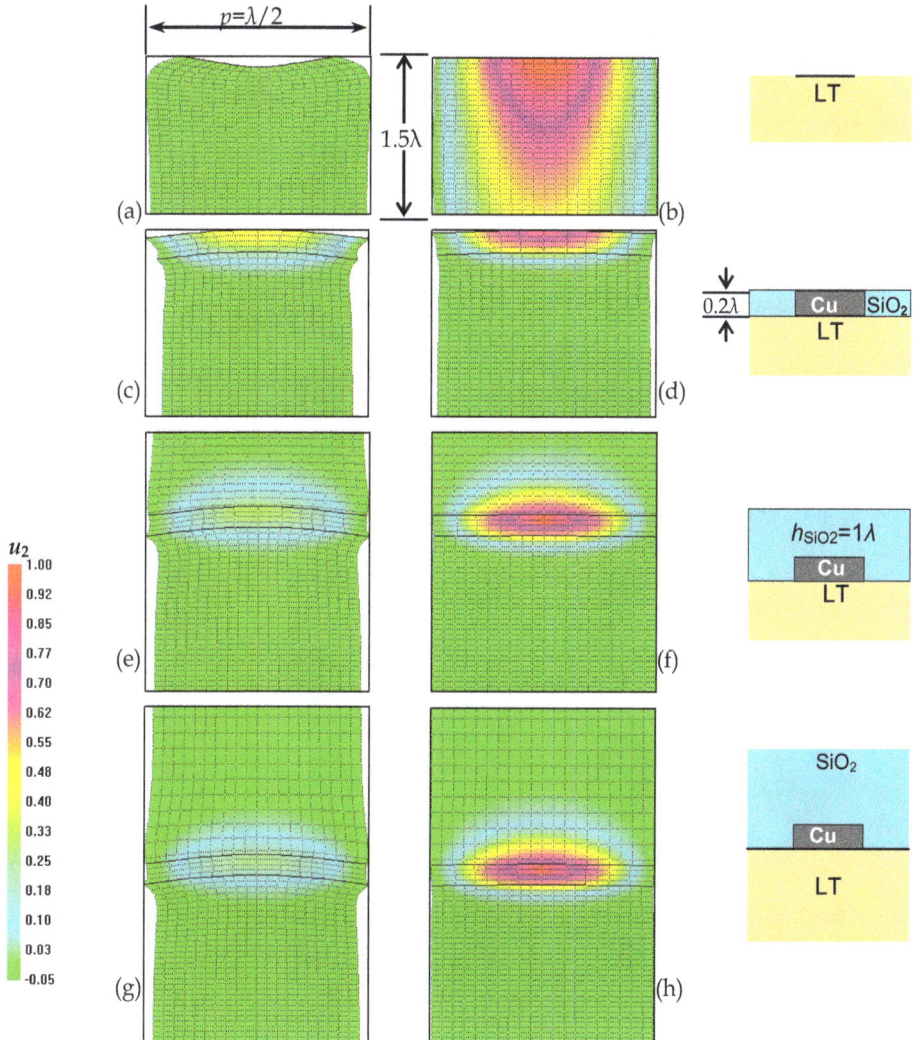

Fig. 11. Transformation of two SAW modes in 42°YX LT with Cu grating with increasing SiO₂ film thickness; (a), (c), (e), (g) SAW1, (b), (d), (f), (h) SAW2

Fig.11 demonstrates how the acoustic fields associated with the modes SAW1 and SAW2 change with increasing SiO₂ film thickness. OC electrical condition is considered, by way of example. The components of mechanical displacements in the sagittal plane are revealed as perturbations of the regular mesh and the values of SH components are presented as colored diagrams. When Cu thickness is small (Fig.11,a,b), SAW1 is nearly perfect Rayleigh wave and LSAW is a quasi-bulk SH-type wave slowly attenuating with depth. When $h_{Cu} > 0.075\lambda$, LSAW transforms into SAW2, which is also SH-type wave. At Cu thicknesses about 0.12λ, both SAW modes are perturbed by interaction between them. In the interval of

Cu thicknesses between 0.12λ and 0.2λ, the modes SAW1 and SAW2, which have been determined as the lower-velocity and higher velocity modes, exchange their polarizations. After the second interaction, which occurs at $h_{SiO2}=h_{Cu}=0.2\lambda$, SAW1 and SAW2 turn back into Rayleigh-type and SH-type waves, respectively. However, at $h_{Cu}=0.2\lambda$ (Fig.11,c,d) both waves still have mixed polarizations. With increasing SiO_2 film thickness, SAW1 and SAW2 transform into the boundary waves BW1 and BW2, respectively (Fig. 11, g, h), with acoustic waves localized in Cu grating and around it. The boundary waves have mixed polarizations, which would be impossible in isotropic substrate with isotropic thin film, but due to specific features of the analyzed LT orientation, BW1 is nearly sagittally polarized wave and BW2 is nearly pure SH wave. BW2 penetrates deeper into SiO_2 film than into LT substrate. A numerical analysis reveals that SiO_2 thickness about 1.5λ is sufficient for transformation of SAW into the boundary wave.

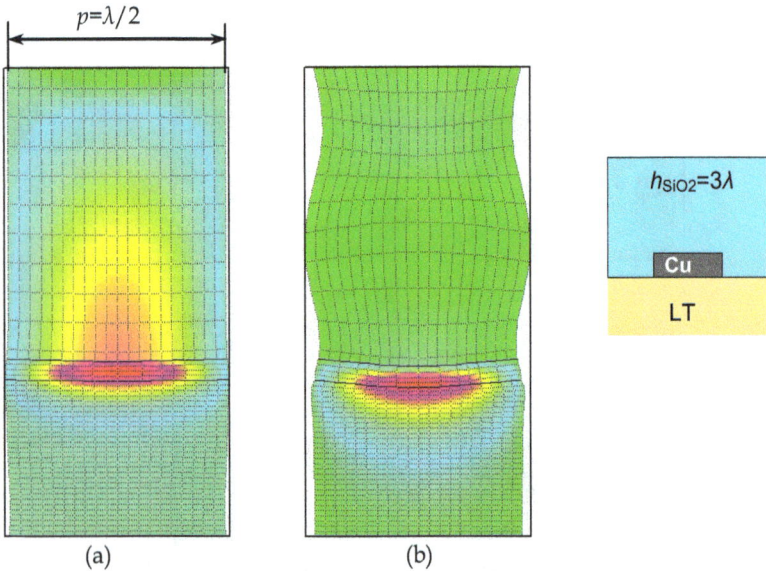

Fig. 12. Acoustic fields associated with two higher-order modes propagating in 42°YX LT with Cu grating and SiO_2 film when $h_{SiO2}=3\lambda$. (a) 1st mode; (b) 2nd mode

The acoustic fields associated with propagation of the two higher-order modes (Fig.10,b) have been also investigated. These modes have leaky wave nature. Fig.12 illustrates the structure of these modes at $h_{SiO2}=3\lambda$. The first mode, which exists when $h_{SiO2}>0.7\lambda$, has SH polarization deeply penetrating into SiO_2 (Fig.12,a). With increasing SiO_2 thickness, this mode degenerates into the SH BAW propagating in SiO_2. The second mode, which exists at $h_{SiO2}>1.6\lambda$, looks as a combination of SH-type SAW in LT substrate with Cu grating and sagittally polarized quasi-bulk wave propagating in SiO_2 (Fig.12,b), with the amplitude of SH polarization component much higher than the amplitudes of two other components. This example demonstrates the effect of anisotropy on the propagation of acoustic waves in multilayered structures and reveals another application of the numerical technique described in this chapter. Similar investigation can be performed for other structures of practical importance or serve to study the wave processes in multilayered structures.

6. Conclusion

In this chapter, some layered and multilayered structures, which look promising as substrates for modern SAW devices developed for applications in cellular phones, communication and navigation systems have been overviewed. A universal numerical technique, which enables fast and accurate analysis of these and other structures have been presented and, by way of example, applied to some multilayered structures of practical interest. The physical insight into the mechanisms of SAW transformation with increasing film thickness in a multilayered structure was provided via simulation of acoustic fields in one of the structures.

7. References

Abbott B.P., Naumenko N.F. & Caron J. (2005). Characterization of bonded wafer for RF filters with reduced TCF, *Proceedings of IEEE Ultrasonics Symposium*, pp.926-929, Rotterdam, the Netherlands, Sept. 2005.

Adler E. L. (1990). Matrix methods applied to acoustic waves in multi-layers, *IEEE Trans. Ultrason., Ferroelectr., Freq. Control*, Vol. 37, No 6, pp. 485-490.

Ballandras S., Reinhardt A., Laude L., Soufyane A., Camou S. & Ventura P. (2004). Simulations of surface acoustic wave devices built on stratified media using mixed finite element/boundary integral formulation, *J. Appl. Phys.*, Vol. 96, No 12, pp.7731-7741.

Benetti M., Cannatta D., Pietrantonio F. D. & Verona E. (2005). Growth of AlN piezoelectric film on diamond for high-frequency surface acoustic wave devices, *IEEE Trans. Ultrason., Ferroelectr., Freq. Control*, Vol. 52 , pp. 1806–1811.

Benetti M., Cannata D., Di Pietrantonio F., Verona E., Almaviva S., Prestopino G., Verona C. & Verona-Rinati G. (2008). Surface acoustic wave devices on AlN/single-crystal diamond for high frequency and high performances operation, *Proceedings of IEEE Ultrasonics Symposium*, pp. 1924-1927, Beijing, China, Nov. 2008.

Blotekjear K., Ingebrigsten K. & Skeie H. (1973). A method for analyzing waves in structures consisting of metal strips on dispersive media, *IEEE Trans. Electronic Devices*, vol. 20, no. 12, pp. 1133–1138.

Brizoual L. L., Sarry F. , Elmazria O. , Alnot P. , Ballandras S. & Pastureaud T. (2008). GHz frequency ZnO/Si SAW device, *IEEE Trans. Ultrason., Ferroelectr., Freq. Control*, Vol. 55, No 2, pp. 442–450.

Didenko I. S., Hickernell F. S. & Naumenko N. F. (2000). The theoretical and experimental characterization of the SAW propagation properties for zinc oxide films on silicon carbide, *IEEE Trans. Ultrason., Ferroelectr., Freq. Control*, Vol. 47, No. 1, pp.179-187.

Eda K., Onishi K., Sato H., Taguchi Y. & Tomita M. (2000). Direct bonding of piezoelectric materials and its applications, *Proceedings of IEEE Ultrasonics Symposium* , pp. 299-309, San Juan, Puerto Rico, Sep. 2000.

Emanetoglu N.W., Patounakis G., Muthukumar S. & Lu Y. (2000). Analysis of temperature compensated modes in ZnO/SiO₂/Si multilayer structures, *Proceedings of IEEE Ultrasonics Symposium*, pp. 325-328, San Juan, Puerto Rico, Sep. 2000.

Endoh G., Hashimoto K. & Yamagouchi M. (1995). SAW propagation characteristics by Finite Element Method and Spectral Domain Analysis, *Jpn. J. Appl. Phys.*, Vol. 34, No 5B, pp. 2638-2641.

Fujii S., Kawano S., Umeda T., Fujioka M. & Yoda M. (2008). Development of a 6GHz resonator by using an AlN diamond structure, *Proceedings of IEEE Ultrasonics Symposium*, pp. 1916-1919, Beijing, China, Nov. 2008.

Hashimoto, K. Y. (2000). *Surface Acoustic Wave Devices in Telecommunications: Modeling and Simulation*, Springer, ISBN 354067232X, USA.

Hashimoto K., Omori T. & Yamaguchi M. (2007). Extended FEM/SDA Software for Characterizing Surface Acoustic Wave Propagation in Multi-layered Structures, *Proceedings of IEEE Ultrasonics Symposium*, pp. 711-714, Rome, Italy, Nov. 2008.

Hashimoto K., Wang Y., Omori T., Yamaguchi M., Kadota M., Kando H. & Shibahara T. (2008). Piezoelectric Boundary Waves: Their Underlying Physics and Applications, *Proceedings of IEEE Ultrasonics Symposium*, pp. 999-1005, New York, USA, Oct. 2007.

Ingebrigtsen K.A. (1969). Surface waves in piezoelectrics, *J. Appl. Phys.*, Vol. 40, No 7, pp.2681-2686.

Kadota M. & Minakata M. (1998). Piezoelectric Properties of ZnO Films on a Sapphire Substrate Deposited by an RF-Magnetron-Mode ECR Sputtering System, *Jpn. J. Appl. Phys.*, Vol.37, pp.2923-2926.

Kadota M. (2007). High Performance and Miniature Surface Acoustic Wave Devices with Excellent Temperature Stability Using High Density Metal Electrodes, *Proceedings of IEEE Ultrasonics Symposium*, pp.496-506, New York, USA, Oct. 2007.

Kadota M., Nakao T., Murata T. & Matsuda K. (2008). Surface Acoustic Wave Filter in High Frequency Range with Narrow Bandwidth and Excellent Temperature Property, *Proceedings of IEEE Ultrasonics Symposium*, pp. 1584-1587, Beijing, China, Nov. 2008.

Kovacs G., Ruile W., Jakob M., Rosler U., Maier E., Knauer U. & Zottl H. (2004). A SAW Duplexer with Superior Temperature Characteristics for US-PCS, *Proceedings of IEEE Ultrasonics Symposium*, pp. 974-977, Montreal, Canada, Aug. 2004.

Milsom R. F., Reilly N.C.H. & Redwood M. (1977). Analysis of generation and detection of surface and bulk acoustic waves by interdigital transducers, *IEEE Trans. Sonics and Ultrasonics*, Vol. 24, pp. 147-166.

Nakahata H., Higaki K., Fujii S. , Hachigo A., Kitabayashi H. , Tanabe K., Seki Y. & Shikata S. (1995). SAW Devices on Diamond, *Proceedings of IEEE Ultrasonics Symposium*, pp.361-370, Seattle, USA, Nov. 1995.

Nakahata H., Hachigo A., Itakura K., Fujii S. & Shikata S. (2000). SAW Resonators on SiO₂/ZnO/Diamond Structure in GHz Range, *Proceedings of IEEE Freq. Contr. Symposium*, pp. 315-320, Kansas City, USA, June 2000.

Nakai Y., Nakao T. , Nishiyama K. & Kadota M. (2008). Surface Acoustic Wave Duplexer composed of SiO₂ film with Convex and Concave on Cu-electrodes/LiNbO3 Structure, *Proceedings of IEEE Ultrasonics Symposium*, pp.1580-1583, Beijing, China, Nov. 2008.

Nakanishi H., Nakamura H. , Hamaoka Y., Kamiguchi H. & Iwasaki Y. (2008). Small-sized SAW Duplexers with Wide Duplex Gap on a $SiO_2/Al/LiNbO_3$ structure by using Novel Rayleigh-mode Spurious Suppression Technique, *Proceedings of IEEE Ultrasonics Symposium*, pp. 1588-1591, Beijing, China, Nov. 2008.

Naumenko N. F. (1996). Application of exceptional wave theory to materials used in surface acoustic wave devices", *J. Appl. Phys.*, Vol.79, pp. 8936-8943.

Naumenko N.F. & Didenko I.S. (1999). High-velocity surface acoustic waves in diamond and sapphire with zinc oxide film, *Appl. Phys. Lett.*, Vol. 75, No 19, pp. 3029-3031.

Naumenko N.F. & Abbott B.P. (2003). Surface Acoustic Wave Devices Using Optimized Cuts of a Piezoelectric substrate, *US Patent #6,556,104 B2*. Int.Cl. H03H 9/64. Apr.29, 2003

Naumenko N.F. & Abbott B.P. (2004). Surface Acoustic Wave Filter , *US Patent 6,833,774 B2*. Int.Cl. H03H 9/16. Dec. 21, 2004.

Naumenko N.F. (2009). Transformation of Surface Acoustic Waves into Boundary Waves in Piezoelectric/Metal/Dielectric Structures, *Proceedings of IEEE Ultrasonics Symposium*, pp.2635-2638, Rome, Italy, Sep. 2009.

Naumenko N.F. (2010). A Universal Technique for Analysis of Acoustic Waves in Periodic Grating Sandwiched Between Multi-Layered Structures and Its Application to Different Types of Waves, *Proceedings of IEEE Ultrasonics Symposium*, San Diego, USA, Oct. 2010.

Naumenko N.F. & Abbott B.P. (2010). Insight into the Wave Transformation Mechanisms via Numerical Simulation of Acoustic Fields in Piezoelectric/Metal Grating/ Dielectric Structures, *Proceedings of IEEE Ultrasonics Symposium*, San Diego, USA, Oct. 2010.

Omori T., Kobayashi A., Takagi Y., Hashimoto K. & Yamaguchi M. (2008). Fabrication of SHF range SAW devices on AlN/Diamond-substrate, *Proceedings of IEEE Ultrasonics Symposium*, pp. 196-200, Beijing, China, Nov. 2008.

Reinhardt A., Pastureaud T., Ballandras S. & Laude L. (2003). Scattering matrix method for modeling acoustic waves in piezoelectric, fluid, and metallic multilayers, *J. Appl. Phys.*, Vol. 94, No 10, pp.6923-6931.

Ruppel C.C.W. & Fjeldy T. A. Eds. (2001). *Advances in Surface Acoustic Wave Technology, Systems and Applications. Volume 2* . ISBN 981-02-4538-6. World Scientific Publishing Co. Pte Ltd, Singapore.

Ruppel, C. C. W., Reindl, L. & Weigel, R. (2002). SAW devices and their wireless communication applications, *IEEE Microwave Magazine*, ISSN 1527-3342 3(2): pp. 65–71.

Stroh A. N. (1962). Steady state problems in anisotropic elasticity, *J. Math. Phys.* Vol. 41., pp. 77-103.

Tan E. L. (2002). A robust formulation of SAW Green's functions for arbitrarily thick multilayeres at high frequencies, *IEEE Trans.Ultrason., Ferroelectr., Freq. Control*, Vol. 49, No 7, pp. 929-936.

Tsutsumi J. , Inoue S., Iwamoto Y., Miura M., Matsuda T., Satoh Y., Nishizawa T., Ueda M. & Ikata O. (2004). A Miniaturized 3x3-mm SAW Antenna Duplexer for the US-PCS band with Temperature-Compensated $LiTaO_3$/Sapphire Substrate, *Proceedings of IEEE Ultrasonics Symposium*, pp. 954-958, Montreal, Canada, Aug. 2004.

Viktorov, I.A. (1967). *Rayleigh and Lamb Waves: physical theory and applications*. Plenum Press, New York.

Viktorov I.A. (1981). *Surface sound waves in solids*. Moscow: Nauka (in Russian).

Zhang Y., Desbois J. & Boyer L. (1993). Characteristic parameters of surface acoustic waves in a periodic metal grating on a piezoelectric substrate, *IEEE Trans.Ultrason., Ferroelectr., Freq. Control*, vol. 40, pp. 183-192, May 1993.

6

Shear Mode Piezoelectric Thin Film Resonators

Takahiko Yanagitani
Nagoya Institute of Technology
Japan

1. Introduction

1.1 Shear mode bulk acoustic wave devices and sensors

Acoustic microsensor technique, well known as QCM (Quartz crystal microbalance) or TSM (Thickness shear mode) sensor, is an effective method to detect small mass loading on the sensor surface. This sensor can be operated even in liquid by using shear mode resonance. Therefore, shear mode piezoelectric film resonators are attractive for liquid microsensor technique such as biosensors and immunosensors.

Shear wave has some unique features compared with the longitudinal wave, for example, it has extremely low velocity in the liquid. Longitudinal wave velocity in the water is 1492.6 m/s, whereas, shear wave velocity in the water is 20-60 m/s at 20-200 MHz (Matsumoto et al., 2000). Therefore, shear mode vibrating solid maintains its vibration even in the liquid, because the difference of acoustic impedance which determines the refection coefficient of solid / liquid interface is very large in the case of shear wave.

The complex refection coefficient Γ of the interface is given as

$$\Gamma = \frac{Z_l - Z_s}{Z_l + Z_s} \tag{1}$$

where Z_s and Z_l are the complex acoustic impedance of solid and liquid.
Complex acoustic impedance can be written as

$$Z = R + jX = \left(\rho (c + j\omega\eta) \right)^{1/2} \tag{2}$$

R and X represent the real part and imaginary part of the acoustic impedance and ρ, c and η represent mass density, stiffness constant and viscosity in the medium, respectively.
Acoustic wave equation gives dispersion relation of

$$\left(\frac{\omega}{v} - j\alpha \right)^2 (c + j\omega\eta) = \rho\omega^2 \tag{3}$$

where v is velocity and α is attenuation factor (B. A. Auld, 1973).
According to (2) and (3), acoustic impedance gives

$$R = \frac{\rho v\omega^2}{\omega^2 + \alpha^2 v^2} \ , \ X = \frac{\rho v^2 \omega\alpha}{\omega^2 + \alpha^2 v^2} \tag{4}$$

Longitudinal and shear wave velocities of water were reported as 1492.6 m/s (Kushibiki et al., 1995) and 35 m/s (Matsumoto et al., 2000), respectively, at 100 MHz. Attenuations of longitudinal and shear wave in the water were also measured to be $\alpha/f^2 = 2.26 \times 10^{-14}$ neper·s^2/m (Kushibiki et al, 1995) and $\alpha/f^2 = 2.12 \times 10^{-9}$ neper·s^2/m, (Matsumoto et al., 2000) respectively. By substituting these values into Eq. (4), the complex longitudinal wave and shear wave acoustic impedance of the water can be estimated to be $1489000+j800$ N·s/m^3 and $14510+j17340$ N·s/m^3 at 100 MHz, respectively.

From these values and Eq. (1), when quartz resonator is immersed in water, the reflection coefficient of acoustic energy $|\Gamma|^2$ in an X-cut quartz vibrating in thickness extensional mode ($Z_s = 15.23 \times 10^{-6}$ N·s/m^3) is estimated to be only 68 % whereas that in an AT-cut quartz vibrating in thickness shear mode ($Z_s = 8.795 \times 10^{-6}$ N·s/m^3) is 98 %. This is because an AT-cut quartz has been used as a QCM or TSM sensor operating in liquid. Sensitivity of the QCM mass sensor is determined by the ratio of the mass and the entire mass of the vibrating part in the sensor, at constant sensor active area (Sauerbrey, 1959). Therefore, it is important to decrease thickness of the vibrating part of sensor. Shear mode thin film is promising for high sensitivity mass sensor.

1.2 Piezoelectric thin film for shear mode excitation

Piezoelectric thin film, which excites shear wave, is expected to provide higher sensitivity and IC compatibility, but it is not straightforward. To excite shear wave by standard sandwiched electrode configuration, polarization axis in the film must be tilted or parallel to the film plane. Although perovskite ferroelectric films have large piezoelectricity, their polarization axis is generally normal to the film surface due to the nature of crystal growth, difficultly of in-plane polarization treatment and domain control. Trigonal piezoelectric material such as LiNbO$_3$, LiTaO$_3$ and quartz are difficult to crystallize (tend to form amorphous structure) or to obtain a strong preferred orientation in polycrystalline film.

6mm wurtzite AlN and ZnO film can be easily crystallized, but they tend to develop their polarization axis (c-axis) perpendicular to the substrate plane. This c-axis oriented film cannot excite shear wave in the case of standard sandwiched electrode structure.

Crystalline orientation control for both in-plane and out-of-plane direction is necessary to excite shear wave. One solution is to use an epitaxial growth technique. However, the combinations of the shear mode piezoelectric film and substrate are limited due to the lattice mismatch. a-plane ZnO or AlN/r-plane sapphire (Mitsuyu et al., 1980; Wittstruck et al., 2003), a-plane ZnO/42° Y-X LiTaO$_3$ (Nakamura et al., 2000) where c-axis in the film is parallel to the substrate plane have been reported.

Ion beam orientation control technique (Yanagitani & Kiuchi, 2007c), which enables in-plane and out-of-plane orientation without use of epitaxial growth, is introduced in the third section. This technique is a good candidate for obtaining c-axis parallel films which excites pure shear wave without any excitation of longitudinal wave.

2. Electromechanical coupling properties of wurtzite crystal

Elastic and piezoelectric properties of wurtzite crystals vary with direction due to the crystal anisotropy. Electromechanical coupling changes as a function of the angle between the c-axis and the applied electric field direction (Foster et al., 1968; Auld, 1973).

The analytical model of a thin film resonator is shown in Fig. 1. The electric field is applied in the x_3 direction. The c-axis is assumed to lie in the x_1-x_3 plane and be inclined at an angle β with respect to the x_3 direction.

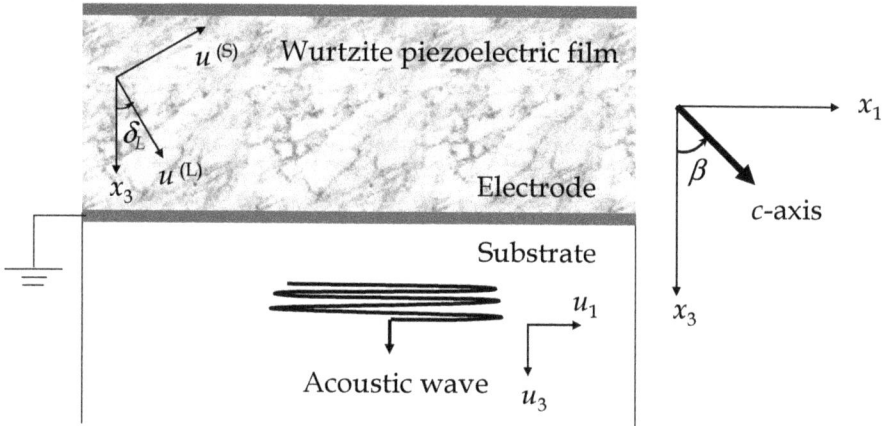

Fig. 1. Analytical model of a thin film resonator

The physical constants of the crystal in each direction are determined by the transformed coordinate of each constant tensor. Bond's method (Bond, 1943) for transforming the elastic and piezoelectric constant tensor is introduced below, which can be applied to the constant tensor with abbreviated subscript notation. For example, the transformation matrix $[a]$ of a clockwise rotation through an angle β about the x_2-axis is described by:

$$[a] = \begin{bmatrix} \cos\beta & 0 & -\sin\beta \\ 0 & 1 & 0 \\ \sin\beta & 0 & \cos\beta \end{bmatrix} \quad (5)$$

The dielectric constant ε' transforms as

$$[\varepsilon'] = [a]\,[\varepsilon]\,[a]^T. \quad (6)$$

The 6×6 transformation matrix of coefficients M is defined as

$$[M] = \begin{bmatrix}
a_{xx}^2 & a_{xy}^2 & a_{xz}^2 & 2a_{xy}a_{xz} & 2a_{xz}a_{xx} & 2a_{xx}a_{xy} \\
a_{yx}^2 & a_{yy}^2 & a_{yz}^2 & 2a_{yy}a_{yz} & 2a_{yz}a_{yx} & 2a_{yx}a_{yy} \\
a_{zx}^2 & a_{zx}^2 & a_{zz}^2 & 2a_{zy}a_{zz} & 2a_{zz}a_{zx} & 2a_{zx}a_{zy} \\
a_{yx}a_{zx} & a_{yy}a_{zy} & a_{yz}a_{zz} & a_{yy}a_{zz}+a_{yz}a_{zy} & a_{yx}a_{zz}+a_{yz}a_{zx} & a_{yy}a_{zx}+a_{yx}a_{zy} \\
a_{zx}a_{xx} & a_{zy}a_{xy} & a_{zz}a_{xz} & a_{xy}a_{zz}+a_{xz}a_{zy} & a_{xz}a_{zx}+a_{xx}a_{zz} & a_{xx}a_{zy}+a_{xy}a_{zx} \\
a_{xx}a_{yx} & a_{xy}a_{yy} & a_{xz}a_{yz} & a_{xy}a_{yz}+a_{xz}a_{yy} & a_{xz}a_{yx}+a_{xx}a_{yz} & a_{xx}a_{yy}+a_{xy}a_{yx}
\end{bmatrix} \quad (7)$$

Finally, using the above transformation matrix, transformed elastic constant and piezoelectric constant tensors c' and e' are obtained:

$$[c'] = [M] \, [c] \, [M]^T \, , \ [e'] = [M] \, [e] \, [M]^T \tag{8}$$

In the x_2 axis rotation of a hexagonal (6mm) crystal, the transformed stiffness and piezoelectric constant tensors c' and e' are given by

$$[c'] = \begin{bmatrix} c'_{11} & c'_{12} & c'_{13} & 0 & c'_{15} & 0 \\ c'_{12} & c'_{22} & c'_{23} & 0 & c'_{25} & 0 \\ c'_{13} & c'_{23} & c'_{33} & 0 & c'_{35} & 0 \\ 0 & 0 & 0 & c'_{44} & 0 & c'_{46} \\ c'_{15} & c'_{25} & c'_{35} & 0 & c'_{55} & 0 \\ 0 & 0 & 0 & c'_{46} & 0 & c'_{66} \end{bmatrix}, \ [e'] = \begin{bmatrix} e'_{11} & e'_{12} & e'_{13} & 0 & e'_{15} & 0 \\ 0 & 0 & 0 & e'_{24} & 0 & e'_{26} \\ e'_{31} & e'_{32} & e'_{33} & 0 & e'_{35} & 0 \end{bmatrix} \tag{9}$$

In case, wave propagation toward x_3 direction is only focused, the term of $\partial/\partial x_1$ and $\partial/\partial x_2$ can be ignored. Thus, the wave motion equation for the x_3 direction is given by mechanical displacement component u_1, u_2 and u_3:

$$\frac{\partial T_{31}}{\partial x_3} = \rho \frac{\partial^2 u_1}{\partial t^2} \tag{10a}$$

$$\frac{\partial T_{33}}{\partial x_3} = \rho \frac{\partial^2 u_3}{\partial t^2} \tag{10b}$$

$$\frac{\partial T_{32}}{\partial x_3} = \rho \frac{\partial^2 u_2}{\partial t^2} \tag{10c}$$

where

$$T_{31} = c'^E_{55} \frac{\partial u_1}{\partial x_3} + c'^E_{35} \frac{\partial u_3}{\partial x_3} + e'_{35} \frac{\partial \varphi}{\partial x_3} \tag{11a}$$

$$T_{33} = c'^E_{35} \frac{\partial u_1}{\partial x_3} + c'^E_{33} \frac{\partial u_3}{\partial x_3} + e'_{33} \frac{\partial \varphi}{\partial x_3} \tag{11b}$$

$$T_{32} = c'^E_{44} \frac{\partial u_2}{\partial x_3} \tag{11c}$$

As div D = 0, the electrostatic equation is given by

$$\frac{\partial D_3}{\partial x_3} = e'_{35} \frac{\partial^2 u_1}{\partial x_3^2} + e'_{33} \frac{\partial^2 u_3}{\partial x_3^2} - \varepsilon'^S_{33} \frac{\partial^2 \varepsilon}{\partial x_3^2} = 0 \tag{12}$$

In Eqs. (10)-(12), T_{31} and T_{33} are stress components, D_3 is the electric displacement, $c_{33}{}^E$, $c_{35}{}^E$ and $c_{55}{}^E$ are the stiffness constants with constant electric field, e_{33} and e_{35} are piezoelectric constants, $\varepsilon_{33}{}^S$ and $\varepsilon_{35}{}^S$ are dielectric constants with constant strain, and φ is the electric potential.

Equation (10c) describes a pure shear wave with a u_2 displacement component in the x_2 direction and propagates along the x_3 direction with a phase velocity of $\sqrt{c_{44}/\rho}$. Eqs. (10a) and (10b) represent a quasi-longitudinal wave and quasi-shear wave. These waves incorporate u_1, u_3, and φ, which are coupled with each other. It is well known that Eqs (10a), and (10b) have plane-wave solutions:

$$\begin{pmatrix} u_1 \\ u_3 \\ \varphi \end{pmatrix} = \begin{pmatrix} A \\ B \\ C \end{pmatrix} \exp\left\{ j\omega\left(t - \frac{x_3}{v}\right)\right\} \tag{13}$$

Substituting Eq. (13) into Eqs. (11) and (12), the simultaneous equations are obtained

$$\begin{pmatrix} c_{55}^D - \rho v^2 & c_{35}^D & 0 \\ \overline{c}_{35} & c_{33}^D - \rho v^2 & 0 \\ -e_{35} & -e_{33} & \varepsilon_{33}^S \end{pmatrix} \begin{pmatrix} A \\ B \\ C \end{pmatrix} = 0, \tag{14}$$

where

$$c_{33}^D = c_{33}'^E + \left(e_{33}'\right)^2 / \varepsilon_{33}'^S,$$

$$c_{35}^D = c_{35}'^E + \left(e_{33}' e_{35}'\right) / \varepsilon_{33}'^S, \tag{15}$$

$$c_{55}^D = c_{55}'^E + \left(e_{35}'\right)^2 / \varepsilon_{33}'^S.$$

A, B and C are all nonzero when the coefficient matrix in Eq. (14) is zero. From this condition, we obtain the phase velocity $v^{(L,S)}$ of a quasi-longitudinal wave and quasi-shear wave:

$$v^{(L,S)} = \left[\frac{c_{33}^D + c_{55}^D}{2\rho} \pm \sqrt{\left(\frac{c_{33}^D - c_{55}^D}{2\rho}\right)^2 + \left(\frac{c_{35}^D}{\rho}\right)^2} \right]^{\frac{1}{2}} \tag{16}$$

Figure 2 shows the calculated results of phase velocity of a quasi-longitudinal wave and quasi-shear wave for a ZnO crystal as function of the angle β between the c-axis and x_3 direction. Physical constants in a ZnO single crystal reported by Smith were used in the calculation (Smith, 1969).

The general solutions for u_1, u_3 and φ are given by

$$\begin{pmatrix} u_1 \\ u_3 \\ \varphi \end{pmatrix} = \begin{pmatrix} A_1 \\ B_1 \\ C_1 \end{pmatrix} \exp\left\{ j\omega\left(t - \frac{x_3}{V^{(+)}}\right)\right\} + \begin{pmatrix} A_2 \\ B_2 \\ C_2 \end{pmatrix} \exp\left\{ j\omega\left(t - \frac{x_3}{V^{(-)}}\right)\right\} \tag{17}$$

and

$$\frac{B_1}{A_1} = -\frac{A_2}{B_2} \tag{18}$$

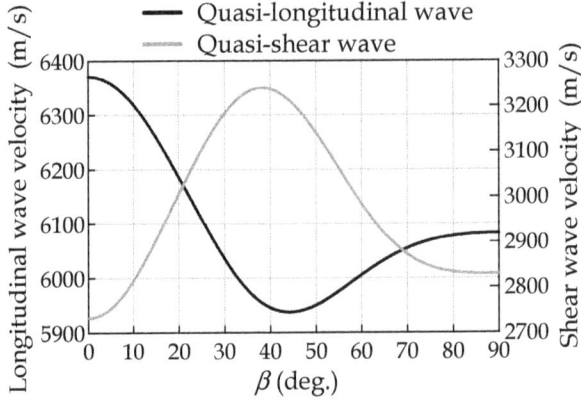

Fig. 2. Phase velocity of quasi-longitudinal wave and quasi-shear wave for a ZnO crystal as function of the angle β between the c-axis and x_3 direction

is derived from Eqs. (14) and (16). It can be seen that the displacement components of the quasi-longitudinal wave and quasi-shear wave are perpendicular to each other. From Eqs. (14) and (16), the angle δ_L between the quasi-longitudinal wave displacement u_3 and the x_3 direction and the angle δ_S between the quasi-shear wave displacement u_1 and the x_1 direction are given by

$$\delta_L = \tan^{-1}\left(\frac{A_1}{B_1}\right), \ \delta_S = \tan^{-1}\left(\frac{B_2}{A_2}\right) \tag{19}$$

The extensional and shear effective piezoelectric constants $e^{(L)}_{eff}$ and $e^{(S)}_{eff}$ are defined as

$$e^{(L)}_{eff} = e'_{35}\sin\delta_L + e'_{33}\cos\delta_L, \ e^{(S)}_{eff} = e'_{35}\cos\delta_S - e'_{33}\sin\delta_S \tag{20}$$

Thus, the quasi-longitudinal and quasi-shear-mode electromechanical coupling coefficients $k^{(L)}$ (transformed k_{33}) and $k^{(S)}$ (transformed k_{15}) are

$$\left(k^{(S)}\right)^2 = \left(e^{(S)}\right)^2 \Big/ e'_{33}\rho \left(V^{(-)}\right)^2, \ \left(k^{(L)}\right)^2 = \left(e^{(L)}\right)^2 \Big/ e'_{33}\rho \left(V^{(+)}\right)^2 \tag{21}.$$

Finally, Figs. 3 (a) and (b) show the calculated angle δ and the electromechanical coupling coefficients (k values) of the quasi-longitudinal and quasi-shear waves for the ZnO crystal as function of the angle β (Foster et al., 1968)

From these figures, we can see a relatively large shear-mode electromechanical coupling k_{15} = 0.39 at c-axis tilt angle of β = 28°. Several author reported FBAR (film bulk acoustic resonator)-type viscosity sensor and biosensor, consisting of c-axis tilted wurtzite films (Weber et al., 2006; Link et al., 2007; Wingqvist et al., 2007, 2009, 2010; Yanagitani, 2010, 2011a). However, the thickness extensional mode (longitudinal wave mode) also has the coupling of k_{33} = 0.155 and the displacement inclination angle of δ_S = 4.1° at angle of β = 28°. This indicates that the resonator excites both thickness extensional and shear mode (longitudinal and shear wave modes), and the shear displacement direction is not perpendicular to the propagation direction. Larger δ_S values may result in energy leakage

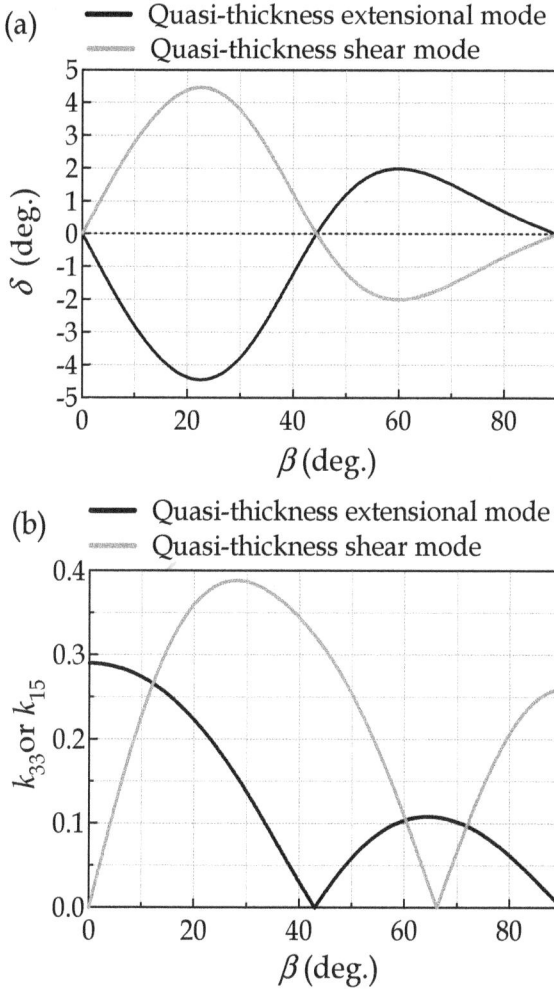

Fig. 3. (a) Angle δ between the wave displacement u and the x direction and
(b) electromechanical coupling coefficient of the quasi-longitudinal and quasi-shear waves
for the ZnO crystal as function of the angle β between the c-axis and x_3 direction

due to mode conversion in the reflection plane. This induces the decrease of Q value. Both of
the no extensional mode coupling and small δ_S values of 0.38° can be obtained at $\beta = 43°$,
however, it is difficult to adjust such as large c-axis tilt angle in a large area deposition. One
option is to use a pure-shear-mode ($\beta = 90°$) resonator to satisfy both the conditions of no
extensional coupling and $\delta_S = 0°$. Pure shear mode excitation can be achieved by two electric
field-orientation combination. One is to apply the cross-electric field to c-axis parallel film
by sandwiched electrode (Yanagitani et al., 2007d), and the other is to apply the in-plane
electric field to c-axis normal film by IDT electrode (Corso et al., 2007; Milyutin et al., 2008,
2010). Of course, the latter is the easiest way to obtain pure shear mode because deposition

technique of c-axis normal film has been well established, but effective electrometrical coupling is weak (k_{eff}=0.04-0.06) (Corso et al., 2007; Milyutin et al., 2008). The former has large electrometrical coupling (k_{15}=0.24) (Yanagitani et al., 2007a), and recently the c-axis parallel oriented film can be easily obtained by using ion beam orientation control technique (presented in next section), even in a large area (Kawamoto et al., 2010).

3. Ion beam orientation control technique for shear mode piezoelectric films

3.1 Ion beam orientation control of wurtzite thin film by ion beam irradiation

Polycrystalline films tend to grow in their most densely packed plane parallel to the substrate plane. Bravais proposed the empirical rule that the growth rate of the crystal plane is proportional to the surface atomic density. Namely, the lattice plane with higher surface atomic density grows more rapidly. Curie argued that the growth rate perpendicular to a plane is proportional to the surface free energy (Curie, 1885).

Ion bombardment during film deposition can modify this preferred orientation of the films. This is usually explained by a change in anisotropy of the growing rate of the crystal plane in the grain, which is reflected by the difference in the degree of the ion channeling effect or ion-induced damage in the crystal plane (Bradley et al., 1986; Ensinger, 1995; Ressler et al., 1997; Dong & Srolovitz, 1999). For example, during ion beam irradiation, the commonly observed <111> preferred orientation in a face-centered cubic film changes to a <110> preferred orientation, which corresponds to the easiest channeling direction (Van Wyk & Smith, 1980; Dobrev, 1982). In-plane texture controls have also been achieved by optimizing the incident angle of the ion beam (Yu et al., 1985; Iijima et al., 1992; Harper et al., 1997; Kaufman et al., 1999; Dong et al., 2001; Park et al., 2005).

In wurtzite films, for example, the surface energy densities of the (0001), (11$\bar{2}$0) and (10$\bar{1}$0) planes of the ZnO crystal are estimated to be 9.9, 12.3, 20.9 eV/nm^2, respectively (Fujimura et al., 1993). The (0001) plane has the lowest surface density. Thus, the ZnO film tends to grow along the [0001] direction. When wurtzite crystal is irradiated with ion beam, the most densely packed (0001) plane should incur more damage than the (10$\bar{1}$0) and (11$\bar{2}$0) planes, which correspond to channeling directions toward the ion beam irradiation. We can therefore expect that the thermodynamically preferred (0001) oriented grain growth will be disturbed by ion damage so that the damage-tolerant (10$\bar{1}$0) or (11$\bar{2}$0) orientated grains (c-axis parallel oriented grain) will preferentially develop instead.

On this basis, in-plane and out-of-plane orientation control of AlN and ZnO films by means of ion beam-assisted deposition technique, such as evaporation (Yanagitani & Kiuchi, 2007c) and sputtering (Yanagitani & Kiuchi, 2007e, 2011b) was achieved. c-axis parallel oriented can be obtained even in a conventional magnetron sputtering technique using a low pressure discharge (<0.1 Pa) (Yanagitani et al., 2005) or RF substrate bias (Takayanagi, 2011), which leads ion bombardment on the substrate. Figure 4 shows the XRD patterns of the ZnO films deposited with various ion energy and amount of flux in ion beam assisted evaporation (Yanagitani & Kiuchi, 2007c). Table 1 shows the ion current densities in the case of "Large ion flux" and "Small ion flux" in Fig 4. The tendency of the (10$\bar{1}$0) orientation is enhanced with increasing ion energy and amount of ion irradiation, demonstrating that the ion bombardment induced the (0001) orientation to change into a (10$\bar{1}$0) orientation, which corresponds to the ion channeling direction.

Ion energy	A: Large ion flux	B: Small ion flux
0.05 keV	0-5 μA/cm^2	
0.25 keV	30-50 μA/cm^2	
0.5 keV	190 μA/cm^2	140 μA/cm^2
0.75 keV	220 μA/cm^2	130 μA/cm^2
1.0 keV	240 μA/cm^2	120 μA/cm^2

Table 1. Ion current densities in "Large ion flux" and "Small ion flux"

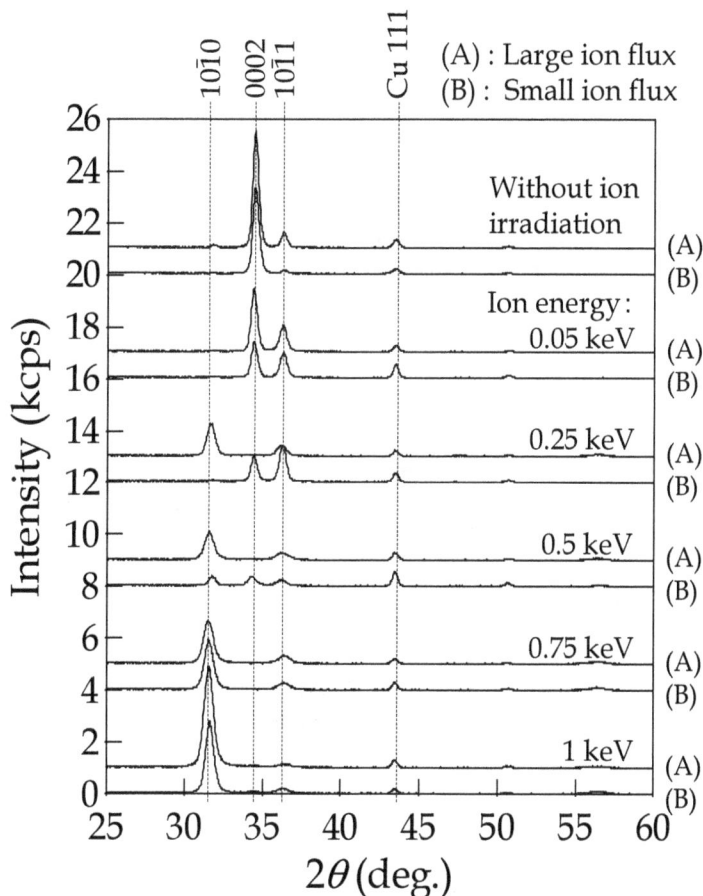

Fig. 4. 2θ-ω scan XRD patterns of the ZnO films deposited without ion irradiation, and with ion irradiation of 0-1 keV with "Large ion flux" and "Small ion flux" (Yanagitani & Kiuchi, 2007c)

Figure 5 shows the XRD patterns of the samples deposited under the conditions that various RF and DC bias are applied to the substrate. Although any dramatic change in usual (0001) preferred orientation is not occurred in the case of positive or negative DC bias, (0001) orientation changed to (11$\bar{2}$0) and (10$\bar{1}$0) orientation with the increase of RF bias power which induces the bombardment of positive ion on substrate. Interestingly, the order of the appearance of the (0001) to (11$\bar{2}$0) and (10$\bar{1}$0) corresponds to the order of increasing surface atomic density, which may be the order of damage tolerance against ion bombardment.

In order to excite shear wave in the c-axis parallel film, c-axis is required to orient not only in out-of-plane direction but also in in-plane direction. The ion beam orientation control technique allows us to control even in in-plane c-axis direction and polarization by the direction of beam incident direction (Yanagitani et al., 2007d).

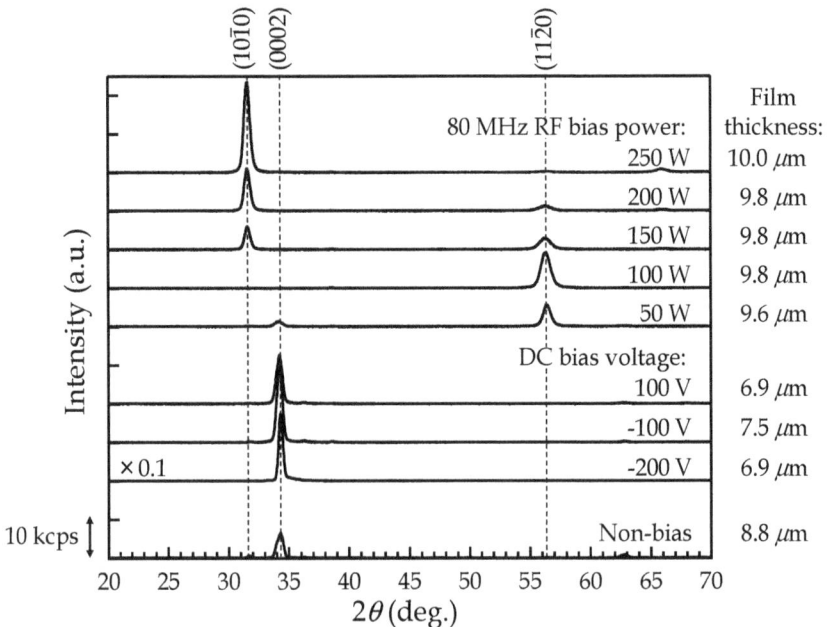

Fig. 5. 2θ-ω scan XRD patterns of the samples deposited without bias, with 80 MHz RF bias of 50 to 250 W, or with -200 to 100 DC bias. All samples were measured at the center of the bias electrode (Takayanagi et al., 2011)

4. Method for determining *k* values in piezoelectric thin films

4.1 *k* value determination using as-deposited structure (HBAR structure)

A method for determining piezoelectric property in thin films is described in this section. In general, electromechanical coupling coefficient (*k* value) in thin film can be easily determined by series and parallel resonant frequency of a FBAR consisting of top electrode layer/piezoelectric layer/bottom electrode layer or SMR (Solidly mounted resonator) consisting of top electrode layer/piezoelectric layer/bottom electrode layer/Bragg reflector. In case thickness of electrode film is negligible small compared with that of piezoelectric film. *k* of the piezoelectric film can be written as follows (Meeker, 1996):

$$k^2 = \frac{\pi}{2}\frac{f_s}{f_p}\tan\left(\frac{\pi}{2}\frac{f_p - f_s}{f_p}\right) \tag{22}$$

where f_p and f_s are the parallel resonant frequency and series resonant frequency, respectively.

However, it takes considerable time and effort to fabricate FBAR structure which have self-standing piezoelectric layer. It is convenient if k value can be determined from as deposited structure, namely so-called an HBAR (high-overtone bulk acoustic resonator) or composite resonator structure consisting of top electrode layer/piezoelectric layer/bottom electrode layer/thick substrate. Methods for determining the k value of the films from HBAR structure are more complex than that for the self-supported single piezoelectric film structure (FBAR structure). Several groups have investigated methods for the determination of k_t value from the HBAR structure (Hickernell, 1996; Naik, et al., 1998; Zhang et al., 2003). One of the easiest ways of k determination is to use a conversion loss characteristic of the HBAR structure. When the thickness of electrode layers is negligible small compared with that of piezoelectric layer, capacitive impedance of resonator is equal to the electrical source impedance, and k value of the piezoelectric layer is smaller than 0.3, conversion loss CL is approximately represented by k value at parallel resonant frequency (Foster et al., 1968):

$$CL \approx 10\log_{10}\frac{\pi}{8k^2}\cdot\frac{Z_s}{Z_p} \tag{23}$$

where, Z_s and Z_p is acoustic impedance of the substrate and piezoelectric layer, respectively. However, various inhomogeneities sometimes exist in the film resonator, such as non-negligible thick and heavy electrode layers, thickness taper, or the piezoelectrically inactive layer composed of randomly oriented gains growing in the initial stages of the deposition. In this case, the k values of the film can be determined so as to match the experimentally measured conversion losses (CL) of the resonators with theoretical minimum CL by taking k value as adjustable parameter. The theoretical CL in this case can be calculated by Mason's equivalent circuit model including electrode layer, film thickness taper and piezoelectrically inactive layer. This method allows various inhomogeneous effect of film to be taken into account (Yanagitani et al., 2007b, 2007c).

4.2 Experimental method to estimate conversion loss of HBAR structure

The experimental CL of HBAR can be determined from reflection coefficients (S_{11}) of the resonators, which can be obtained using a network analyzer with a microwave probe. The inverse Fourier transform of S_{11} frequency response of the resonator gives the impulse response of the resonator in the time domain. In the HBAR structure, the impulse response is expected to include echo pulse trains reflected from the bottom surface of the substrate, and the insertion loss of resonator can be obtained from the Fourier transform of the first echo in this impulse response. This experimental insertion loss $IL_{experiment}$ includes doubled CL in the piezoelectric film and round-trip diffraction loss DL and round-trip propagation loss PL in the silica glass substrate. Therefore, CL can be expressed as

$$CL = \frac{1}{2}\left(IL_{experiment} - DL - PL\right), \tag{24}$$

where diffraction loss DL can be calculated according to the method reported by Ogi *et al.* (Ogi et al., 1995). This method is based on integration of the velocity potential field in the divided small transducer elements, which allows calculation of the DL with electrode areas of various shapes. The round-trip propagation loss PL is given as

$$PL = 2d_s \frac{\alpha_s}{f^2},$$ (25)

where d_s is the thickness of the substrate, α_s represents the shear wave attenuation in the substrate, for example, $\alpha_s / f^2 = 19.9 \times 10^{-16}$ (dB s^2/m) in silica glass substrate (Fraser, 1967).

4.3 Conversion loss simulation in HBAR by Mason's equivalent circuit model

Electromechanical coupling coefficient k can be estimated by comparing an experimental CL with a theoretical CL of the HBAR. One-dimensional Mason's equivalent circuit model is convenient tool for simulating theoretical CL of the resonator. Generally, in case non-piezoelectric elastic solid vibrates in thickness mode, its can be described as T-type equivalent circuit (Fig. 6 (a)) where F_1 and F_2 are force and v_1 and v_2 are particle velocity acting on each surface of elastic solid. Piezoelectric elastic solid can be represented as the Mason's three ports equivalent circuit which includes additional electric terminal concerning electric voltage V and current I (Fig. 6 (b)) (Mason, 1964). Here, γ is propagation constant, Z is acoustic impedance and d_p is thickness of elastic solid. To take account of attenuation of vibration, mechanical quality factor Q_m is defined as $Q_m = c_r / c_i$ where c_r and c_i are real part and imaginary part of elastic constant, respectively. Using mechanical quality factor Q_m, propagation constant γ and acoustic impedance Z are given as:

$$\gamma = j\omega \sqrt{\frac{\rho}{c_r \left\{ 1 + j\ (1/Q_m) \right\}}} , \quad Z = S \sqrt{\rho c_r \left\{ 1 + j\ (1/Q_m) \right\}}$$ (26)

where ρ is density of the elastic solid and S is electrode area of the resonator.
Static capacitance C_0 and ratio of transformer ϕ_0 in the circuit are given as:

$$C_0 = \varepsilon_{11}^S \frac{S}{d_p} , \quad \phi_0 = \left[\frac{C_0 v_p Z_p}{d_p} \left(\frac{k_{15}^2}{1 - k_{15}^2} \right) \right]^{\frac{1}{2}},$$ (27),

where d is the thickness of the layers, ε_{11}^S is permittivity, and v is the velocity of the shear wave. Subscript p, $e1$, $e2$ and s respectively represent piezoelectric layer, top electrode layer, bottom electrode layer and substrate. k value affects the equivalent circuit through the ratio of transformer ϕ_0.

Equivalent circuit for the over-moded resonator structure is given in Fig. 7 by cascade arranging non-piezoelectric and piezoelectric part as described in Figs. 6 (a) and (b). Substrate thickness is assumed infinite to ignore reflection waves from bottom surface of the substrate in this case. When the surface of the top electrode is stress-free, the acoustic input port is shorted. As top electrode part circuit can be simplified, three-port circuit in Fig. 7 is transformed to the two-ports circuit shown in Fig. 8 (Rosenbaum, 1988).

(a)

v_1 $Z_p \tanh (\gamma_p d_p / 2)$ $Z_p \tanh (\gamma_p d_p / 2)$ v_2

$Z_p / \sinh (\gamma_p d_p)$

F_1 F_2

(b)

v_1 $Z_p \tanh (\gamma_p d_p / 2)$ $Z_p \tanh (\gamma_p d_p / 2)$ v_2

$Z_p / \sinh (\gamma_p d_p)$

F_1 i $-C_0$ F_2

V C_0

$1 : \phi_0$

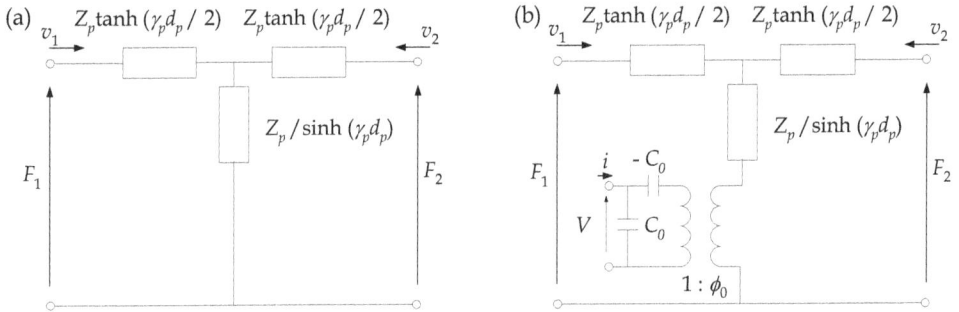

Fig. 6. Equivalent circuit model of (a) non-piezoelectric (b) piezoelectric elastic solid

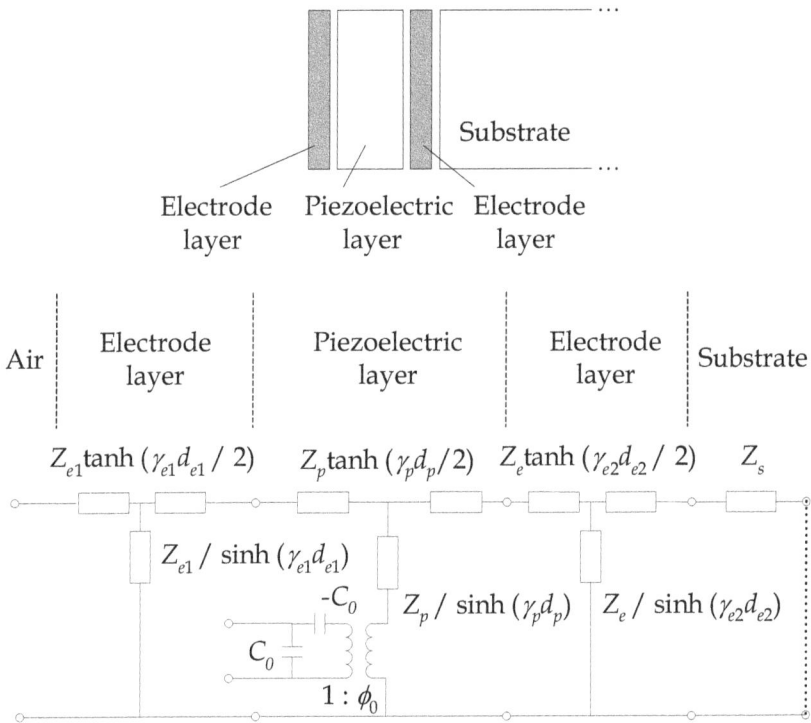

Electrode layer   Piezoelectric layer   Electrode layer   Substrate

Air | Electrode layer | Piezoelectric layer | Electrode layer | Substrate

$Z_{e1} \tanh (\gamma_{e1} d_{e1} / 2)$ $Z_p \tanh (\gamma_p d_p / 2)$ $Z_e \tanh (\gamma_{e2} d_{e2} / 2)$ Z_s

$Z_{e1} / \sinh (\gamma_{e1} d_{e1})$

$-C_0$ $Z_p / \sinh (\gamma_p d_p)$ $Z_e / \sinh (\gamma_{e2} d_{e2})$

C_0

$1 : \phi_0$

Fig. 7. Equivalent circuit model of the over-moded resonator structure

It is convenient to derive whole impedance of the circuit by using $ABCD$-parameters (Paco et al., 2008) As shown in Eqs. (28)-(32), $ABCD$-parameters of whole circuit is derived multiplying each circuit element.

$$F_{Transformer} = \begin{bmatrix} 1/\phi_0 & 0 \\ 0 & \phi_0 \end{bmatrix}, \quad F_{Electric\ port} = \begin{bmatrix} 1 & 0 \\ j\omega C_0 & 1 \end{bmatrix} \cdot \begin{bmatrix} 1 & -1/j\omega C_0 \\ 0 & 1 \end{bmatrix}, \quad F_{Substrate} = \begin{bmatrix} 1 & Z_s \\ 0 & 1 \end{bmatrix} \quad (28)$$

$$F_{Piezo+Electrode\ layer} = \begin{bmatrix} 1 & Z_p / \sinh(\gamma_p d_p) \\ 0 & 1 \end{bmatrix} \cdot \begin{bmatrix} 1 & 0 \\ 1 / \left\{ Z_{e1} \tanh(\gamma_{e1} d_{e1} / 2) + Z_p \tanh(\gamma_p d_p / 2) \right\} & 1 \end{bmatrix}$$
$$\cdot \begin{bmatrix} 1 & Z_p \tanh(\gamma_p d_p / 2) \\ 0 & 1 \end{bmatrix} \tag{29}$$

$$F_{Counter-electrode} = \begin{bmatrix} 1 & Z_{e2} \tanh(\gamma_{e2} d_{e2} / 2) \\ 0 & 1 \end{bmatrix} \cdot \begin{bmatrix} 1 & 0 \\ \sinh(\gamma_{e2} d_{e2}) / Z_{e2} & 1 \end{bmatrix} \cdot \begin{bmatrix} 1 & Z_{e2} \tanh(\gamma_{e2} d_{e2} / 2) \\ 0 & 1 \end{bmatrix} \tag{30}$$

$$F_{Substrate} = \begin{bmatrix} 1 & Z_s \\ 0 & 1 \end{bmatrix} \tag{31}$$

$$F_{Over-moded\ resonator} = F_{Electric\ port} \cdot F_{Transformer} \cdot F_{Piezo+Electrode\ layer} \cdot F_{Counterelectrode} \cdot F_{Substrate} \tag{32}$$

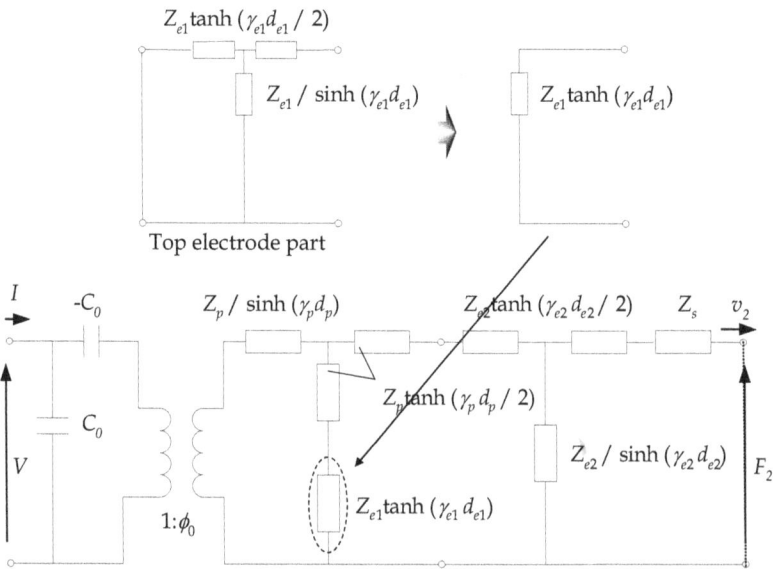

Fig. 8. Simplification of equivalent circuit model for over-moded resonator structure

Insertion loss IL is expressed as the ratio of the signal power delivered from a source into load resistance to the power delivered from a source into the inserted network. IL of the resonators can be calculated with the following equation using conductance of the electrical source G_S (0.02 S), input conductance G_f, and susceptance B_f of the circuit model, which can be derived from ABCD-parameter to Y-parameter conversion of eq. (32):

$$IL = 20 \log_{10} \frac{\Re \left\{ (G_f + jB_f) \frac{G_S^2}{(G_S + G_f)^2 + B_f^2} \right\}}{G_S / 4}. \tag{33}$$

Hence the CL is

$$CL = \frac{IL}{2} = 10\log_{10}\frac{4G_SG_f}{\left(G_S+G_f\right)^2+B_f^2}.$$ (34)

4.4 *k* value determination from conversion loss curves

Figure 9 (a) shows the pure shear mode theoretical and experimental CL curves of the c-axis parallel film HBAR as an example. By comparing experimental curve with theoretical curves

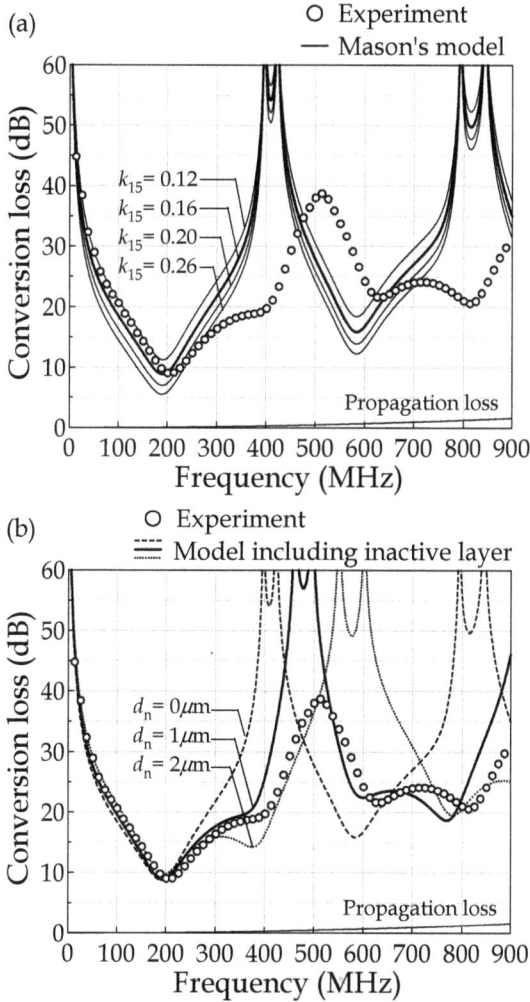

Fig. 9. Frequency response of the experimental shear mode CL (open circles). (a) The simulated shear mode CL curves (solid line) in various k_{15} values and (b) the curve simulated by the model including various thickness of piezoelectrically inactive layer (Yanagitani & Kiuchi, 2007c)

at minimum CL point (near the parallel resonant frequency), we can determine the k_{15} value of the film. As shown in Fig. 9 (b), effective thickness of the piezoelectrically inactive layer d_n in the initial stages of the deposition also can be estimated from comparison of the curves. Figure 10 shows the correlation between k_{15} value and crystalline orientation of the film. FWHM values of ψ-scan and ϕ-scan curve of the XRD (X-ray diffraction) pole figure show the degree of crystalline orientation in out-of plane and in-plane, respectively. Thicker films tend to have large k_{15} values even though they have same degree of crystalline orientation as thinner one. This kind of correlations and inhomogeneities characterization in wafer can be easily obtained from as-deposited film structure, by using present k value determination method.

4.5 Conclusion
In this chapter, shear mode piezoelectric thin film resonators, which is promising for the acoustic microsensors operating in liquid, were introduced. Theoretical predictions of electromechanical coupling and tilt of wave displacement as functions of c-axis tilt angle showed that pure shear mode excitation by using c-axis parallel oriented wurtzite piezoelectric films expected to achieve high-Q and high-coupling sensor. Fabrication of c-axis parallel oriented films by ion beam orientation control technique and characterization of the film by a conversion loss of the as-deposited resonator structure were discussed.

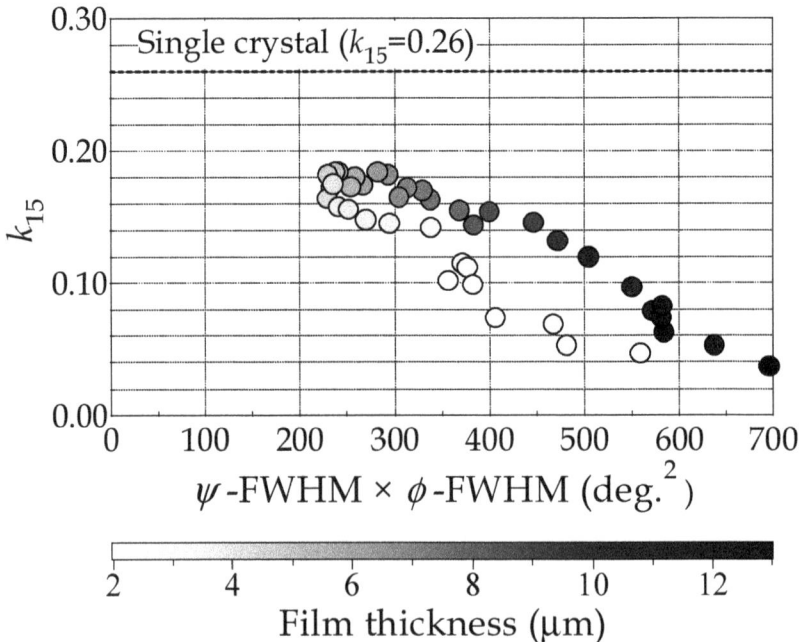

Fig. 10. k_{15} values of the ZnO piezoelectric layers as a function of multiplication of ψ-scan and ϕ-scan profile curve FWHM values extracted from XRD pole figure (indicating the degree of crystalline orientation in out-of-plane and in-plane) (Yanagitani et al., 2007b)

5. References

Auld, B. A. Acoustic Fields and Waves in Solid. (1973). Vol. 1, pp. 73–76, A Wiley-Interscience Publication.

Bond, W. (1943). The Mathmatics of the Physical Properties of Crystals. *Bell System Technical Journal*, Vol. 22, pp. 1–72.

Bradley R. M.; Harper J. M. E. & Smith, D. A. (1986). Theory of Thin–film Orientation by Ion bombardment during deposition. *J. Appl. Phys.*, Vol. 60, No. 12, pp. 4160–4164.

Corso, C. D.; Dickherber, A. & Hunt, W. D. (2007). Lateral Field Excitation of Thickness Shear Mode Waves in a Thin Film ZnO Solidly Mounted Resonator. *J. Appl. Phys.* Vol. 101, pp. 054514-1–054514-7.

Curie, P. (1885). *Bull. Soc. Franc. Miner. Crist.* Vol. 8 p. 145.

Dobrev, D. (1982). Ion-beam-induced Texture Formation in Vacuum-condensed Thin Metal Films. *Thin Solid Films*, Vol. 92, No. 1-2, pp. 41–53.

Dong, L. & Srolovitz, D. J. (1999). Mechanism of Texture Development in Ion-beam-assisted Deposition. *J. Appl. Phys. Lett.*, Vol. 75, No. 4, pp. 584–586.

Dong, L.; Srolovitz, D. J.; Was, G. S.; Zhao, Q. & Rollett, A. D. (2001). Combined Out-of-plane and In-plane Texture Control in Thin Films using Ion Beam Assisted Deposition. *J. Mater. Res.*, Vol. 16, No. 1, pp. 210–216.

Ensinger, W. (1995). On the Mechanism of Crystal Growth Orientation of Ion Beam Assisted Deposited Thin Films. *Nucl. Instrum. Meth. Phys. Res. B*, Vol. 106, pp. 142–146.

Foster, N. F.; Coquin, G. A.; Rozgonyi, G. A. & Vannatta, F. A. (1968). Cadmium Sulphide and Zinc Oxide Thin-Film Transducers. *IEEE Trans. Sonic. Ultrason.*, Vol. 15, pp. 28–41.

Fraser, D. B.; Krause, J. T. & Meitzler, A. H.; (1967). Physical Limitations on the Performance of Vitreous Silica in High–frequency Ultrasonic and Acousto–optical Devices. *Appl. Phys. Lett.* Vol. 11, No. 10, pp. 308–310.

Fujimura, N.; Nishihara, T.; Goto, S.; Xu, J. & Ito, T. (1993). Control of Preferred Orientation for ZnOx Films: Control of Self-texture. *J. Crystal Growth*, Vol. 130, pp. 269–279.

Harper, J. M. E.; Rodbell, K. P.; Colgan, E. G. & Hammond, R. H. (1997). Control of In-plane Texture of Body Centered Cubic Metal Thin Films. *J. Appl. Phys.* Vol. 82, pp. 4319–4326.

Hickernell, F. S. (1996). Measurement Techniques for Evaluating Piezoelectric Thin Films. *Proc. IEEE Ultrason. Symp.*, pp. 235–242.

Iijima, Y.; Tanabe, N.; Kohno, O. & Ikeno, Y. (1992). In-plane Aligned YBa$_2$Cu$_3$O$_{7-x}$ Thin Films Deposited on Polycrystalline Metallic Substrates. *Appl. Phys. Lett.*, Vol. 60, No. 6, pp. 769–771.

Kaufman, D. Y.; DeLuca, P. M.; Tsai, T. & Barnett, S. A. (1999). High-rate Deposition of Biaxially Textured Yttria-stabilized zirconia by Dual Magnetron Oblique Sputtering. *J. Vac. Sci. Technol. A*, Vol. 17, pp. 2826–2829.

Kawamoto, T.; Yanagitani T.; Matsukawa, M.; Watanabe, Y.; Mori1, Y.; Sasaki, S. & Oba M. (2010). Large-Area Growth of In-Plane Oriented (11-20) ZnO Films by Linear Cathode Magnetron Sputtering. *Jpn. J. Appl. Phys.* Vol. 49 pp. 07HD16-1–07HD16-4.

Kushibiki, J-I.; Akashi, N.; Sannomiya, T.; Chubachi, N.; & Dunn, F. (1995). VHF/UHF Range Bioultrasonic Spectroscopy System and Method. *IEEE Trans. Ultrason., Ferroelect., Freq. Contr.*, Vol. 42 No. 6, pp. 1028–1039

Link, M.; Weber, J.; Schreiter, M.; Wersing, W.; Elmazria, O. & Alonot, P. (2007). Sensing Characteristic of High-freqency Shear Mode Resonators in Glycerol Solutions. *Sens. Actuators B*, Vol. 121, No.2, pp. 372–378.

Matsumoto, Y.; Ujiie, T. & Kushibiki, J-I. (2000). Measurement of Shear Acoustic Properties of Water using the Ultrasonic Reflectance Method in Pulse-mode Operation. *Spring Meeting of Acoustical Society of Japan* (in Japanese)

Meeker, T. R. (1996). IEEE Stabdard on Piezoelectricity (ANSI/IEEE Std. 176-1987). *IEEE Trans. Ultrason., Ferroelect., Freq. Contr.*, Vol. 43, No. 5, pp. 719–772..

Milyutin, E. & Mural, P. (2010). Electro-Mechanical Coupling in Shear-Mode FBAR with Piezoelectric Modulated Thin Film. *IEEE Trans. Ultrason., Ferroelect., Freq. Contr.*, Vol. 58, No. 4, pp. 685–688.

Milyutin, E.; Gentil, S. & Mural, P. (2008). Shear Mode Bulk Acoustic Wave Resonator Based on c-axis Oriented AlN Thin Film. *J. Appl. Phys.* Vol. 104, pp. 084508-1–084508-6.

Mitsuyu, T.; Ono S. & Wasa, K. (1980). Structures and SAW Properties of Rf-sputtered Single-crystal Films of ZnO on Sapphire. *J. Appl. Phys.*, Vol. 51, No. 5, pp. 2464–2470.

Naik, B. S.; Lutsky, J. J.; Rief R.; & Sodini, C. D. (1998). Electromechanical Coupling Constant Extraction of Thin-film Piezoelectric Materials Using a Bulk Acoustic Wave Resonator. *IEEE Trans. Ultrason., Ferroelect., Freq. Contr.*, Vol.45, No.1, pp. 257–263.

Nakamura, K.; Shoji, T. & Kang, H. (2000). ZnO Film Growth on $(01\bar{1}2)$ LiTaO₃ by Electron Cyclotron Resonance-assisted Molecular Beam Epitaxy and Determination of Its Polarity. *Jpn. J. Appl. Phys.*, Vol. 39, No. 6A, pp. L534–L536.

Ogi, H.; Hirao, M.; Honda T. & Fukuoka, H. (1995). Ultrasonic Diffraction from a Transducer with Arbitrary Geometry and Strength Distribution. *J. Acoust. Soc. Am.* Vol. 98, No. 2, pp. 1191–1198.

Paco, P.; Menéndez, Ó. & Corrales E. (2008) Equivalent Circuit Modeling of Coupled Resonator Filters. *IEEE Trans. Ultrason., Ferroelect., Freq. Contr.*, Vol. 55, No. 9, pp. 2030–2037.

Park, S. J.; Norton, D. P. & Selvamanickam, V. (2005). Ion-beam Texturing of Uniaxially Textured Ni Films. *Appl. Phys. Lett.*, Vol. 87, pp. 031907–031909.

Ressler, K. G.; Sonnenberg, N. & Cima, M. J. (1997). Mechanism of Biaxial Alignment of Oxide Thin Films during Ion-Beam-Assisted Deposition. *J. Am. Ceram. Soc.*, Vol. 80, No. 10, pp. 2637–2648.

Rosenbaum, J. F. (1988). Bulk Acoustic Waves: Theory and Devices. Artech House Boston London.

Sauerbrey, G. (1959) Verwendung von Schwingquarzen zur Wägung Dünner Schichten und zur Mikrowägung. *Z. Physik*, Vol. 155, pp. 206–222.

Smith, R. T. & Stubblefield, V. E. (1969). Temperature Dependence of the Electroacoustical Constants of Li-doped ZnO Single Crystals. *J. Acoust. Soc. Am.*, Vol. 46, pp. 105.

Takayanagi, S.; Yanagitani, T.; Matsukawa, M. & Watanabe Y. (2010). A Simple Technique for Obtaining (11-20) or (10-10) Textured ZnO Films by RF Bias Sputtering," *Proc. 2010 IEEE Ultrason. Symp.* pp. 1060–1063.

Van Wyk, G. N. & Smith, H. J. (1980). Crystalline Reorientation Due to Ion Bombardment. *Nucl. Instrum. Meth.*, Vol. 170, pp. 433–439.

Weber, J.; Albers, W. M.; Tuppurainen, J.; Link, M.; Gabl, R.; Wersing, W. & Schreiter, M. (2006). Shear Mode FBARs as Highly Sensitive Liquid Biosensors. *Sens. Actuators A*, Vol. 128, No. 2, pp. 84–88.

Wingqvist G.; Anderson H.; Lennartsson C.; Weissbach T.; Yantchev V. & Lloyd Spetz A. (2009). On the Applicability of High Frequency Acoustic Shear Mode Biosensing in View of Thickness Limitations Set by The Film Resonance. *Biosens. Bioelectron.*, Vol. 24 No. 11, pp. 3387–3390.

Wingqvist, G. (2010). AlN-based Sputter-deposited Shear Mode Thin Film Bulk Acoustic Resonator (FBAR) for Biosensor Applications. *Surf. Coat. Tech.*, Vol. 205, No. 5, pp. 1279–1286

Wingqvist, G.; Bjurstrom, J.; Liljeholm, L.; Yantchev, V. & Katardjiev, I. (2007). Shear mode AlN Thin Film Electro-acoustic Resonant Sensor Operation in Viscous Media. *Sens. Actuators B*, Vol. 123, No. 1, pp. 466–473.

Wittstruck, R. H.; Tong, X.; Emanetoglu, N. W.; Wu, P.; Chen, Y.; Zhu, J.; Muthukumar, S.; Lu, Y.; & Ballato, A. (2003) Characteristic of $Mg_xZn_{1-x}O$ Thin Film Bulk Acoustic Wave Devices. *IEEE Trans. Ultrason., Ferroelect., Freq. Contr.*, Vol. 50, pp. 1272–1277.

Yanagitani, T. & Kiuchi, M. (2007c). Control of In-plane and Out-of-plane Texture in Shear Mode Piezoelectric ZnO Films by Ion-beam Irradiation. *J. Appl. Phys.*, Vol. 102, pp. 044115-1–044115-7.

Yanagitani, T. & Kiuchi, M. (2007e). Highly Oriented ZnO Thin Films Deposited by Grazing Ion-beam Sputtering: Application to Acoustic Shear Wave Excitation in the GHz Range. *Jpn. J. Appl. Phys.* Vol. 46, pp. L1167–L1169.

Yanagitani, T. & Kiuchi, M. (2011b). Texture Modification of Wurtzite Piezoelectric Films by Ion Beam Irradiation. *Surf. Coat. Technol.*, in press.

Yanagitani, T., Matsukawa, M., Watanabe, Y., Otani, T. (2005). Formation of Uniaxially (11-2 0) Textured ZnO Films on Glass Substrates. *J. Cryst. Growth*, Vol. 276, No. 3-4, pp. 424-430.

Yanagitani, T.; Arakawa, K.; Kano, K.; Teshigahara, A.; Akiyama, M. (2010). Giant Shear Mode Electromechanical Coupling Coefficient k_{15} in c-axis Tilted ScAlN Films. *Proc. 2010 IEEE Ultrason. Symp.*, pp. 2095–2098.

Yanagitani, T.; Kiuchi, M.; Matsukawa, M. & Watanabe, Y. (2007b). Shear Mode Electromechanical Coupling Coefficient k_{15} and Crystallites Alignment of (11-20) Textured ZnO Films. *J. Appl. Phys.*, Vol. 102, pp. 024110-1–024110-7.

Yanagitani, T.; Kiuchi, M.; Matsukawa, M. & Watanabe, Y. (2007d). Characteristics of Pure-shear Mode BAW Resonators Consisting of (11-20) Textured ZnO Films. *IEEE Trans. Ultrason., Ferroelect., Freq. Contr.*, Vol. 54, No. 8, pp. 1680–1686.

Yanagitani, T.; Mishima, N.; Matsukawa, M. & Watanabe, Y. (2007a). Electromechanical Coupling Coefficient k_{15} of Polycrystalline ZnO Films with the c-axes Lie in the Substrate Plane. *IEEE Trans. Ultrason., Ferroelect., Freq. Contr.*, Vol. 54, No. 4, pp. 701–704.

Yanagitani, T.; Morisato, N.; Takayanagi, S.; Matsukawa, M. & Watanabe, Y. (2011a) c-axis Zig-Zag ZnO Film Ultrasonic Transducers for Designing Longitudinal and Shear

Wave Resonant Frequencies and Modes. *IEEE Trans. Ultrason., Ferroelect., Freq. Contr.*, Vol. 58, No. 5, pp. 1062–1068.

Yu, L. S.; Harper, J. M. E.; Cuomo, J. J. & Smith, D. A. (1985). Alignment of Thin Films by Glancing Angle Ion Bombardment During Deposition. *Appl. Phys. Lett.*, Vol. 47, No. 9, pp. 932–933.

Zhang, Y.; Wang Z. & Cheeke, J. D. N. (2003). Resonant Spectrum Method to Characterize Piezoelectric Films in Composite Resonators. *IEEE Trans. Ultrason., Ferroelect., Freq. Contr.*, Vol. 50, No. 3, pp. 321–333.

7

SAW Parameters Analysis and Equivalent Circuit of SAW Device

Trang Hoang

Faculty of Electrical-Electronics Engineering, University of Technology, HoChiMinh City
VietNam

1. Introduction

Surface Acoustic Wave (SAW) devices, using interdigital electrodes, play a key role in today's telecommunication systems and are widely used as electronic filters, resonators, delay lines, convolvers or wireless identification systems (ID tags).

During the last three decades, demands set by the expansion of the telecommunication industry and many applications in sensor have resulted in the introduction of a new generation of the SAW devices. Consequently, the design of high performance SAW devices requires precise and efficient models, simulation tools. Several methods have been proposed for modeling, analyzing SAW devices. These include the impulse model, the equivalent circuit models, the coupling-of-mode (COM) model, P-matrix model, angular spectrum of waves models [1] and the Scattering Matrix approach that was presented by Coldren and Rosenberg [2]. While the impulse model is only a first order model, the other models include second order effects, e.g. reflections, dispersion, and charge distribution effects. Purely numerical methods have also been and are being developed by many authors [3]-[35], [41].

In this chapter, the method for calculating the SAW parameters, including modeling and simulation, is given.

Section 2 gives the calculation of SAW properties and analyses of different SAW device structures.

Section 3 presents the equivalent circuit of SAW delay line based on Mason model.

The equivalent circuit of SAW delay line based on Couple-Of-Mode theory is presented in section 4.

Based on section 3 and 4, section 5 shows comparison between using the equivalent circuit of SAW delay line device based on Mason model and COM theory. This model is useful and fast model for designing the SAW device.

2. Calculation of SAW parameters

2.1 SAW parameters

The most important parameter for SAW device design is the center frequency, which is determined by the period of the IDT fingers and the acoustic velocity. The governing equation that determines the operation frequency is:

$$f_0 = v_{SAW}/ \lambda \tag{1}$$

where
λ is the wavelength, determined by the periodicity of the IDT and v_{SAW} is the acoustic wave velocity . For the technology being used in this research:

$$\lambda = p = \text{finger width} \times 4 \tag{2}$$

with the finger width (as shown in Figure 1) is determined by the design rule of the technology which sets the minimum metal to metal distance.
v_{SAW} is surface acoustic wave velocity.

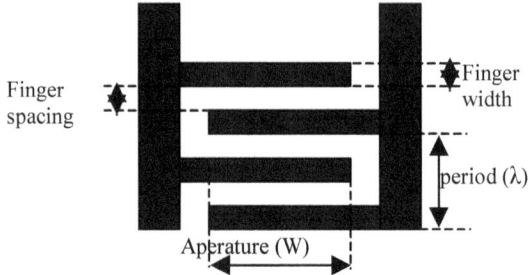

Fig. 1. IDT parameters

By using matrix method or Finite Element Method (FEM) in section 2.2, the velocity v of acoustic wave is derived in two cases:
-    Wave velocity V_0 is velocity in case of free surface.
-    Wave velocity Vs is velocity in case of short-circuit surface.
Therefore, the electromechanical coupling coefficient K is calculated approximately by Ingebrigtsen [54] as:

$$K^2 = 2\frac{V_o - V_s}{V_o} \tag{3}$$

By using the matrix method or FEM and approximation of coupling factor as in (3), the SAW parameters in different structures AlN/Si, AlN/SiO$_2$/Si and AlN/Mo/Si are calculated and analysed in three next sections.

2.2 Matrix method and Finite Element Method (FEM). The choice between them

Matrix method

The SAW propagation properties on one layer or multilayer structure are obtained by using matrix approach, proposed by J.J.Campbell and W.R.Jones [50], K.A.Ingebrigtsen [54], and then developed by Fahmy and Adler [31], [32], [33] and other authors [51], [52], [53]. The numerical solution method is based on characterizing each layer by means of a transfer matrix relating the mechanical and electrical field variables at the boundary planes. The boundary conditions for multilayer are based on the mechanical and electrical field variables those quantities that must be continuous at material interfaces. This matrix method is used to calculate the wave velocity and therefore, the electromechanical coupling factor. A general view and detail of this approach are given as follows and also presented in [50]-[53].

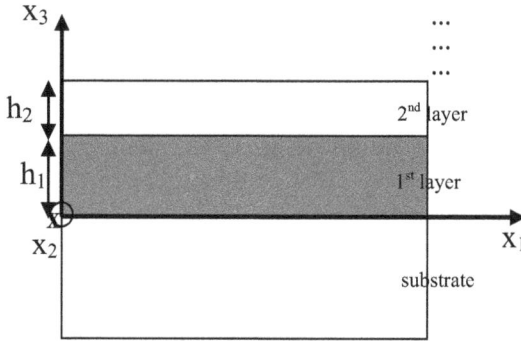

Fig. 2. Multilayer structure

Constitutive equations:

$$c_{ijkl}\frac{\partial^2 u_l}{\partial x_j \partial x_k} + e_{kij}\frac{\partial^2 \phi}{\partial x_j \partial x_k} = \rho\frac{\partial^2 u_i}{\partial t^2} \tag{4}$$

$$e_{jkl}\frac{\partial^2 u_l}{\partial x_j \partial x_k} - \varepsilon_{jk}\frac{\partial^2 \phi}{\partial x_j \partial x_k} = 0 \tag{5}$$

where

$c_{ijkl}, e_{ijk}, \varepsilon_{jk}, \rho$ are elastic tensor, piezoelectric tensor, dielectric tensor and mass density, respectively, of the considered material.
U is the particle displacement.
ϕ is the scalar electric potential.
The boundary conditions are shown in Table 1

Position	Mechanical conditions	Electrical conditions
$x_3 = 0$	$U_i^S = U_i^{1^{st}}$ $T_{3i}^S = T_{3i}^{1^{st}}$	Boundary is open $\phi^S = \phi^{1^{st}}$, $D^S = D^{1^{st}}$ Boundary is short $\phi^S = \phi^{1^{st}} = 0$
$x_3 = h_1$	$U_i^{1^{st}} = U_i^{2^{nd}}$ $T_{3i}^{1^{st}} = T_{3i}^{2^{nd}}$	Boundary is open $\phi^{1^{st}} = \phi^{2^{nd}}$, $D^{1^{st}} = D^{2^{nd}}$ Boundary is short $\phi^{1^{st}} = \phi^{2^{nd}} = 0$
$x_3 = h_1 + h_2$	$T_{3i}^{2^{nd}} = 0$	Boundary is open $D^{2^{nd}} = -\varepsilon.k.\phi^{2^{nd}}$ Boundary is short $\phi^{2^{nd}} = 0$

Table 1. Boundary conditions

where D: electronic displacement,

$$D_k = \frac{\partial \phi}{\partial x_k} \tag{6}$$

The general solution for U_l and ϕ (1) and (2) may be written as follows:

$$U_l = \sum_{m=1}^{n} C_m A_l^{(m)} \exp[ik(b^{(m)}x_3 + x_1 - vt)] \tag{7}$$

where $l = 1, 2, 3$

$$\phi = \sum_{m=1}^{n} C_m A_4^{(m)} \exp[ik(b^{(m)}x_3 + x_1 - vt)] \tag{8}$$

The coefficients C_m are determined from boundary conditions.
By substituting (7) and (8) in every layer into the boundary conditions, we have general form

$$[H] \begin{bmatrix} C_1^S \\ \cdots \\ C_4^S \\ C_1^{1st} \\ \cdots \\ C_8^{1st} \\ C_1^{2nd} \\ \cdots \\ C_8^{2nd} \end{bmatrix} = 0 \tag{9}$$

Phase velocity is determined from the condition:

$$\text{Det}(H) = 0 \tag{10}$$

(use approximation to solve (10))
Figure 3 shows the wave velocity of structure AlN/SiO$_2$(1.3μm)/Si(4μm).

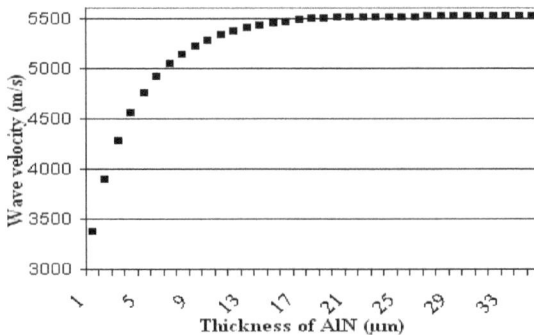

Fig. 3. Wave velocity in structure AlN/SiO$_2$(1.3μm)/Si(4μm) with different thicknesses of AlN

Finite Element Method (FEM)

In the design procedure of SAW devices, simple models like Equivalent Circuit Model coming from Smith Model and COM Model as presented above are used to achieve short calculation time and to get a general view of response of SAW devices. They are a good approach for designing SAW devices, for getting the frequency response, impedance parameters and transfer characteristics of SAW device. They could allow the designer to determine the major dimensions and parameters in number of fingers, finger width, and aperture. However, they are subjected to some simplifications and restrictions.

Field theory is the most appropriate theory for the design SAW devices as it involves the resolution of all the partial differential equations for a given excitation. The Finite Element Model (FEM) is the most appropriate numerical representation of field theory where the piezoelectric behaviour of the SAW devices can be discretized [45], [46]. Besides, nowadays, FEM tools also provide 3D view for SAW device, such as COMSOL® [47], Coventor® [48], ANSYS® [49].

The typical SAW devices can include a lot of electrodes (hundreds or even thousands of electrodes). In fact, we would like to include as many IDT finger pairs as possible in our FEM simulations. This would however significantly increase the scale of the device. Typically finite element models of SAW devices require a minimum of 20 mesh elements per wavelength to ensure proper convergence. A conventional two-port SAW devices consisting of interdigital transducers (IDT) may have – especially on substrate materials with low piezoelectric coupling constants - a length of thousands of wavelengths and an aperture of hundred wavelengths. Depending on the working frequency, the substrate which carries the electrode also has a depth of up to one hundred wavelengths. Taking into account that FEM requires a spatial discretization with at least twenty first order finite elements per wavelength and that an arbitrary piezoelectric material has at least four degrees of freedom, this leads to 8×10^8 unknowns in the three dimensional (3-D) case. Hence, the 3-D FEM representation of SAW device with hundreds of IDT fingers would require several million elements and nodes. The computational cost to simulate such a device is extremely high, or the amount of elements could not be handled by nowadays computer resources.

Fortunately, SAW devices consist of periodic section. M.Hofer et al proposed the Periodic Boundary Condition (PBC) in the FEM that allows the reduction of size of FE model tremendously [45], [46].

A good agreement between FEM and analytic method is obtained via the results in case of SAW with AlN thickness of 4μm, wavelength of 8μm presented in Table 2.

	Matrix method	FEM	Difference between Matrix method and FEM (%)
f_0 (MHz)	771.13	775.48	0.56
f_s (MHz)	770.26	774.57	0.56
v_0 (m/s)	6169.02	6203.87	0.56
v_s (m/s)	6162.07	6196.54	0.56
K (%)	4.74	4.86	2.4

Table 2. Comparison between matrix method and FEM

From this table, matrix method and FEM give the same results. However, FEM would takes a long time and require a trial and error to find the results. Consequently, to reduce time, the matrix method proposed to be used to extract the parameters of SAW devices; FEM is used to get a 3D view and explain some results that can not be explained by equivalent circuit. This point will be presented in next sections.

The three next sections present and analyse SAW parameters in different structures AlN/Si, AlN/SiO₂/Si and AlN/Mo/Si.

2.3 Wave velocity, coupling factor in AlN/Si structure

Figure 4 shows the dependence of Rayleigh wave velocity V_0 and Vs on the normalized thickness as respect to the wavelength, khAlN of AlN layer in SAW device AlN/Si substrate, where normalized thickness is defined by:

$$kh = \frac{2\pi h}{\lambda} \qquad (11)$$

In this graph, when the normalized thickness of AlN, khAlN is larger than 3, the wave velocity reaches the velocity of the Rayleigh wave in AlN substrate $v_{(AlN\ substrate)}$=6169 (m/s). This could be explained that the wave travels principally in AlN layer when khAlN is larger than 3, because for low frequency the wave penetrates inside the other layer and this work is in the case where the wave are dispersive. It is better to be in the frequency range where the Rayleigh wave is obtained to have a constant velocity.

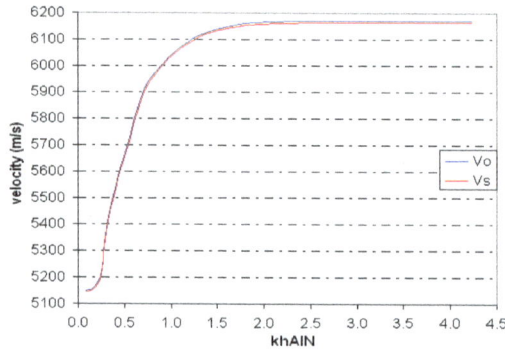

Fig. 4. Calculated values of wave velocity V_0 and Vs in SAW device AlN/Si substrate depend on the normalized thickness khAlN of AlN layer

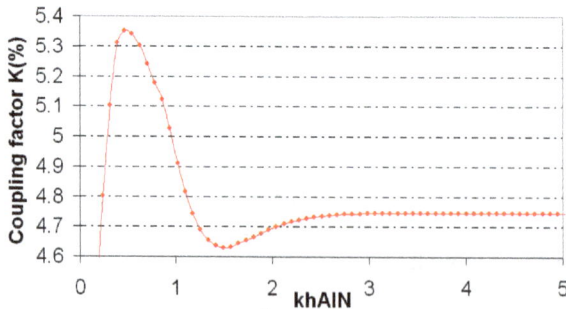

Fig. 5. Calculated values of coupling factor K(%) in SAW device AlN/Si substrate depends on the normalized thickness khAlN of AlN layer

The coupling factor K for this kind of device is shown in Figure 5. When normalized thickness of AlN layer is larger than 3, the coupling factor K still remain at 4.74% by that the wave travels principally in AlN layer.

In this configuration, K is at its maximum value of 5.34% when khAlN=0.55.

2.4 Wave velocity, coupling factor in AlN/SiO$_2$/Si structure

Wave velocity and coupling factor in structure AlN/SiO$_2$/Si are also presented in Figure 6 and Figure 7, respectively.

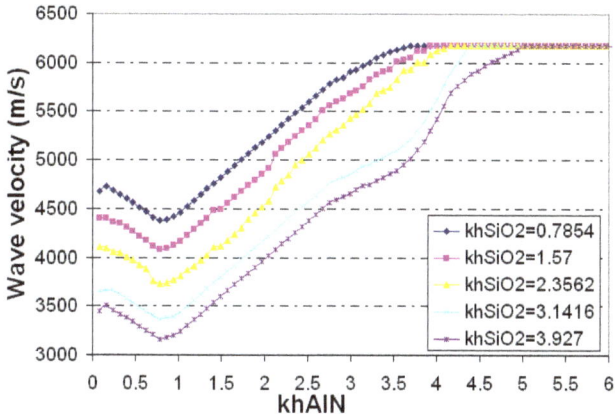

Fig. 6. Dependence of wave velocity in SAW device AlN/SiO$_2$/Si substrate on the normalized thickness khAlN of AlN layer and khSiO$_2$

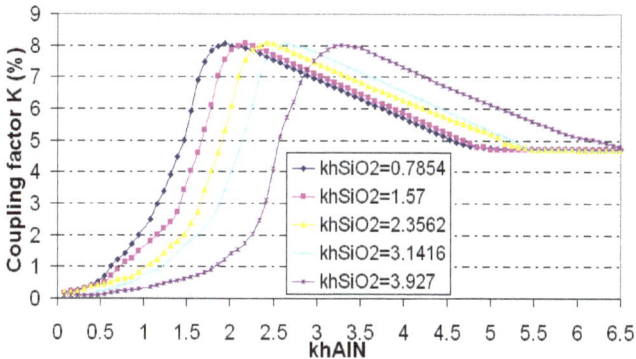

Fig. 7. Dependence of coupling factor K(%) in SAW device AlN/ SiO$_2$/Si substrate on the normalized thickness khAlN of AlN layer and khSiO$_2$

In this configuration, as results in Figure 6, when $khAlN < 6$, with the same thickness of AlN layer, an increase in thickness of SiO$_2$ would decrease the wave velocity. When $khAlN > 6$, the wave velocity reaches the velocity of the Rayleigh wave in AlN substrate v(AlN substrate)=6169 (m/s). A same conclusion is formulated also for coupling factor for this kind of structure, AlN/SiO$_2$/Si, in Figure 7; when $khAlN > 6$, K remains at the value of 4.7%.

To understand the above behavior, we use FEM method to display displacement profile along the depth of multilayer AlN/SiO$_2$/Si. These results obtained from FEM method in case of khSiO$_2$=0.7854, khAlN=5 and khAlN=0.2 are compared as in Figure 8.

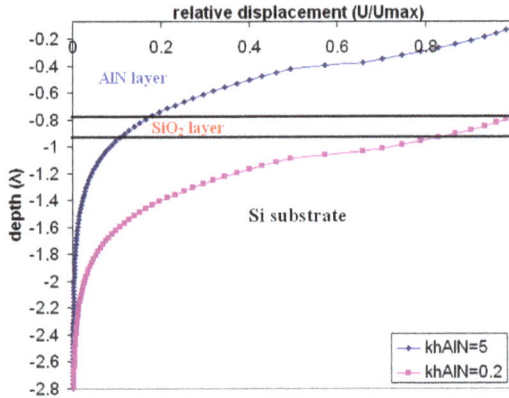

Fig. 8. Displacement profile along the depth of the multilayer AlN/SiO$_2$/Si, khSiO$_2$=0.7854

From Figure 8, we note that wave travels principally in AlN layer for a khAlN value of 5. By this reason, from a khAlN value of larger than 5, the coupling factor K remains at 4.7% and wave velocity remains at 6169m/s. For khAlN=0.2, wave travels principally in SiO$_2$ layer and Si substrate that are not piezoelectric layer. Consequently, the coupling factor K reaches the value of 0%.

In conclusion, the values of wave velocity and coupling factor depend on wave propagation medium, in which constant values of wave velocity and coupling factor indicate a large contribution of AlN layer, and coupling factor value of near 0% indicates a large contribution of SiO$_2$ layer and Si substrate.

2.5 Wave velocity, coupling factor in AlN/Mo/Si structure

For our devices, a thin Mo layer will be also deposited below the AlN layer to impose the crystal orientation of AlN. Besides this dependence, the Mo layer also has influences on

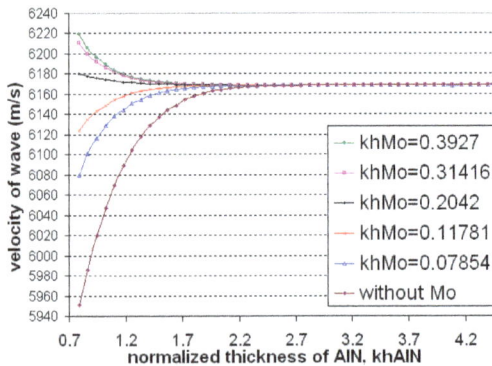

Fig. 9. Wave velocity AlN/Mo/Si substrate depends on the normalized thickness khAlN and khMo

wave velocity and coupling factor K. These influences are shown in Figure 9 and Figure 10, respectively.

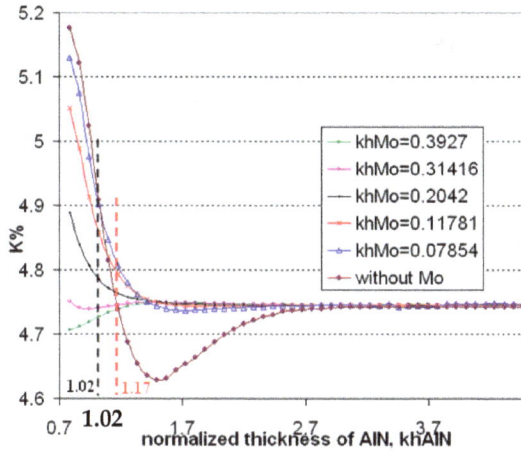

Fig. 10. Coupling factor K(%) in SAW device AlN/Mo/Si substrate depends on the normalized thickness khAlN and khMo

From Figure 9, the use of Mo layer would increase the wave velocity with any thickness of AlN layer and Mo layer. In case of coupling factor K as in Figure 10, the Mo layer, however, could decrease K when the khAlN is less than 1.02. When the normalized thickness of AlN layer khAlN is in the range from 1.17 to 2.7, the Mo layer would increase the coupling factor K. And when the khAlN is larger than 2.7, the Mo has no influence on wave velocity and coupling factor. The reason of this effect could be explained by the displacement profile in AlN/Mo/Si structure, as shown in Figure 11 for thickness AlN value of khAlN=2.7. We could note that when $khAlN \geq 2.7$, the first interesting point is that the wave travels principally in AlN layer and Si substrate, the second one is that the relative displacement U/Umax in Mo layer will be smaller than 0.5. These points would explain the reason why when $khAlN \geq 2.7$ the use of Mo has no influence on wave velocity and coupling factor.

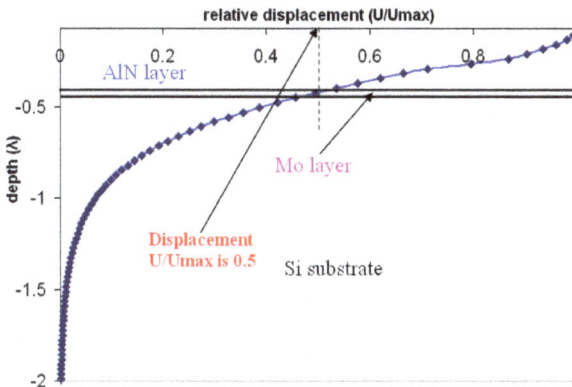

Fig. 11. Displacement profile along the depth of the multilayer AlN/Mo/Si, khAlN=2.7

3. Equivalent circuit for SAW delay line based on Mason model

3.1 Why Equivalent Circuit model is chosen?

Actual devices exist in a three-dimensional physical continuum. Their behaviour is governed by the laws of physics, chemistry, biology, and electronics. From a general point of view, the analysis of devices can be carried out by using some equations of laws of physics, chemistry ... For example; the analysis of piezoelectric resonators or transducers and their application to ultrasonic system can be solved by using the wave equation [36],[37]. But through analysis, equivalent electrical circuit representations of devices can be extracted. So, they can be readily expressible with Equivalent Electric Circuit. Below is the presentation of advantages and disadvantages of equivalent circuit.

Advantages:
- There are an immensely powerful set of intellectual tools to understand electric circuits.
- The equivalent circuit approach has distinct advantages over the direct physical, chemical equations approach (such as direct wave equations approach).
- Many theories, problems of electric circuits have already been solved such as microwave network theory, integrated circuit etc.
- Electric circuit approach is intrinsically correct from an energy point of view [56].
- A further advantage of electric circuit model is that it permits efficient modelling of the interaction between the electric and non-electric components of a microsystem. Both the electrical and mechanical portions of a system are represented by the same means. With software like Simulink, the block diagram is easily constructed, easily to build a more complex system but when we would like to connect a mechanical element to electrical circuits, Simulink can not do that. The analogies between electrical and mechanical elements are presented clearly by Warren P.Mason [57], [58].

Disadvantages:
- Care must be taken to make sure whether the boundary conditions are compatible with those used in the original derivation of the equivalent circuit [58].

In many systems, both commercial and industrial, pressure measurement plays a key role. Since pressure is a normal stress (force per unit area), pressure measurement can be done by using piezoelectric material which can convert stress into voltage. Equivalent circuits such as Mason's model [36] provide a powerful tool for the analysis and simulation of piezoelectric transducer elements. Most of the analogous circuits which have appeared in the literature implement transducers as the circuit elements. This model simulates both the coupling between the mechanical and electrical systems and the coupling between the mechanical and acoustical systems [39]. The mechanical, electrical and acoustic parts of piezoelectric transducer can be varied and analysed about behaviour by implementing equivalent circuits on computer tools such as Ansoft®, Spice, ADS, etc. For IDT composing of N periodic sections, Smith et al [41] developed the equivalent circuit model based on Berlincourt et al [40] work about equivalent circuit for Length Expander Bar with parallel electric field and with perpendicular electric field and based on the equivalent circuit for electromechanical transducer presented by Mason [36]. "Smith model" henceforth will be used to indicate this model. From this model, some models for SAW device in literature have been implemented. However, these models would include only IDTs [42], [43]. In SAW pressure sensor, one of sensitive parts is propagation path. It should be included in the model. The hybrid model based on Smith model for SAW pressure sensor which includes the IDTs and propagation path have been constructed.

Another equivalent model is based on the Coupling-Of-Modes (COM) theory. An excellent recent review of COM theory used in SAW devices was written by K.Hashimoto [10]. Based on the COM equations, as the force and voltage analogy can be used, the relationships between the terminal quantities at the one electrical port and two acoustic ports for an IDT have been done. K.Nakamura [44] introduced a simple equivalent circuit for IDT based on COM approach that is developed in section 4.

In conclusion, the equivalent-circuit model is chosen because it can allow fast design. This allows the designer to determine the major dimensions and parameters in number of fingers, fingers width, aperture, delay line distance, frequency response, impedance parameters and transfer characteristics of SAW device.

3.2 Equivalent circuit for IDT including N periodic sections

Based on Berlincourt et al [39] about equivalent circuit for Length Expander Bar with parallel electric field and with perpendicular electric field and based on the equivalent circuit for electromechanical transducer presented by Mason [36], Smith and al [41] have developed the equivalent circuit for IDT composed of N periodic sections of the form shown in Figure 12.

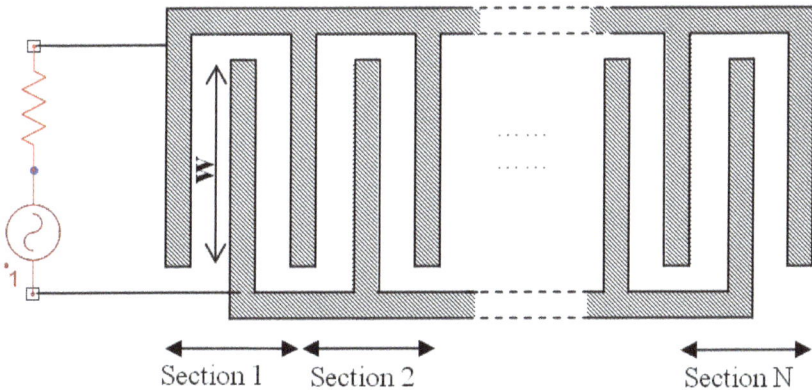

Fig. 12. Interdigital transducer diagram

One periodic section as shown in Figure 13 (a) can be presented by analogous one-dimensional configurations: "crossed-field" model as in Figure 13 (b), and "in-line" model as in Figure 13 (c). In "crossed-field" model, the applied electric field is normal to the acoustic propagation vector; while in "in-line field" model, the electric field is parallel to the propagation vector.

The important advantage of two one-dimensional models is that each periodic section can be represented by equivalent circuit of Mason, as shown in Figure 14 for "crossed-field" model and Figure 15 for "in-line field" model. The difference between these two equivalent circuits is that in "crossed-field" model, the negative capacitors are short-circuited.

Where:

$$\alpha = \frac{\theta}{4} = \frac{\pi}{2}\frac{\omega}{\omega_0}$$

(12)

With periodic section transit angle

$$\theta = 2\pi \frac{\omega}{\omega_0}$$

$$R_0 = \frac{2\pi}{\omega_0 C_s k^2} \tag{13}$$

R_0 is electrical equivalent of mechanical impedance Z_0 [59]
k: electromechanical coupling coefficient
$C_0 = C_s/2$ with Cs : electrode capacitance per section
ω_0 is center angular frequency

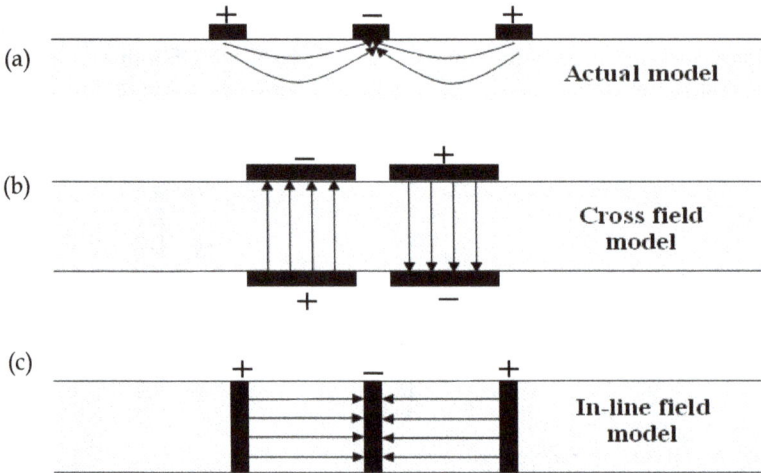

Fig. 13. Side view of the interdigital transducer and 2 analogous one-dimensional configurations (a) Actual model, (b) "crossed-field" model, (c) "in-line field" model

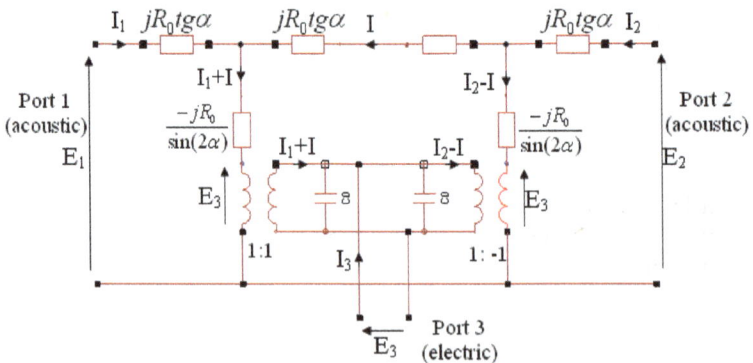

Fig. 14. Mason equivalent circuit for one periodic section in "crossed-field" model

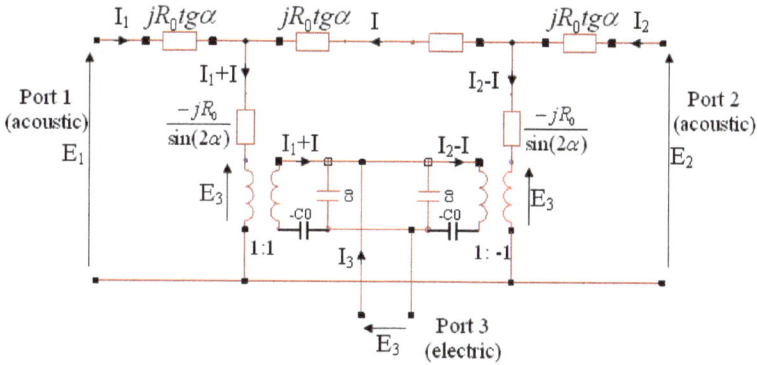

Fig. 15. Mason equivalent circuit for one periodic section in "in-line field" model

One periodic section can be represented by the 3-port network [y] matrix. The [y] matrix of one periodic section for 2 models as follows (see Appendix, section Appendix 1), with $G_0 = R_0^{-1}$, R_0 is expressd by (13):

- for the "crossed-field" model:

$$y_{11} = -jG_0 \cot g(4\alpha)$$

$$y_{12} = \frac{jG_0}{\sin(4\alpha)}$$

$$y_{13} = -jG_0 tg\alpha$$ \hfill (14)

$$y_{33} = j(2\omega C_0 + 4G_0 tg\alpha)$$

- for the "in-line field" model:

$$y_{11} = -jG_0 \cot g\alpha \left(\frac{G_0}{\omega C_0} - \cot g(2\alpha)\right)\left[2 - \frac{\left(\frac{G_0}{\omega C_0} - \frac{1}{\sin(2\alpha)}\right)^2}{\left(\frac{G_0}{\omega C_0} - \cot g(2\alpha)\right)^2}\right]$$

$$y_{12} = jG_0 \frac{\cot g\alpha \left(\frac{G_0}{\omega C_0} - \frac{1}{\sin(2\alpha)}\right)^2}{2\left(\frac{2G_0}{\omega C_0} - \cot g\alpha\right)\left(\frac{G_0}{\omega C_0} - \cot g(2\alpha)\right)}$$ \hfill (15)

$$y_{13} = -jG_0 \frac{tg\alpha}{1 - \frac{2G_0}{\omega C_0} tg\alpha}$$

$$y_{33} = \frac{j2\omega C_0}{1 - \frac{2G_0}{\omega C_0} tg\alpha}$$

In IDT including N periodic sections, the N periodic sections are connected acoustically in cascade and electrically in parallel as represented in Figure 16.

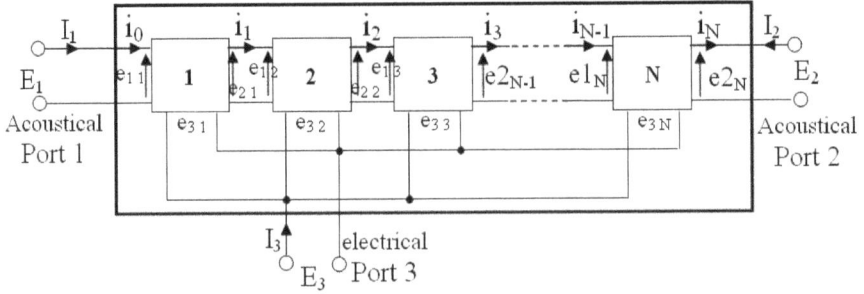

Fig. 16. IDT including the N periodic sections connected acoustically in cascade and electrically in parallel

Matrix [Y] representation of N-section IDT for two models, "crossed-field" model and "in-line" model are in (16) and (17), respectively (the calculation development is presented in Appendix, section Appendix 1):
-   In "crossed-field" model:

$$Y_{11} = -jG_0 \cot g(4N\alpha)$$

$$Y_{12} = \frac{jG_0}{\sin(4N\alpha)}$$

$$Y_{13} = -jG_0 tg\alpha$$

$$Y_{33} = jN(2\omega C_0 + 4G_0 tg\alpha)$$

(16)

-   In "in-line field" model:

$$Y_{11} = -\frac{Q_{11}}{Q_{12}}$$

$$Y_{12} = \frac{1}{Q_{12}}$$

$$Y_{13} = -jG_0 \frac{tg\alpha}{1 - \frac{2G_0}{\omega C_0} tg\alpha}$$

$$Y_{33} = \frac{j2\omega N C_0}{1 - \frac{2G_0}{\omega C_0} tg\alpha}$$

(17)

It was shown in the literature that the crossed field model yielded better agreement than the experiment when compared to the in-line model when K is small. In section 2, K is always smaller than 7.2%. Besides, in section stated above, the "crossed-field" model is simpler than "in-line field" model in term of equations of all element of [Y] matrix. Consequently, the "crossed-field" model is selected henceforth for the calculating, modeling the devices.

3.3 Equivalent circuit for propagation path
The delay line SAW device could be used for pressure sensor application. The sensitive part of this kind of device will be the propagation path. To model the pressure sensor using

SAW, it is necessary to construct the model for propagation path. Based on the equivalent circuit for electromechanical transducer presented by Mason [36], equivalent circuit of propagation path is presented as in Figure 17.

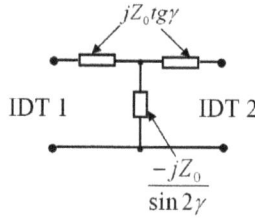

Fig. 17. Equivalent circuit of propagation path, based on Mason model

Where

$$\gamma = \frac{\omega l}{2v} \tag{18}$$

with v is SAW velocity, l is propagation length.

3.4 Equivalent circuit for SAW delay line

Due to the piezoelectric effect, an RF signal applied at input IDT stimulates a micro-acoustic wave propagating on its surface. These waves propagate in two directions, one to receiving IDT and another to the medium. The approximations as follows are assumed to construct the equivalent circuit for SAW delay line:

- Assume that the IDT radiates the wave into a medium of infinite extent. Experimentally, an infinite medium is approximated either by using absorber, such as wax, polyimide to provide acoustic termination, or by using a short RF pulse measurement. The condition of infinite medium means that no wave reflects back to input IDT. This is created for SAW device model by connecting the acoustic characteristic admittance Y_0 to one terminal of IDT.
- Assume that the wave propagating to receiving IDT has no attenuation during propagation way between two IDTs. So, the propagation path between two IDTs can be expressed as the no-loss transmission line.

Based on these two approximations, the [Y] matrix representation of IDT in section 3.2, and propagation path representation in section 3.3, the SAW delay line can be expressed as equivalent circuit as in Figure 18.

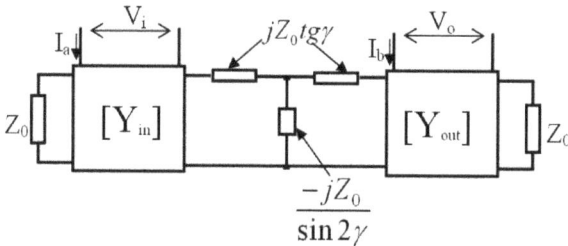

Fig. 18. Equivalent circuit of SAW delay line, based on Mason model

Section 3 gives the equivalent circuit of SAW delay line, including IDT input, IDT output and propagation path. All of calculation developments are presented in appendix, section 2. In this appendix, a new equivalent circuit of IDT including N periodic section plus one finger, which we call it "N+1/2", also are developed and presented. Another representation of SAW delay line is [ABCD] matrix representation which also proposed in appendix, section Appendix 4. [ABCD] matrix representation has one interesting property that in cascaded network, the [ABCD] matrix of total network can be obtained easily by multiplying the matrices of elemental networks.

4. Equivalent circuit for IDT based on the Coupling-Of-Mode theory

The Coupling-Of-Modes formalism is a branch of the highly developed theory of wave propagation in periodic structure, which has an history of more than 100 years. This theory covers a variety of wave phenomena, including the diffraction of EM waves on periodic gratings, their propagation in periodic waveguides and antennas, optical and ultrasonic waves in multi-layered structures, quantum theory of electron states in metal, semiconductors, and dielectrics.... Theoretical aspects of the wave in periodic media and applications were reviewed by C.Elachi [4], in which it included theories of waves in unbounded and bounded periodic medium, boundary periodicity, source radiation in periodic media, transients in periodic structures, active and passive periodic structures, waves and particles in crystals. An excellent recent review of COM theory used in SAW devices was written by K.Hashimoto [10].

A simple equivalent circuit for IDT based on COM approach was proposed by K.Nakamura [29]. This model would be useful to analyze and design SAW devices. Based on the COM equations, the relationships between the terminal quantities at the one electrical port and two acoustic ports for an IDT have been done.

4.1 COM equation for particle velocities

Consider an IDT including N periodic sections with periodic length of L as shown in Figure 19.

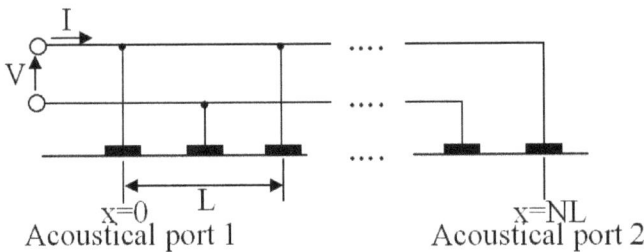

Fig. 19. IDT including N periodic sections

The particle velocities v+(x) and v-(x) of the wave propagating in the +x and –x directions in the periodic structure can be expressed as follows with the time dependence exp(jωt) term:

$$v^+(x) = A^+(x)e^{-jkx} \tag{19}$$

$$v^-(x) = A^-(x)e^{jkx} \tag{20}$$

Where k is the wave number

$$k = \omega / V_{SAW} \tag{21}$$

The amplitude A$^+$(x) and A$^-$(x) obey the following coupled-mode equations [60]:

$$\frac{dA^+(x)}{dx} = -jK_{11}A^+(x) - jK_{12}e^{j2\delta x}A^-(x) + j\zeta e^{j\delta x}V \tag{22}$$

$$\frac{dA^-(x)}{dx} = jK_{12}e^{-j2\delta x}A^+(x) + jK_{11}A^-(x) - j\zeta e^{-j\delta x}V \tag{23}$$

Where V is the voltage applied to the IDT,
ζ is the constant associated with the convention from electrical to SAW quantities,
K_{11} and K_{12} are coupling coefficients, sum of the coupling coefficient coming from the piezoelectric perturbation and that coming from the mechanical perturbation.

$$\delta = k - k_0 \text{ , with } k_0 = \frac{2\pi}{L} \tag{24}$$

The solution to (22) and (23) can be expressed as

$$v^+(x) = \left(h_1 e^{-j\beta_1 x} + ph_2 e^{-j\beta_2 x} + q\zeta V\right)e^{-jk_0 x} \tag{25}$$

$$v^-(x) = \left(ph_1 e^{-j\beta_1 x} + h_2 e^{-j\beta_2 x} + q\zeta V\right)e^{jk_0 x} \tag{26}$$

Where the subscripts 1 and 2 indicate the elementary waves with wavenumbers $k_0 + \beta_1$ and $k_0 + \beta_2$ in the +x direction, and the magnitudes h$_1$ and h$_2$, respectively.

$$\beta_1, \beta_2 = \pm\sqrt{(\delta + K_{11})^2 - K_{12}^2} \tag{27}$$

$$p = \frac{\beta_1 - \delta - K_{11}}{K_{12}} \tag{28}$$

$$q = \frac{1}{\delta + K_{11} + K_{12}} \tag{29}$$

4.2 Equivalent circuit for IDT based on COM theory
From the equations (25) and (26), the particle velocities at the both ends of the IDT can be expressed as:

$$v^+(0) = h_1 + ph_2 + q\zeta V \tag{30}$$

$$v^-(0) = ph_1 + h_2 + q\zeta V \tag{31}$$

$$v^+(NL) = \pm\left(e^{-j\beta_1 NL}h_1 + e^{j\beta_1 NL}ph_2 + q\zeta V\right) \tag{32}$$

$$v^-(NL) = \pm\left(e^{-j\beta_1 NL}ph_1 + e^{j\beta_1 NL}h_2 + q\zeta V\right) \tag{33}$$

The upper and lower signs in (32) and (33) correspond to the cases N=i and N=i+0.5, respectively, where i is an integer. Consequently, the total particle velocities at the two acoustical ports can be expressed as:
-    Particle velocity at port 1 (x=0):

$$v_1 = v^+(0) + v^-(0) = (1+p)(h_1 + h_2) + 2q\zeta V \tag{34}$$

-    Particle velocity at port 2 (x=NL):

$$v_2 = -\left[v^+(NL) + v^-(NL)\right] = \mp\left[(1+p)(e^{-j\beta_1 NL}h_1 + e^{j\beta_1 NL}h_2) + 2q\zeta V\right] \tag{35}$$

The two forces at two acoustic ports are considered to be proportional to the difference of v^+ and v^-. For the simplicity, these forces can be expressed as follows:

$$F_1 = v^+(0) - v^-(0) = (1-p)(h_1 - h_2) \tag{36}$$

$$F_2 = v^+(NL) - v^-(NL) = \pm\left[(1-p)(e^{-j\beta_1 NL}h_1 - e^{j\beta_1 NL}h_2)\right] \tag{37}$$

From these equations, h_1 and h_2 are the terms of F_1 and F_2 as follows:

$$h_1 = \frac{e^{j2\beta_1 NL}}{(1-p)(e^{j2\beta_1 NL}-1)}F_1 \mp \frac{e^{j\beta_1 NL}}{(1-p)(e^{j2\beta_1 NL}-1)}F_2 \tag{38}$$

$$h_2 = \frac{1}{(1-p)(e^{j2\beta_1 NL}-1)}F_1 \mp \frac{e^{j\beta_1 NL}}{(1-p)(e^{j2\beta_1 NL}-1)}F_2 \tag{39}$$

The current I at the electrical ports can be expressed as:

$$I = \eta\int_0^{NL}\left[(1+p)(h_1 e^{-j\beta_1 x} + h_2 e^{-j\beta_2 x}) + 2q\zeta V\right]dx + j\omega NC_s V$$
$$= j\eta\left\{(1+p)\left[\frac{h_1}{\beta_1}(e^{-j\beta_1 NL}-1) + \frac{h_2}{\beta_2}(e^{j\beta_1 NL}-1)\right]\right\} + 2q\zeta\eta NLV + j\omega NC_s V \tag{40}$$

where η is the constant associated with the convention from SAW to electrical quantities, therefore associated with the coupling factor K.
C_s is the capacitance for one electrode pair.
By substituting equations (38) and (39) in (34), (35) and (40), the following equations can be obtained:

$$I = (j\omega NC_s + 2q\zeta\eta NL)V + \frac{\eta(1+p)}{j\beta(1-p)}F_1 \mp \frac{\eta(1+p)}{j\beta(1-p)}F_2 \tag{41}$$

$$v_1 = 2q\zeta V + \frac{1+p}{1-p}\frac{1}{j\tan 2\theta}F_1 \mp \frac{1+p}{1-p}\frac{1}{j\sin 2\theta}F_2 \tag{42}$$

$$v_2 = \mp 2q\zeta V \mp \frac{1+p}{1-p}\frac{1}{j\sin 2\theta}F_1 + \frac{1+p}{1-p}\frac{1}{j\tan 2\theta}F_2 \tag{43}$$

Where

$$\theta = \beta NL / 2 \tag{44}$$

$$\beta \equiv \beta_1 = -\beta_2 \tag{45}$$

From these equations, the matrix as follows can be obtained:

$$\begin{bmatrix} I \\ v_1 \\ v_2 \end{bmatrix} = \begin{bmatrix} (j\omega NC_s + 2q\zeta\eta NL) & \dfrac{\eta(1+p)}{j\beta(1-p)} & \mp\dfrac{\eta(1+p)}{j\beta(1-p)} \\ 2q\zeta & \dfrac{1+p}{1-p}\dfrac{1}{j\tan 2\theta} & \mp\dfrac{1+p}{1-p}\dfrac{1}{j\sin 2\theta} \\ \mp 2q\zeta & \mp\dfrac{1+p}{1-p}\dfrac{1}{j\sin 2\theta} & \dfrac{1+p}{1-p}\dfrac{1}{j\tan 2\theta} \end{bmatrix} \begin{bmatrix} V \\ F_1 \\ F_2 \end{bmatrix} \tag{46}$$

In the acoustic wave transducer using piezoelectric effect, the force and voltage analogy can be used. Therefore, the COM-based circuit of IDT as matrix in (46) can be considered as the reciprocal circuit. The reciprocity theorem states that if a voltage source E acting in one branch of a network causes a current I to flow in another branch of the network, then the same voltage source E acting in the second branch would cause an identical current I to flow in the first branch. By using this theorem in this case, replacing V and F₁ together, the same value I requirement leads the following equations:

$$2q\zeta = \frac{\eta(1+p)}{j\beta(1-p)} \tag{47}$$

$$\eta = 2j\zeta \tag{48}$$

From (46), (47), and (48), the matrix as in (46) becomes:

$$\begin{bmatrix} I \\ v_1 \\ v_2 \end{bmatrix} = \begin{bmatrix} j\omega C_T + \dfrac{\phi^2}{j2\theta Z_0} & \dfrac{\phi}{j2\theta Z_0} & \mp\dfrac{\phi}{j2\theta Z_0} \\ \dfrac{\phi}{j2\theta Z_0} & \dfrac{1}{jZ_0\tan 2\theta} & \mp\dfrac{1}{jZ_0\sin 2\theta} \\ \mp\dfrac{\phi}{j2\theta Z_0} & \mp\dfrac{1}{jZ_0\sin 2\theta} & \dfrac{1}{jZ_0\tan 2\theta} \end{bmatrix} \begin{bmatrix} V \\ F_1 \\ F_2 \end{bmatrix} \tag{49}$$

Where

$$Z_0 = \frac{1-p}{1+p} = \frac{1}{q\beta} \tag{50}$$

$$\phi = \eta NL = 2j\zeta NL \tag{51}$$

$$C_T = NC_s \tag{52}$$

Consequently, the simple equivalent circuit obtained for IDT with N electrode pairs is shown in Figure 20:

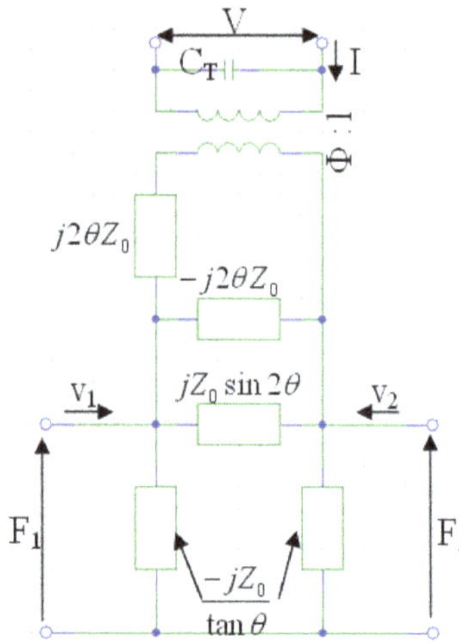

Fig. 20. Equivalent circuit IDT based on COM theory

4.3 Equivalent circuit for propagation path based on COM theory

In SAW devices, the propagation path should be taken into account. It is necessary to determine the equivalent circuit for a propagation path of distance l between 2 IDTs. This propagation path is a uniform section of length l, with a free surface or a uniformly metallized surface. In this case, $K_{11}=K_{12}=0$, and $\beta=\delta$.

Consequently, from equations (38) and (39), h_1 and h_2 can be expressed as:

$$h_1 = \frac{e^{j2\beta l}}{e^{j2\beta l}-1}F_1 - \frac{e^{j\beta l}}{e^{j2\beta l}-1}F_2 \tag{53}$$

$$h_2 = \frac{1}{e^{j2\beta l}-1}F_1 - \frac{e^{j\beta l}}{e^{j2\beta l}-1}F_2 \tag{54}$$

And, the particle velocities are expressed as:

$$v^+(x) = h_1 e^{-j(k_0+\beta_1)x} = h_1 e^{-jkx} \tag{55}$$

$$v^-(x) = h_2 e^{j(k_0+\beta_1)x} = h_2 e^{jkx} \tag{56}$$

If the v_1, v_2, F_1, and F_2 are defined as:

$$v_1 = v^+(0) + v^-(0) \tag{57}$$

$$v_2 = -[v^+(NL) + v^-(NL)] \tag{58}$$

$$F_1 = v^+(0) - v^-(0) \tag{59}$$

$$F_2 = v^+(NL) - v^-(NL) \tag{60}$$

Then, by expressing h_1 and h_2 in terms of F_1 and F_2 based on equations (53) and (54), the v_1 and v_2 become as follows:

$$v_1 = h_1 + h_2 = \frac{e^{j2\beta l}+1}{e^{j2\beta l}-1}F_1 - \frac{2e^{j\beta l}}{e^{j2\beta l}-1}F_2 \tag{61}$$

$$v_2 = h_1 e^{-jkl} + h_2 e^{jkl} = \frac{2e^{j\beta l}}{e^{j2\beta l}-1}F_1 - \frac{e^{j2\beta l}+1}{e^{j2\beta l}-1}F_2 \tag{62}$$

Using the relation between complex number and trigonometry, the v_1 and v_2 can be expressed as follows:

$$v_1 = \frac{1}{jZ_0' \tan 2\theta'}F_1 - \frac{1}{jZ_0' \sin 2\theta'}F_2 \tag{63}$$

$$v_2 = -\frac{1}{jZ_0' \sin 2\theta'}F_1 + \frac{1}{jZ_0' \tan 2\theta'}F_2 \tag{64}$$

Consequently, the equivalent circuit for propagation path can be represented by the π-circuit of Figure 21:

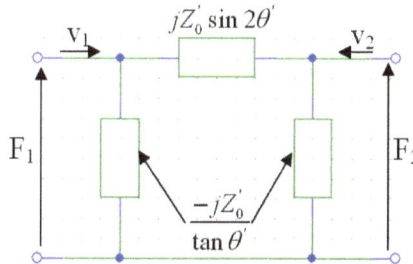

Fig. 21. Equivalent circuit of propagation path based on COM theory

Based on Mason model, the equivalent circuit of propagation path was presented in Figure 17, which has star form. In Figure 21, the circuit has triangle form. By using triangles and stars transformation theory published by A.E. Kennelly, equivalent circuit of propagation in these two figures is the same. Consequently, the approachs that are based on Mason model and COM theory can get the same equivalent circuit of propagation path.

4.4 Equivalent circuit for SAW delay line based on COM theory
Based on section 4.2 and 4.3, equivalent circuit of SAW delay line based on COM theory is presented in Figure 22.
In this model, some parameters must to be calculated or extracted. SAW velocity v, piezoelectric coupling factor K could be calculated from section 2. The periodic length L (or wavelength λ) is determined by design and fabrication.
The parameters K_{11} and K_{12} are coupling coefficients. They are sum of the coupling coefficient coming from the piezoelectric perturbation and that coming from the mechanical perturbation, and their equations for calculation are complicated [55]. Exact equations for

K_{11} and K_{12} were given by Y.Suzuki et al [55], but it seems so complex that their usefulnesses could be limited. However, from this work of Y.Suzuki et al [55], we propose the K_{11} and K_{12} could be expressed as follows:

$$K_{11} = O_{11}K^2 k_0 \tag{65}$$

$$K_{12} = O_{12}K^2 k_0 \tag{66}$$

Where k_0 is stated by (24) $k_0 = \dfrac{2\pi}{L}$ and K is piezoelectric coupling factor.

O_{11} is so-called self-coupling constant of finger, and O_{12} is so-called coupling constant between fingers. O_{12} could also presents the reflective wave between two fingers.

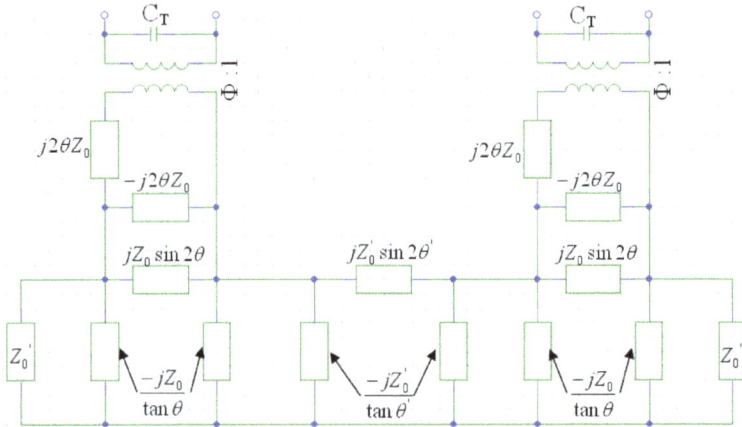

Fig. 22. Equivalent circuit of SAW delay line based on COM theory

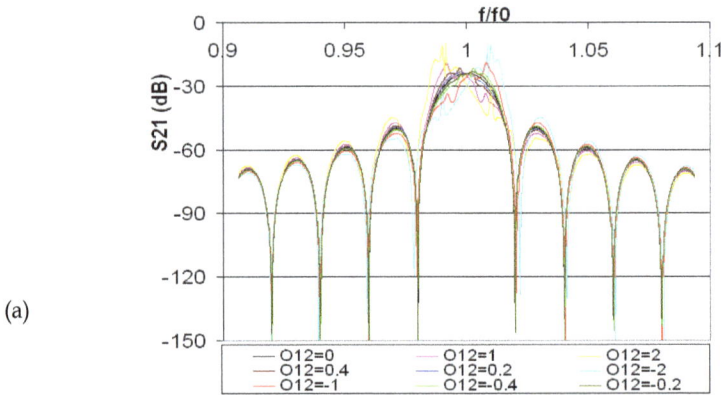

(a)

Fig. 23. Effect of O_{12} on S21(dB), N=50, v_{SAW}=5120m/s, λ=8µm, K=0.066453, O_{11}=0

Figure 23 shows the effects of O_{12} on S21(dB) of SAW device N=50, v_{SAW}=5120m/s, λ=8µm, K=0.066453 when O_{11}=0. S_{21} is the transmission coefficient in the scattering matrix representation [28].

Figure 24 shows the effects of O_{11} on S21(dB) of SAW device N=50, v_{SAW}=5120m/s, λ=8μm, K=0.066453 when O_{12}=0. So, O_{11} coefficient shifts the center frequency of SAW device, the positive value of O_{11} reduces the center frequency f_0 of device, the negative on will increase the f_0.

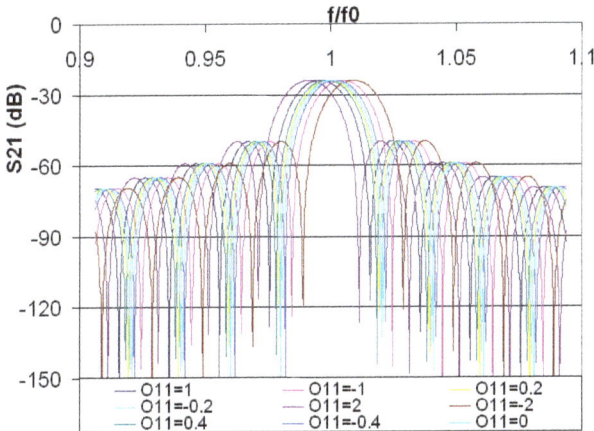

Fig. 24. Effect of O_{11} on S21(dB), N=50, v_{SAW}=5120m/s, λ=8μm, K=0.066453, O_{12}=0

The effect of K_{11} and K_{12} could be explained by their measurement method [61]. K_{11} could be derived from the measurement of frequency response, therefore the usefulness of its calculation could be limited. Meanwhile, K_{12} can be extracted from FEM. It is shown in literature that K_{12} depends on the thickness of finger with respect to the wavelength. In our work, the ratio thickness/wavelength (its maximum value is 300nm/8μm) is too small that its effect can be ignored. In conclusion, in our work, value of K_{11} and K_{12} are 0.

5. Comparison of equivalent circuit of SAW device based on Mason model and COM thoery

Figure 25 presents the comparison between hybrid model and COM model in that O_{11}= O_{12}=0, distance between 2 IDTs is 50λ. These models could be the same, except that a

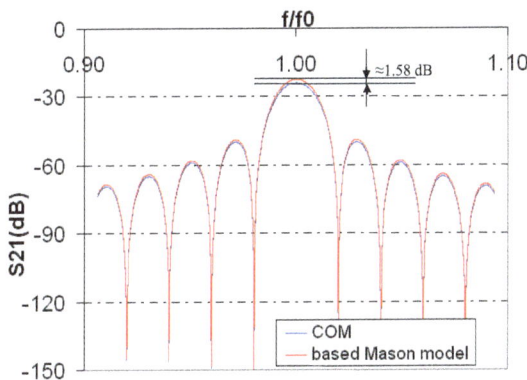

Fig. 25. Comparison between Hybrid model and COM model (O_{11}=O_{12}=0)

small difference in the peak value of S21 (dB) occurs. This difference could be explained by using "crossed-filed" model instead of actual model as in Figure 13.

6. Conclusion

The model used for SAW pressure sensor based on delay line are presented. For usefulness and reduction of time in design process, the equivalent circuit based on COM model, in which K_{11}, K_{12}=0 is proposed to be used.

Acoustic wave properties in different structures of AlN/SiO$_2$/Si, AlN/Si, and AlN/Mo/Si are analyzed. The wave velocity, coupling factor could depend on the wave propagation medium.

From analyses of these structures, the range in which there is a weak dependence of the wave velocity, coupling factor on the AlN layer thickness could be known. The SAW devices should be fabricated in this range to facilitate manufacturing.

For AlN/Si structure, this range is $khAlN \geq 3$.

For AlN/Mo/Si, if this kind of SAW device is fabricated in the range from $khAlN \geq 2.7$ to facilitate manufacturing, the use of Mo layer is useless. Consequently, to take full advantage of using Mo layer in term of wave velocity and coupling factor, it should be required to control the fabrication process carefully to obtain the required AlN thickness from khAlN=1.02 to khAlN=2.7.

For AlN/SiO$_2$/Si, this range is $khAlN \geq 5$ for khSiO$_2$=0.7854, for thicker SiO$_2$ layer, this range changes based on Figure 6 and Figure 7. Besides, using SiO$_2$ layer would reduce temperature dependence of frequency. To choose the thickness of SiO$_2$ layer, it would consider the effect of temperature dependence and analyses of wave velocity, coupling factor.

7. Appendix: Development of calculation for equivalent circuit of SAW device

7.1 Appendix 1. Equivalent circuit for normal IDT including N periodic sections

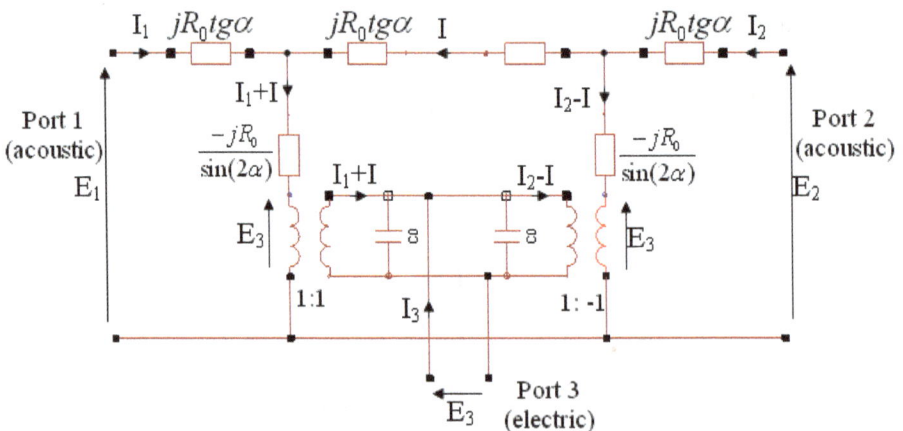

Fig. Appendix.1. Mason equivalent circuit for one periodic section in "crossed-field" model

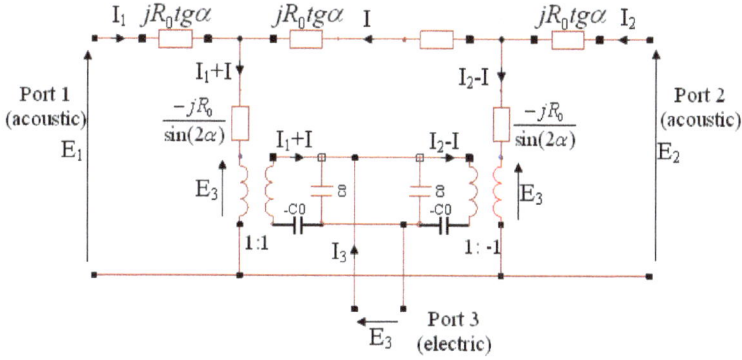

Fig. Appendix.2. Mason equivalent circuit for one periodic section in "in-line field" model

One periodic section can be expressed by the 3-port network as follows:

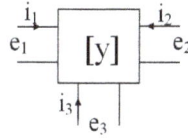

Fig. Appendix.3. One periodic section represented by 3-port network, admittance matrix [y]

$$\begin{bmatrix} i_1 \\ i_2 \\ i_3 \end{bmatrix} = \begin{bmatrix} y_{11} & y_{12} & y_{13} \\ y_{21} & y_{22} & y_{23} \\ y_{31} & y_{32} & y_{33} \end{bmatrix} \begin{bmatrix} e_1 \\ e_2 \\ e_3 \end{bmatrix}$$ (Appendix.1)

By the symmetrical properties of one periodic section (the voltage applied at port 3 will result in stress of the same value at port 1 and 2), the [y] matrix in (Appendix.1) becomes (Appendix.2) for Figure Appendix.4 and becomes (Appendix.3) for Figure Appendix.5.

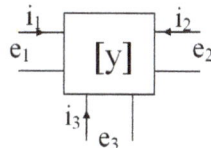

Fig. Appendix.4. 3-port network representation of one periodic section, with the change of sign between Y_{13} and Y_{23} to ensure that acoustic power flows symmetrically away from transducer

$$\begin{bmatrix} i_1 \\ i_2 \\ i_3 \end{bmatrix} = \begin{bmatrix} y_{11} & y_{12} & y_{13} \\ y_{12} & y_{11} & -y_{13} \\ y_{13} & -y_{13} & y_{33} \end{bmatrix} \begin{bmatrix} e_1 \\ e_2 \\ e_3 \end{bmatrix}$$ (Appendix.2)

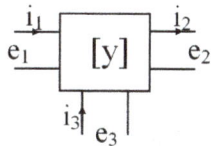

Fig. Appendix.5. 3-port network representation of one periodic section, with the no change of sign between Y_{13} and Y_{23}

$$\begin{bmatrix} i_1 \\ i_2 \\ i_3 \end{bmatrix} = \begin{bmatrix} y_{11} & y_{12} & y_{13} \\ -y_{12} & -y_{11} & y_{13} \\ y_{13} & -y_{13} & y_{33} \end{bmatrix} \begin{bmatrix} e_1 \\ e_2 \\ e_3 \end{bmatrix}$$ (Appendix.3)

Applying circuit theory, definitions of [y] matrix elements are presented:

$$y_{11} = \left.\frac{i_1}{e_1}\right|_{\substack{e_2=0 \\ e_3=0}} \qquad ; \qquad y_{12} = \left.\frac{i_1}{e_2}\right|_{\substack{e_1=0 \\ e_3=0}}$$

$$y_{13} = \left.\frac{i_1}{e_3}\right|_{\substack{e_1=0 \\ e_2=0}} \qquad ; \qquad y_{33} = \left.\frac{i_3}{e_3}\right|_{\substack{e_1=0 \\ e_2=0}}$$

(Appendix.4)

And using trigonometric functions as follows:

$$tg\alpha - \frac{2}{\sin(2\alpha)} = -\cot g\alpha$$

$$tg\alpha - \frac{1}{\sin(2\alpha)} = \frac{1}{2}\left(tg\alpha - \cot g\alpha\right)$$

(Appendix.5)

$$tg\alpha\frac{3\cos(2\alpha)+1-tg\alpha\sin(4\alpha)-tg\alpha\sin(2\alpha)}{tg\alpha\sin(4\alpha)+tg\alpha\sin(2\alpha)-\cos(2\alpha)} = -tg(4\alpha)$$

The [y] matrix can be obtained for 2 models as follows:
-      for the "crossed-field" model:

$$y_{11} = -jG_0 \cot g(4\alpha)$$

$$y_{12} = \frac{jG_0}{\sin(4\alpha)}$$

(Appendix.6)

$$y_{13} = -jG_0 tg\alpha$$
$$y_{33} = j(2\omega C_0 + 4G_0 tg\alpha)$$

-      for the "in-line field" model:

$$y_{11} = -jG_0 \cot g\alpha\left(\frac{G_0}{\omega C_0} - \cot g(2\alpha)\right)\left[2 - \frac{\left(\dfrac{G_0}{\omega C_0} - \dfrac{1}{\sin(2\alpha)}\right)^2}{\left(\dfrac{G_0}{\omega C_0} - \cot g(2\alpha)\right)^2}\right]$$

$$y_{12} = jG_0\frac{\cot g\alpha\left(\dfrac{G_0}{\omega C_0} - \dfrac{1}{\sin(2\alpha)}\right)^2}{2\left(\dfrac{2G_0}{\omega C_0} - \cot g\alpha\right)\left(\dfrac{G_0}{\omega C_0} - \cot g(2\alpha)\right)}$$

(Appendix.7)

$$y_{13} = -jG_0\frac{tg\alpha}{1 - \dfrac{2G_0}{\omega C_0}tg\alpha}$$

$$y_{33} = \frac{j2\omega C_0}{1 - \dfrac{2G_0}{\omega C_0}tg\alpha}$$

In IDT including N periodic sections, the N periodic sections are connected acoustically in cascade and electrically in parallel as Figure Appendix.6.

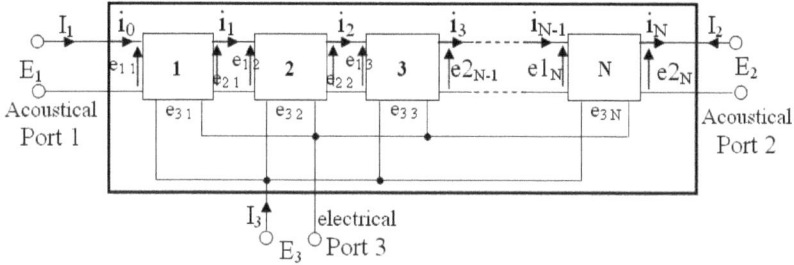

Fig. Appendix.6. IDT including the N periodic sections connected acoustically in cascade and electrically in parallel

Because the symmetric properties of the IDT including N section like these of one periodic section, and from (Appendix.2), (Appendix.3), Figure Appendix.4 and Figure Appendix.5, the [Y] matrices of N-section IDT are represented as follows:

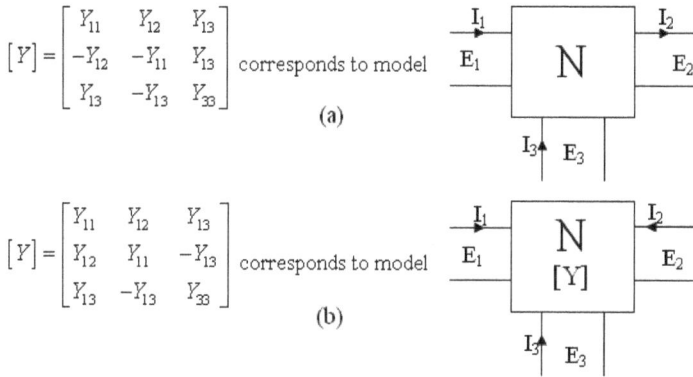

$$[Y] = \begin{bmatrix} Y_{11} & Y_{12} & Y_{13} \\ -Y_{12} & -Y_{11} & Y_{13} \\ Y_{13} & -Y_{13} & Y_{33} \end{bmatrix} \text{corresponds to model}$$

(a)

$$[Y] = \begin{bmatrix} Y_{11} & Y_{12} & Y_{13} \\ Y_{12} & Y_{11} & -Y_{13} \\ Y_{13} & -Y_{13} & Y_{33} \end{bmatrix} \text{corresponds to model}$$

(b)

Fig. Appendix.7. The [Y] matrices and the model corresponsive models

Since the periodic sections are identical, the recursion relation as follows can be obtained:

$$e_{1\,m} = e_{2\,m-1} \qquad\qquad \text{(Appendix.8)}$$

$$e_{3\,N} = e_{3\,N-1} = e_{3\,N-2} = ... = e_{3\,2} = e_{3\,1} = E_3 \qquad\qquad \text{(Appendix.9)}$$

$$i_{1\,m} = i_{2\,m-1} \qquad\qquad \text{(Appendix.10)}$$

With m is integer number, m=1,2, …, N-1, N
The total transducer current is the sum of currents flowing into the N sections.

$$I_3 = i_{31} + i_{32} + ... + i_{3N-1} + i_{3N}$$
$$= (y_{13}e_{11} - y_{13}e_{21} + y_{33}e_{31}) + (y_{13}e_{12} - y_{13}e_{22} + y_{33}e_{32}) + ... \qquad \text{(Appendix.11)}$$
$$+ (y_{13}e_{1N-1} - y_{13}e_{2N-1} + y_{33}e_{3N-1}) + (y_{13}e_{1N} - y_{13}e_{2N} + y_{33}e_{3N})$$

By applying (Appendix.8), (Appendix.9) and boundary conditions (e_{11} = E_1, e_{2N}=E_2), (Appendix.11) becomes:

$$I_3 = y_{13}e_{11} - y_{13}e_{2N} + Ny_{33}E_3 = y_{13}E_1 - y_{13}E_2 + Ny_{33}E_3 \qquad \text{(Appendix.12)}$$

From Figure Appendix.7, the Y_{13} and Y_{33} can be expressed as:

$$Y_{13} = y_{13} \qquad \text{(Appendix.13)}$$

$$Y_{33} = Ny_{33} \qquad \text{(Appendix.14)}$$

Because the N periodic sections are connected acoustically in cascade and electrically in parallel, the model as in Figure Appendix.5 should be used to obtain the [Y] matrix of N-section IDT.

From (Appendix.3) for one section, the i_1 and i_2 can be expressed

$$i_1 = y_{11}e_1 + y_{12}e_2 + y_{13}e_3, \quad i_2 = -y_{12}e_1 - y_{12}e_2 + y_{13}e_3 \qquad \text{(Appendix.15)}$$

Equations (Appendix.15) can be represented in matrix form like [ABCD] form in electrical theory as follows:

$$\begin{bmatrix} e_2 \\ i_2 \end{bmatrix} = [K]\begin{bmatrix} e_1 \\ i_1 \end{bmatrix} + [L]e_3 \qquad \text{(Appendix.16)}$$

Where

$$[K] = \begin{bmatrix} -\dfrac{y_{11}}{y_{12}} & \dfrac{1}{y_{12}} \\[2ex] \dfrac{y_{11}^2 - y_{12}^2}{y_{12}} & -\dfrac{y_{11}}{y_{12}} \end{bmatrix} \qquad \text{(Appendix.17)}$$

$$[L] = \begin{bmatrix} -\dfrac{y_{13}}{y_{12}} \\[2ex] \dfrac{y_{11}y_{13} + y_{12}y_{13}}{y_{12}} \end{bmatrix} \qquad \text{(Appendix.18)}$$

By applying (Appendix.16) into N-section IDT as in Figure Appendix.6 and using (Appendix.9), the second recursion relation is obtained as follows:

$$\begin{bmatrix} e_m \\ i_m \end{bmatrix} = [K]\begin{bmatrix} e_{m-1} \\ i_{m-1} \end{bmatrix} + [L]E_3 \qquad \text{(Appendix.19)}$$

Where m is integer number, m=1,2, ..., N-1, N
Starting (Appendix.19)(Appendix.19) by using with m=N, and reducing m until m=1 gives the expression:

$$\begin{bmatrix} e_N \\ i_N \end{bmatrix} = [Q]\begin{bmatrix} e_0 \\ i_0 \end{bmatrix} + [X]E_3 \qquad \text{(Appendix.20)}$$

Where

$$[Q] = [K]^N \qquad \text{(Appendix.21)}$$

$$[X] = \begin{bmatrix} X_1 \\ X_2 \end{bmatrix} = \sum_{n=0}^{N-1} [K]^n [L] \qquad \text{(Appendix.22)}$$

Solving (Appendix.20) and using the boundary conditions ($e_0 = E_1$, $i_0 = I_1$) gives:

$$I_1 = -\frac{Q_{11}}{Q_{12}} E_1 + \frac{1}{Q_{12}} E_2 - \frac{X_1}{Q_{12}} E_3 \qquad \text{(Appendix.23)}$$

Consequently,

$$Y_{11} = -\frac{Q_{11}}{Q_{12}} \qquad \text{(Appendix.24)}$$

$$Y_{12} = \frac{1}{Q_{12}} \qquad \text{(Appendix.25)}$$

$$Y_{13} = -\frac{X_1}{Q_{12}} \qquad \text{(Appendix.26)}$$

The Y_{13} is known by (Appendix.13), so (Appendix.26) and matrix [X] don't need to be solved.

To solve (Appendix.24) and (Appendix.25), matrix [Q] should be solved.

In "crossed-field" model, matrix [Q] can be represented in a simple form as follows:

$$[K] = \begin{bmatrix} \cos(4\alpha) & -jR_0\sin(4\alpha) \\ -jG_0\sin(4\alpha) & \cos(4\alpha) \end{bmatrix} \qquad \text{(Appendix.27)}$$

$$[K]^2 = \begin{bmatrix} \cos(8\alpha) & -jR_0\sin(8\alpha) \\ -jG_0\sin(8\alpha) & \cos(8\alpha) \end{bmatrix} \qquad \text{(Appendix.28)}$$

$$[K]^3 = \begin{bmatrix} \cos(12\alpha) & -jR_0\sin(12\alpha) \\ -jG_0\sin(12\alpha) & \cos(12\alpha) \end{bmatrix} \qquad \text{(Appendix.29)}$$

...... etc. Consequently, matrix [Q] will be given:

$$[Q] = [K]^N = \begin{bmatrix} \cos(N4\alpha) & -jR_0\sin(N4\alpha) \\ -jG_0\sin(N4\alpha) & \cos(N4\alpha) \end{bmatrix} \qquad \text{(Appendix.30)}$$

From (Appendix.24) and (Appendix.35), Y_{11} and Y_{12} in "cross-field" model can be expressed:

$$Y_{11} = -jG_0 \cot g(N4\alpha) \qquad \text{(Appendix.31)}$$

$$Y_{12} = \frac{jG_0}{\sin(N4\alpha)} \qquad \text{(Appendix.32)}$$

In conclusion, matrix [Y] representation of N-section IDT is:

-   In "crossed-field" model, from (Appendix.6), (Appendix.13), (Appendix.14), (Appendix.31) and (Appendix.32):

$$Y_{11} = -jG_0 \cot g(4N\alpha)$$
$$Y_{12} = \frac{jG_0}{\sin(4N\alpha)}$$
$$Y_{13} = -jG_0 tg\alpha \qquad \text{(Appendix.33)}$$
$$Y_{33} = jN(2\omega C_0 + 4G_0 tg\alpha)$$

-   In "in-line field" model, from (Appendix.7), (Appendix.13), (Appendix.14), (Appendix.24) and (Appendix.25):

$$Y_{11} = -\frac{Q_{11}}{Q_{12}}$$

$$Y_{12} = \frac{1}{Q_{12}}$$

$$Y_{13} = -jG_0 \frac{tg\alpha}{1 - \frac{2G_0}{\omega C_0} tg\alpha} \qquad \text{(Appendix.34)}$$

$$Y_{33} = \frac{j2\omega NC_0}{1 - \frac{2G_0}{\omega C_0} tg\alpha}$$

Where [Q] can be calculated from (Appendix.17) and (Appendix.21).

7.2 Appendix 2: Equivqlent circuit for "N+1/2" model IDT

In case IDT includes N periodic sections (like in section 3.2 plus one finger (in color red) as shown in Figure Appendix.8 that we call "N+1/2" model IDT.

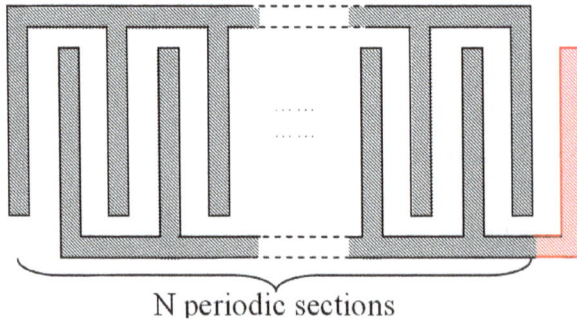

Fig. Appendix.8. "N+1/2" model IDT

The equivalent circuit for this model is shown in Figure Appendix.9 and the matrix [Yd] representation is shown as in Figure Appendix.10 (letter "d" stands for different from model [Y] in section 3.2.

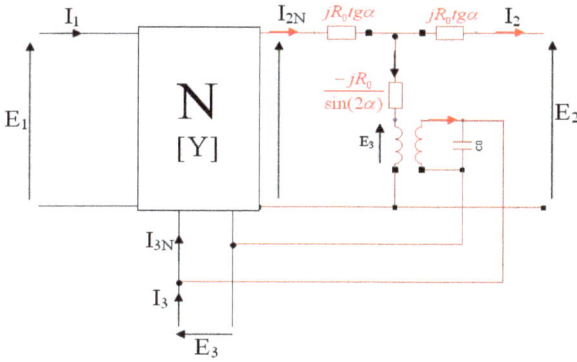

Fig. Appendix.9. Equivalent circuit of "N+1/2" model IDT

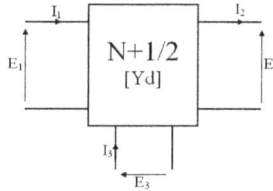

Fig. Appendix.10. [Yd] matrix representation of "N+1/2" model IDT

The form of matrix [Yd] is:

$$[Yd] = \begin{bmatrix} Yd_{11} & Yd_{12} & Yd_{13} \\ Yd_{21} & Yd_{22} & Yd_{23} \\ Yd_{31} & Yd_{32} & Yd_{33} \end{bmatrix}$$ (Appendix.35)

The elements of [Yd] matrix for "crossed-field" model are given as follows:

$$Yd_{11} = jG_0 \left\{ \frac{1}{\sin^2(4N\alpha)(\cot g(2\alpha) + \cot g(4N\alpha))} - \cot g(4N\alpha) \right\}$$ (Appendix.36)

$$Yd_{12} = \frac{jG_0}{\sin(4N\alpha)} \left\{ \cos(2\alpha) - \frac{\sin(2\alpha)[\cot g(4N\alpha)\cos(2\alpha) - \sin(2\alpha)]}{\cos(2\alpha) + \cot g(4N\alpha)\sin(2\alpha)} \right\}$$ (Appendix.37)

$$Yd_{13} = jG_0 \left\{ \frac{(-tg\alpha + 2\cot g(4N\alpha)\sin^2\alpha + \sin(2\alpha))\sin(2\alpha)}{\sin(4N\alpha)(\cos(2\alpha) + \cot g(4N\alpha)\sin(2\alpha))} + \frac{2\sin^2\alpha}{\sin(4N\alpha)} - tg\alpha \right\}$$ (Appendix.38)

$$Yd_{21} = -jG_0 \frac{1}{\sin(4N\alpha)(\cos(2\alpha) + \cot g(4N\alpha)\sin(2\alpha))}$$ (Appendix.39)

$$Yd_{22} = jG_0 \frac{\cot g(4N\alpha)\cos(2\alpha) - \sin(2\alpha)}{\cos(2\alpha) + \cot g(4N\alpha)\sin(2\alpha)}$$ (Appendix.40)

$$Yd_{23} = jG_0 \frac{-tg\alpha + 2\cot g(4N\alpha)\sin^2(2\alpha) + \sin(2\alpha)}{\cos(2\alpha) + \cot g(4N\alpha)\sin(2\alpha)} \qquad \text{(Appendix.41)}$$

$$Yd_{31} = -jG_0 tg\alpha \qquad \text{(Appendix.42)}$$

$$Yd_{32} = -jG_0 \sin(2\alpha) \qquad \text{(Appendix.43)}$$

$$Yd_{33} = j\omega C_0(2N-1) + jG_0\{\sin(2\alpha) + (4N+1)tg\alpha\} \qquad \text{(Appendix.44)}$$

7.3 Appendix 3: Scattering matrix [S] for IDT

The scattering matrix [S] of a three-port network characterized by its admittance matrix [Y] is given by [3]:

$$S = \Pi_3 - 2Y(\Pi_3 + Y)^{-1} \qquad \text{(Appendix.45)}$$

Where Π_3 is the 3x3 identity matrix.

After expanding this equation, the scattering matrix elements for a general three-port network are given by the following expressions:

$$S_{11} = \frac{1}{M}\{(1+Y_{33})(1-Y_{11}+Y_{22}-Y_{11}Y_{22}+Y_{12}Y_{21}) + \\ + Y_{13}[Y_{31}(1+Y_{22})-Y_{21}Y_{32}] + Y_{23}[Y_{32}(Y_{11}-1)-Y_{12}Y_{31}]\} \qquad \text{(Appendix.46)}$$

$$S_{12} = -\frac{2}{M}[Y_{12}(1+Y_{33})-Y_{13}Y_{32}] \qquad \text{(Appendix.47)}$$

$$S_{13} = -\frac{2}{M}[Y_{13}(1+Y_{22})-Y_{12}Y_{23}] \qquad \text{(Appendix.48)}$$

$$S_{21} = -\frac{2}{M}[Y_{21}(1+Y_{33})-Y_{23}Y_{31}] \qquad \text{(Appendix.49)}$$

$$S_{22} = \frac{1}{M}\{(1+Y_{33})(1+Y_{11}-Y_{22}-Y_{11}Y_{22}+Y_{12}Y_{21}) + \\ + Y_{13}[Y_{31}(Y_{22}-1)-Y_{21}Y_{32}] + Y_{23}[Y_{32}(Y_{11}+1)-Y_{12}Y_{31}]\} \qquad \text{(Appendix.50)}$$

$$S_{23} = -\frac{2}{M}[Y_{23}(1+Y_{11})-Y_{13}Y_{21}] \qquad \text{(Appendix.51)}$$

$$S_{31} = -\frac{2}{M}[Y_{31}(1+Y_{22})-Y_{21}Y_{32}] \qquad \text{(Appendix.52)}$$

$$S_{32} = -\frac{2}{M}[Y_{32}(1+Y_{11})-Y_{12}Y_{31}] \qquad \text{(Appendix.53)}$$

$$S_{33} = \frac{1}{M}\{(1-Y_{33})(1+Y_{11}+Y_{22}+Y_{11}Y_{22}-Y_{12}Y_{21}) + \\ + Y_{13}[Y_{31}(Y_{22}+1)-Y_{21}Y_{32}] + Y_{23}[Y_{32}(Y_{11}+1)-Y_{12}Y_{31}]\} \qquad \text{(Appendix.54)}$$

where

$$M = \det(\Pi_3 + Y)$$
$$= (1 + Y_{33})[(1 + Y_{11})(1 + Y_{22}) - Y_{12}Y_{21}] - Y_{23}[Y_{32}(1 + Y_{11}) - Y_{12}Y_{31}] - \text{(Appendix.55)}$$
$$- Y_{13}[Y_{31}(1 - Y_{22}) - Y_{21}Y_{32}]$$

For model IDT including N identical sections, these equations can be further simplified. In case of Figure Appendix.7 (b):

$$\begin{aligned} Y_{11} &= Y_{22} \\ Y_{21} &= Y_{12} \\ Y_{31} &= Y_{13} \\ Y_{23} &= Y_{32} = -Y_{13} \end{aligned} \qquad \text{(Appendix.56)}$$

Therefore, S_{ij}'s take the following form

$$S_{11} = S_{22} = \frac{1}{M}\{(1 + Y_{33})(1 - Y_{11}^2 + Y_{12}^2) + 2Y_{13}^2(Y_{11} + Y_{12})\} \qquad \text{(Appendix.57)}$$

$$S_{12} = S_{21} = -\frac{2}{M}\left[Y_{12}(1 + Y_{33}) + Y_{13}^2\right] \qquad \text{(Appendix.58)}$$

$$S_{13} = S_{31} = -\frac{2}{M}Y_{13}(1 + Y_{11} + Y_{12}) \qquad \text{(Appendix.59)}$$

$$S_{23} = S_{32} = -S_{13} \qquad \text{(Appendix.60)}$$

$$S_{33} = \frac{1}{M}\{(1 - Y_{33})[(1 + Y_{11})^2 - Y_{12}^2] + 2Y_{13}^2(1 + Y_{11} + Y_{12})\} \qquad \text{(Appendix.61)}$$

Where

$$M = (1 + Y_{33})[(1 + Y_{11})^2 - Y_{12}^2] - 2Y_{13}^2(1 + Y_{12}) \qquad \text{(Appendix.62)}$$

7.4 Appendix 4: Equivalent circuit for SAW device base on Mason model, [ABCD] Matrix representation
7.4.1 Appendix 4.1: [ABCD] Matrix representation of IDT

In SAW device, each input and output IDTs have one terminal connected to admittance G_0. Therefore, one IDT can be represented as two-port network. [ABCD] matrix (as in Figure Appendix.11) is used to represent each IDT, because [ABCD] matrix representation has one interesting property that in cascaded network, the [ABCD] matrix of total network can be obtained easily by multiplying the matrices of elemental networks.

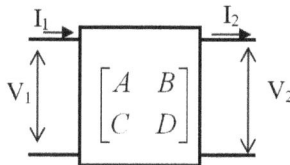

Fig. Appendix.11. [ABCD] representation of two-port network for one IDT

To find the [ABCD] matrix for one IDT in SAW device, the condition that no reflected wave at one terminal of IDT, and the current-voltage relations by [Y] matrix in section are used as follows:

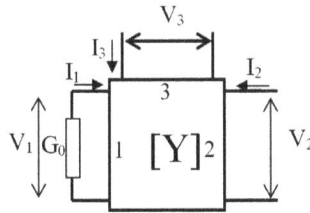

Fig. Appendix.12. Two-port network for one IDT

$$\begin{bmatrix} I_1 \\ I_2 \\ I_3 \end{bmatrix} = \begin{bmatrix} Y_{11} & Y_{12} & Y_{13} \\ Y_{12} & Y_{11} & -Y_{13} \\ Y_{13} & -Y_{13} & Y_{33} \end{bmatrix} \begin{bmatrix} V_1 \\ V_2 \\ V_3 \end{bmatrix}$$

(Appendix.63)

And
$$I_1 = -G_0 V_1$$
(Appendix.64)

From these current-voltage relations, the V_3 and I_3 are given:

$$V_3 = \frac{Y_{11}^2 - Y_{12}^2 + Y_{11}G_0}{Y_{12}Y_{13} + Y_{11}Y_{13} + Y_{13}G_0} V_2 - \frac{G_0 + Y_{11}}{Y_{12}Y_{13} + Y_{11}Y_{13} + Y_{13}G_0} I_2$$
(Appendix.65)

$$I_3 = \frac{-(Y_{13}Y_{12} + Y_{13}Y_{11} + Y_{13}G_0)^2 + (Y_{11}Y_{33} - Y_{13}^2 + Y_{33}G_0)(Y_{11}^2 - Y_{12}^2 + Y_{11}G_0)}{(G_0 + Y_{11})(Y_{12}Y_{13} + Y_{11}Y_{13} + Y_{13}G_0)} V_2 -$$
$$- \frac{Y_{11}Y_{33} - Y_{13}^2 + Y_{33}G_0}{Y_{12}Y_{13} + Y_{11}Y_{13} + Y_{13}G_0} I_2$$

(Appendix.66)

From (Appendix.65) and (Appendix.66), equivalence between port 3 in Figure Appendix.12 equals to port 1 in Figure Appendix.11, and consideration of direction of current I_2 in Figure Appendix.11 and Figure Appendix.12, [ABCD] matrix representation for two-port network of IDT in obtained:

$$A = \frac{Y_{11}^2 - Y_{12}^2 + Y_{11}G_0}{Y_{12}Y_{13} + Y_{11}Y_{13} + Y_{13}G_0}$$
(Appendix.67)

$$B = \frac{G_0 + Y_{11}}{Y_{12}Y_{13} + Y_{11}Y_{13} + Y_{13}G_0}$$
(Appendix.68)

$$C = \frac{-(Y_{13}Y_{12} + Y_{13}Y_{11} + Y_{13}G_0)^2 + (Y_{11}Y_{33} - Y_{13}^2 + Y_{33}G_0)(Y_{11}^2 - Y_{12}^2 + Y_{11}G_0)}{(G_0 + Y_{11})(Y_{12}Y_{13} + Y_{11}Y_{13} + Y_{13}G_0)}$$
(Appendix.69)

$$D = \frac{Y_{11}Y_{33} - Y_{13}^2 + Y_{33}G_0}{Y_{12}Y_{13} + Y_{11}Y_{13} + Y_{13}G_0}$$
(Appendix.70)

In case of "crossed-field" model, the [ABCD] can be further simplified:

$$A = \frac{\sin(4N\alpha) - j\cos(4N\alpha)}{tg\alpha\left[1 - \cos(4N\alpha) - j\sin(4N\alpha)\right]}$$ (Appendix.71)

$$B = \frac{A}{G_0}$$ (Appendix.72)

$$D = \frac{\sin(4N\alpha)}{1 - \cos(4N\alpha) - j\sin(4N\alpha)}\left[N(2\omega C_0 Z_0 \cot\alpha + 4)(\cot(4N\alpha) + j) + tg\alpha\right]$$ (Appendix.73)

$$C = -\frac{1}{B} + G_0 D$$ (Appendix.74)

One interesting property of [ABCD] of "crossed-field" mode is:

$$AD-BC=1$$ (Appendix.75)

This means [ABCD] matrix is reciprocal.
In SAW device, the ouput IDT is inverse of input IDT. By the reciprocal property of [ABCD], the [ABCD] matrix of output IDT can be easily obtained:

$$A_{output} = D_{input}$$ (Appendix.76)

$$B_{output} = B_{input}$$ (Appendix.77)

$$C_{output} = C_{input}$$ (Appendix.78)

$$D_{output} = A_{input}$$ (Appendix.79)

in which N is replaced by M (number of periodic sections in output IDT)
Consequently, the [ABCD] matrix of output IDT is:

$$A_{out} = \frac{\sin(4M\alpha)}{1 - \cos(4M\alpha) - j\sin(4M\alpha)}\left[M(2\omega C_0 Z_0 \cot\alpha + 4)(\cot(4M\alpha) + j) + tg\alpha\right]$$ (Appendix.80)

$$B_{out} = \frac{1}{G_0}\frac{\sin(4M\alpha) - j\cos(4M\alpha)}{tg\alpha\left[1 - \cos(4M\alpha) - j\sin(4M\alpha)\right]}$$ (Appendix.81)

$$D_{out} = \frac{\sin(4M\alpha) - j\cos(4M\alpha)}{tg\alpha\left[1 - \cos(4M\alpha) - j\sin(4M\alpha)\right]}$$ (Appendix.82)

$$C_{out} = -\frac{1}{B_{out}} + G_0 A_{out}$$ (Appendix.83)

At the center frequency f_0, the [ABCD] matrix becomes infinite since $\alpha=0.5\pi(f/f_0)= 0.5\pi$. However, [ABCD] elements may be calculated by expanding for frequency very near frequency f_0.
By setting:

$$\alpha = \frac{\pi}{2}\frac{f - f_0}{f_0} + \frac{\pi}{2} = \frac{x}{2N} + \frac{\pi}{2}$$ (Appendix.84)

Where
$$x = N\pi \frac{f - f_0}{f_0}$$
(Appendix.85)

By using the limit of some functions as follows:

$$\lim_{x \to 0}[\sin(4N\alpha)] = \lim_{x \to 0}[\sin(2x)] \approx 2x$$
(Appendix.86)

$$\lim_{x \to 0}[\cos(4N\alpha)] = \lim_{x \to 0}[\cos(2x)] \approx 1$$
(Appendix.87)

$$\lim_{x \to 0}[tg\alpha] = \lim_{x \to 0}[-\cot(\frac{x}{2N})] \approx -\frac{2N}{x}$$
(Appendix.88)

The [ABCD] matrix of input IDT is obtained:

$$A \approx \frac{2x - j}{j4N}$$
(Appendix.89)

$$B \approx \frac{1}{G_0} \frac{2x - j}{j4N}$$
(Appendix.90)

$$C \approx 2\pi f C_0 x - 4NG_0 - j\left(\pi f C_0 + \frac{4NG_0}{2x - j}\right)$$
(Appendix.91)

$$D \approx 2\pi f C_0 Z_0 x - 4N - j\pi f C_0 Z_0$$
(Appendix.92)

7.4.2 Appendix 4.2: [ABCD] matrix representation of propagation path
Based on equivalent circuit star model of propagation path in section 3.3, [ABCD] matrix representation of propagation way can be obtained clearly:

$$A_{path} = D_{path} = \cos 2\theta$$
(Appendix.93)

$$B_{path} = C_{path} = j \sin 2\theta$$
(Appendix.94)

With
$$\theta = \frac{\pi f l}{v}$$
(Appendix.95)

Where l is the length of propagation path between two IDTs.
So, [ABCD] matrix representations of input IDT, propagation way and output IDT are obtained. They are cascaded as Figure Appendix.13:

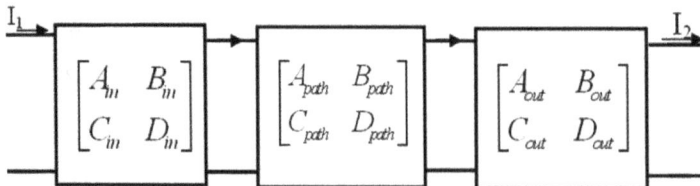

Fig. Appendix.13. Cascaded [ABCD] matrices of input IDT, propagation way and output IDT

And the [ABCD] equivalent matrix of SAW device is shown in Figure Appendix.14

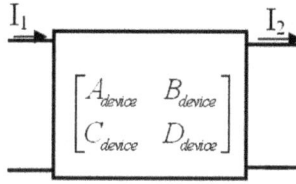

Fig. Appendix.14. [ABCD] matrix of SAW device

[ABCD] matrix of delay line SAW is

$$\begin{bmatrix} A_{device} & B_{device} \\ C_{device} & D_{device} \end{bmatrix} = \begin{bmatrix} A_{in} & B_{in} \\ C_{in} & D_{in} \end{bmatrix}\begin{bmatrix} A_{path} & B_{path} \\ C_{path} & D_{path} \end{bmatrix}\begin{bmatrix} A_{out} & B_{out} \\ C_{out} & D_{out} \end{bmatrix} \qquad \text{(Appendix.96)}$$

$$A_{device} = A_{in}A_{path}A_{out} + B_{in}C_{path}A_{out} + A_{in}B_{path}C_{out} + B_{in}D_{path}C_{out} \qquad \text{(Appendix.97)}$$

$$B_{device} = A_{in}A_{path}B_{out} + B_{in}C_{path}B_{out} + A_{in}B_{path}D_{out} + B_{in}D_{path}D_{out} \qquad \text{(Appendix.98)}$$

$$C_{device} = C_{in}A_{path}A_{out} + D_{in}C_{path}A_{out} + C_{in}B_{path}C_{out} + D_{in}D_{path}C_{out} \qquad \text{(Appendix.99)}$$

$$D_{device} = C_{in}A_{path}B_{out} + D_{in}C_{path}B_{out} + C_{in}B_{path}D_{out} + D_{in}D_{path}D_{out} \qquad \text{(Appendix.100)}$$

Where [ABCD]$_{in}$ is calculated from (Appendix.71), (Appendix.72), (Appendix.73) and (Appendix.74).
[ABCD]$_{out}$ is calculated from (Appendix.80), (Appendix.81), (Appendix.82) and (Appendix.83).
[ABCD]$_{path}$ is calculated from (Appendix.93) and (Appendix.94).

8. References

[1] C.C.W.Ruppel, W.Ruile, G.Scholl, K.Ch.Wagner, and O.Manner, Review of models for low-loss filter design and applications, IEEE Ultransonics Symposium, pp.313-324, 1994.
[2] L.A.Coldren, and R.L.Rosenberg, Scattering matrix approach to SAW resonators, IEEE Ultrasonics Symposium, 1976, pp.266-271.
[3] R.W.Newcomb, Linear Multiport Synthesis, McGraw Hill, 1966.
[4] C.Elachi, Waves in Active and Passive Periodic Structures: A Review, Proceedings of the IEEE, vol.64, No.12, December 1976, pp.1666-1698
[5] M.Hikita, A.Isobe, A.Sumioka, N.Matsuura, and K.Okazaki, Rigorous Treatment of Leaky SAW's and New Equivalent Circuit Representation for Interdigital Transducers, IEEE Transactions on Ultrasonics, Ferroelectrics, and Frequency Control, Vol.43, No.3, May 1996.
[6] L.F.Brown, and D.L.Carlson, Ultrasound Transducer Models for Piezoelectric Polymer Films, IEEE Transactions on Ultrasonics, Ferroelectrics, and Frequency Control, Vol.36, No.3, May 1989.
[7] K.Hashimoto, and M.Yamaguchi, Precise simulation of surface transverse wave devices by discrete Green function theory, IEEE Ultrasonics Symposium, 1994, pp.253-258.

[8] K.Hashimoto, G.Endoh, and M.Yamaguchi, Coupling-of-modes modelling for fast and precise simulation of leaky surface acoustic wave devices, *IEEE Ultrasonics Symposium*, 1995, pp.251-256.

[9] K.Hashimoto, and M.Yamaguchi, General-purpose simulator for leaky surface acoustic wave devices based on Coupling-Of-Modes theory, *IEEE Ultrasonics Symposium*, 1996, pp.117-122.

[10] K.Hashimoto, Surface Acoustic Wave Devices in Telecommunications, *Modelling and Simulation*, Springer, 2000, ISBN: 9783540672326.

[11] P.M.Smith, and C.K.Campbell, A Theoretical and Experimental Study of Low-Loss SAW Filters with Interdigitated Interdigital Transducers, *IEEE Transactions on Ultrasonics, Ferroelectrics, and Frequency Control*, Vol.36, No.1, January 1989, pp.10-15.

[12] C.K.Campbell, Modelling the Transverse-Mode Response of a Two-Port SAW Resonator, *IEEE Transactions on Ultrasonics, Ferroelectrics, and Frequency Control*, Vol.38, No.3, May 1991, pp.237-242.

[13] C.K.Campbell, P.M.Smith, and P.J.Edmonson, Aspects of Modeling the Frequency Response of a Two-Port Waveguide-Coupled SAW Resonator-Filter, *IEEE Transactions on Ultrasonics, Ferroelectrics, and Frequency Control*, Vol.39, No.6, November 1992, pp.768-773.

[14] C.K.Campbell, Longitudinal-Mode Leaky SAW Resonator Filters on 64⁰ Y-X Lithium Niobate, *IEEE Transactions on Ultrasonics, Ferroelectrics, and Frequency Control*, Vol.42, No.5, September 1995, pp.883-888.

[15] C.K.Campbell, and P.J.Edmonson, Conductance Measurements on a Leaky SAW Harmonic One-Port Resonator, *IEEE Transactions on Ultrasonics, Ferroelectrics, and Frequency Control*, Vol.47, No.1, January 2000, pp.111-116.

[16] C.K.Campbell, and P.J.Edmonson, Modeling a Longitudinally Coupled Leaky-SAW Resonator Filter with Dual-Mode Enhanced Upper-Sideband Suppression, *IEEE Transactions on Ultrasonics, Ferroelectrics, and Frequency Control*, Vol.48, No.5, September 2001, pp.1298-1301.

[17] J.Munshi, and S.Tuli, A Circuit Simulation Compatible Surface Acoustic Wave Interdigital Transducer Macro-Model, *IEEE Transactions on Ultrasonics, Ferroelectrics, and Frequency Control*, Vol.51, No.7, July 2004, pp.782-784.

[18] M.P.Cunha, and E.L.Adler, A Network Model For Arbitrarily Oriented IDT Structures, *IEEE Transactions on Ultrasonics, Ferroelectrics, and Frequency Control*, Vol.40, No.6, November 1993, pp.622-629.

[19] D.R.Mahapatra, A.Singhal, and S.Gopalakrishnan, Numerical Analysis of Lamb Wave Generation in Piezoelectric Composite IDT, *IEEE Transactions on Ultrasonics, Ferroelectrics, and Frequency Control*, Vol.52, No.10, October 2005, pp.1851-1860.

[20] A APPENDIX. Bhattacharyya, Suneet Tuli, and S.Majumdar, SPICE Simulation of Surface Acoustic Wave Interdigital Transducers, *IEEE Transactions on Ultrasonics, Ferroelectrics, and Frequency Control*, Vol.42, No.4, July 1995, pp.784-786.

[21] C.M.Panasik, and APPENDIX.J.Hunsinger, Scattering Matrix Analysis Of Surface Acoustic Wave Reflectors And Transducers, *IEEE Transactions On Sonics And Ultrasonics*, Vol.SU-28, No.2, March 1981, pp.79-91.

[22] W.Soluch, Admittance Matrix Of A Surface Acoustic Wave Interdigital Transducer, *IEEE Transactions on Ultrasonics, Ferroelectrics, and Frequency Control*, Vol.40, No.6, November 1993, pp.828-831.

[23] W.Soluch, Scattering Matrix Approach To One Port SAW Resonators, *IEEE Frequency Control Symposium*, 1999, pp.859-862.

[24] W.Soluch, Design of SAW Synchronous Resonators on ST Cut Quartz, *IEEE Transactions on Ultrasonics, Ferroelectrics, and Frequency Control*, Vol.46, No.5, September 1999, pp.1324-1326.

[25] W.Soluch, Scattering Matrix Approach To One-Port SAW Resonators, *IEEE Transactions on Ultrasonics, Ferroelectrics, and Frequency Control*, Vol.47, No.6, November 2000, pp.1615-1618.

[26] W.Soluch, Scattering Matrix Approach To STW Resonators, *IEEE Transactions on Ultrasonics, Ferroelectrics, and Frequency Control*, Vol.49, No.3, March 2002, pp.327-330.

[27] W.Soluch, Scattering Analysis Of Two-Port SAW Resonators, *IEEE Transactions on Ultrasonics, Ferroelectrics, and Frequency Control*, Vol.48, No.3, May 2001, pp.769-772.

[28] W.Soluch, Scattering Matrix Approach To STW Multimode Resonators, Electronics Letters, 6th January 2005, Vol.41, No.1.

[29] K.Nakamura, A Simple Equivalent Circuit For Interdigital Transducers Based On The Couple-Mode Approach, *IEEE Transactions on Ultrasonics, Ferroelectrics, and Frequency Control*, Vol.40, No.6, November 1993, pp.763-767.

[30] K. Nakamura, and K.Hirota, Equivalent circuit for Unidirectional SAW-IDT's based on the Coupling-Of-modes theory, *IEEE Trans on Ultrasonics, Ferroelectrics, and Frequency Control*, Vol.43, No.3, May 1996, pp.467-472.

[31] A.H.Fahmy, and E.L.Adler, Propagation of acoustic surface waves in multilayers: A matrix description. *Applied Physics Letter*, vol. 22, No.10, 1973, pp. 495-497.

[32] E.L.Adler, Matrix methods applied to acoustic waves in multilayers, *IEEE Transactions on Ultrasonics, Ferroelectrics, and Frequency Control*, Vol.37, No.6, November 1990, pp.485-490.

[33] E.L.Adler, SAW and Pseudo-SAW properties using matrix methods, *IEEE Transactions on Ultrasonics, Ferroelectrics, and Frequency Control*, Vol.41, No.5, September 1994, pp.699-705.

[34] G.F.Iriarte, F.Engelmark, I.V.Katardjiev, V.Plessky, V.Yantchev, SAW COM-parameter extraction in AlN/diamond layered structures, *IEEE Transactions On Ultrasonics, Ferroelectrics, And Frequency Control*, Vol. 50, No. 11, November 2003.

[35] M.Mayer, G.Kovacs, A.Bergmann, and K.Wagner, A Powerful Novel Method for the Simulation of Waveguiding in SAW Devices, *IEEE Ultrasonics Symposium*, 2003, pp.720-723.

[36] W.P.Mason, Electromechanical Transducer and Wave Filters, second edition, D.Van Nostrand Company Inc, 1948.

[37] W.P.Mason, Physical Acoustics, Vol 1A, Academic Press, New York 1964.

[38] S.D.Senturia, Microsystem Design, Kluwer Academic Publishers, 2001, ISBN 0-7923-7246-8.

[39] W.Marshall Leach, Controlled-Source Analogous Circuits and SPICE models for Piezoelectric transducers, *IEEE Transactions on Ultrasonics, Ferroelectrics, and Frequency Control*, Vol.41, No.1, January 1994.

[40] D.A.Berlincourt, D.R.Curran and H.Jaffe, Chapter 3, Piezoelectric and Piezomagnetic Materials and Their Function in Tranducers.

[41] W.R.Smith, H.M.Gerard, J.H.Collins, T.M.Reeder, and H.J.Shaw, Analysis of Interdigital Surface Wave Transducers by Use of an Equivalent Circuit Model, *IEEE Transaction on MicroWave Theory and Techniques*, No.11, November 1969, pp.856-864.

[42] C. K. Campbell, Surface acoustic wave devices, in *Mobile and Wireless Communications*, New York: Academic, 1998.

[43] O.Tigli, and M.E.Zaghloul, A Novel Saw Device in CMOS: Design, Modeling, and Fabrication, *IEEE Sensors journal*, vol. 7, No. 2, February 2007, pp.219-227.

[44] K.Nakamura, A Simple Equivalent Circuit For Interdigital Transducers Based On The Couple-Mode Approach, *IEEE Transactions on Ultrasonics, Ferroelectrics, and Frequency Control*, Vol.40, No.6, November 1993, pp.763-767.

[45] M.Hofer, N.Finger, G.Kovacs, J.Schoberl, S.Zaglmayr, U.Langer, and R.Lerch, Finite-Element Simulation of Wave Propagation in Periodic Piezoelectric SAW Structures, *IEEE Transactions on Ultrasonics, Ferroelectrics, and Frequency Control*, Vol.53, No.6, June 2006, pp.1192-1201.

[46] M. Hofer, N. Finger, G. Kovacs, J. Schoberl, U. Langer, and R. Lerch, Finite-element simulation of bulk and surface acoustic wave (SAW) interaction in SAW devices, *IEEE Ultrasonics Symposium*, 2002.

[47] Online: http://www.comsol.com/

[48] Online: http://www.coventor.com/

[49] Online: http://www.ansys.com/

[50] J.J. Campbell, W.R. Jones, A method for estimating optimal crystal cuts and propagation directions for excitation of piezoelectric surface waves, *IEEE Transaction on Sonics Ultrasonics*. SU-15 (4), 1968, pp.209–217.

[51] E.Akcakaya, E.L.Adler, and G.W.Farnell, Apodization of Multilayer Bulk-Wave Transducers, *IEEE Transactions on Ultrasonics Ferroelectrics and Frequency Control*, Vol.36, No.6, November 1989, pp 628-637.

[52] E.L.Adler, J.K.Slaboszewicz, G.W.FARNELL, and C.K.JEN, PC Software for SAW Propagation in Anisotropic Multilayers, *IEEE Transactions on Ultrasonics Ferroelectrics and Frequency Control*, Vol.37, No.2, May 1990, pp.215-223.

[53] E.L.Adler, SAW and Pseudo-SAW Properties Using Matrix Methods, *IEEE Transactions on Ultrasonics Ferroelectrics and Frequency Control*, Vol.41, No.6, pp.876-882, September 1994.

[54] K. A. Ingebrigtsen, Surface waves in piezoelectrics, *Journal of Applied Physics*, Vol.40, No.7, 1969, pp.2681-2686.

[55] Y.Suzuki, H.Shimizu, M.Takeuchi, K.Nakamura, and A.Yamada, Some studies on SAW resonators and multiple-mode filters, *IEEE Ultrasonics Symposium Proceedings*, 1976, pp.297-302.

[56] S.D.Senturia, Microsystem Design, Kluwer Academic Publishers, 2001, ISBN 0-7923-7246-8.

[57] W.P.Mason, Electromechanical Transducer and Wave Filters, second edition, D.Van Nostrand Company Inc, 1948.

[58] W.P.Mason, Physical Acoustics, Vol 1A, Academic Press, New York 1964.

[59] W.R.Smith, H.M.Gerard, J.H.Collins, T.M.Reeder, and H.J.Shaw, Analysis of Interdigital Surface Wave Transducers by Use of an Equivalent Circuit Model, *IEEE Transaction on MicroWave Theory and Techniques*, No.11, November 1969, pp.856-864.

[60] Y.Suzuki, H.Shimizu, M.Takeuchi, K.Nakamura, and A.Yamada, Some studies on SAW resonators and multiple-mode filters, *IEEE Ultrasonics Symposium Proceedings*, 1976, pp.297-302.

[61] K.Nakamura, A Simple Equivalent Circuit For Interdigital Transducers Based On The Couple-Mode Approach, *IEEE Transactions on Ultrasonics, Ferroelectrics, and Frequency Control*, Vol.40, No.6, November 1993, pp.763-767.

8

Sources of Third–Order Intermodulation Distortion in Bulk Acoustic Wave Devices: A Phenomenological Approach

Eduard Rocas and Carlos Collado
Universitat Politècnica de Catalunya (UPC), Barcelona
Spain

1. Introduction

Acoustic devices like Bulk Acoustic Wave (BAW) resonators and filters represent a key technology in modern microwave industry. More specifically, BAW technology offers promising performance due to its good power handling and high quality factors that make it suitable for a wide range of applications. Nevertheless, harmonics and 3IMD arising from intrinsic nonlinear material properties (Collado et al., 2009) and dynamic self-heating (Rocas et al., 2009) could represent a limitation for some applications.

Driven by the need for highly linear devices, there is a demand for further development of accurate models of BAW devices, capable of predicting the nonlinear behavior of the device and its impact on a circuit. Many authors have attempted to model the nonlinearities of BAW devices by using different approaches, mostly involving phenomenological lumped element models. Although these models can be useful because of their simplicity, they are mainly limited to narrow-band operation and they usually cannot be parameterized in terms of device-independent parameters (Constantinescu et al., 2008). Another approach consists of extending all the material properties on the constitutive equations to the nonlinear domain and introducing the nonlinear relations to the model implementation, which leads to several possible nonlinear sources increasing model complexity (Cho et al., 1993; Ueda et al., 2008). On the other hand, (Feld, 2009) presents a one-parameter nonlinear circuit model to account for the intrinsic nonlinearities. Such a model does not include the self-heating mechanism and can underestimate the 3IMD by more than 20 dB.

In this work, a model that uses several nonlinear parameters to predict harmonics and 3IMD distortion is presented. Its novelty lies in its ability to predict the nonlinear effects produced by self-heating in addition to those due to intrinsic nonlinearities in the material properties. The model can be considered an extension of the nonlinear KLM model (originally proposed by Krimholtz, Leedom and Matthaei) (Krimholtz et al., 1970) to include the thermal effects due to self-heating caused by viscous losses and electrode losses. For this purpose a thermal domain circuit model is implemented and coupled to the electro-acoustic model, which allows us to calculate the dynamic temperature variations that change the material properties. In comparison to (Rocas et al., 2009), this work describes the impact that electrode losses produce on the 3IMD, presents closed-form expressions derived from the circuit model and validates the model with extensive measurements that confirm the necessity to include dynamic self-heating to accurately predict the generation of spurious signals in BAW devices.

2. Nonlinear generation mechanisms

The origin of nonlinearities in BAW resonators has been controversial and there still exists no consensus (Nakamura et al., 2010). However, recent results point to several underlying causes which combine in different ways to give rise to a wide range of nonlinear effects (Rocas et al., 2009). We summarize the nonlinear effects of a stiffened elasticity, and then address the nonlinearity due to self-heating caused by viscous losses and electrode losses. We develop a circuit model to describe self-heating effects, and compare the measured results with simulations. Closed-form expressions for a simple one-layer BAW device model are then extracted to better understand the nonlinear generation mechanisms.

2.1 Nonlinear stiffened elasticity

Nonlinear elasticity has been proposed as the predominant contribution to the measured second harmonics and as a potential source of the observed 3IMD products (Collado et al., 2009) in two-tone experiments.

The approach described in (Collado et al., 2009) starts by considering a nonlinear stress-strain relation under electric field described by a nonlinear stiffened elasticity $c^D(T)$ in the form of the polynomial

$$c^D(T) = c_0^D + \Delta c_1^D T + \Delta c_2^D T^2 \tag{1}$$

where T is the stress. As detailed in (Collado et al., 2009), (1) translates into a nonlinear distributed capacitance $C_d(v)$ in the equivalent electric model of the acoustic transmission line (Auld, 1990), in which the voltage v is equivalent to force. In the equivalent electric model (1) transforms into:

$$C_d(v) = C_{d,0} + \Delta C_1 v + \Delta C_2 v^2. \tag{2}$$

Equation (2) leads to the nonlinear acoustic Telegrapher's equations which can be used to obtain the maximum voltage amplitude occurring along a resonating transmission line as shown in (Collado et al., 2009; Collado et al., 2005). When the device is driven by two tones at frequencies ω_1 and ω_2, standing waves with maximum force amplitudes $V_{\omega 1}$ and $V_{\omega 2}$ are trapped in the line. Then, as detailed in (Collado et al., 2009), the nonlinear capacitance (2) is responsible for generating 3IMD signals that result from adding the contributions due to Δc_1^D and Δc_2^D:

$$V_{\omega_{12}} = A_1 Q_L V_{\omega_1}^2 V_{\omega_2}^* \Delta C_1^2 \tag{3}$$

$$V_{\omega_{12}} = A_2 Q_L V_{\omega_1}^2 V_{\omega_2}^* \Delta C_2 \tag{4}$$

where $\omega_{12} = 2\omega_1 - \omega_2$, Q_L is the loaded quality factor and A_1 and A_2 are constants that depend on the geometry of the device and on its materials. Identical results would be obtained for the 3IMD at $2\omega_2 - \omega_1$ (which we will denote as ω_{21}).

2.2 Self-heating

Third-order intermodulation distortion due to dynamic self-heating is a well known process in microwave power amplifiers (Camarchia et al., 2007; Parker et al., 2004; Vuolevi et al., 2001) but has received less attention in passive devices (Rocas et al., 2010). What makes it different from the 3IMD caused by intrinsic nonlinearities is its dependence on the envelope frequency of the signal. For the particular case of a two-tone experiment, in which the envelope is a sinusoid, the thermal generation of 3IMD has a low-pass dependence on the envelope frequency due to the slow dynamics related with heating effects.

Recent results of two-tone 3IMD tests in BAW resonators as a function of the tones spacing reveal the important impact of self-heating effects in thin-Film Bulk Acoustic Resonators (FBAR) (Collado et al., 2009; Feld, 2009; Rocas et al., 2008) and Solidly Mounted Resonators (SMR) (Rocas et al., 2009). Heating produced by viscous damping in the acoustic domain and by ohmic loss in the electric domain produce local temperature oscillations which affect the temperature-dependent material properties.

If $\omega_1 = \omega_0 - \Delta\omega/2$ and $\omega_2 = \omega_0 + \Delta\omega/2$ are the input signals for a two-tone test, dissipation occurs as a result of electric and acoustic losses, and the quadratic dependence of the dissipated power on the signal amplitude

$$P_d \propto \left(V_1 \cos\left(\left(\omega_0 - \frac{\Delta\omega}{2} \right)t \right) + V_2 \cos\left(\left(\omega_0 + \frac{\Delta\omega}{2} \right)t \right) \right)^2 \tag{5}$$

gives rise to several frequency components of the dissipated power:

$$P_d \propto \frac{1}{2}V_1^2 + \frac{1}{2}V_2^2 + \frac{1}{2}V_1^2 \cos(2\omega_0 t - \Delta\omega t) +$$

$$\frac{1}{2}V_2^2 \cos(2\omega_0 t + \Delta\omega t) + V_1 V_2 \cos(2\omega_0 t) \tag{6}$$

$$+ V_1 V_2 \cos(\Delta\omega t).$$

These frequency components produce temperature variations on the device at the same frequencies. These temperature variations $K(\omega)$ can be written in terms of the dissipated power and the thermal impedance as (Parker et al., 2004)

$$K(\omega) = Z_{th}(\omega)P_d(\omega). \tag{7}$$

It is important to point out that the temperature variation at the envelope frequency ($\Delta\omega = \omega_2 - \omega_1$) is the most relevant for the generation of spurious signals because of the low-pass filter character of the thermal impedance $Z_{th}(\omega)$. These slow temperature oscillations induce low frequency changes of the material properties, and consequently, generate undesired 3IMD.

In addition to being able to calculate the temperature oscillations, we also need to determine how these oscillations influence the device performance. For the specific case of BAW devices, there is consensus in assuming that the detuning of BAW devices with temperature is due to the variation of multiple material properties with temperature (Lakin et al., 2000; Ivira et al., 2008; Petit et al., 2007). We reflect this in our model by adding a temperature-dependent term to the stiffened elasticity in (1)

$$c^D(T,K) = c_0^D + \Delta c_1^D T + \Delta c_2^D T^2 + \Delta c_K^D K \tag{8}$$

where K represents the temperature, the equivalent capacitance is

$$C_d(v,K) = C_{d,0} + \Delta C_1 v + \Delta C_2 v^2 + \Delta C_K K, \tag{9}$$

where each of the nonlinear terms ΔC_1, ΔC_2 and ΔC_K are related to their counterparts Δc_1^D, Δc_2^D, Δc_K^D respectively, as detailed in Appendix I.

The term ΔC_K generates 3IMD, whose maximum voltage $V_{\omega_{12}}$ can be found in a similar way as the contribution of ΔC_1 in (3) and ΔC_2 in (4) (see details in Appendix I):

$$V_{\omega_{12}} = A_T Q_L \Delta C_K P_{d\Delta\omega} Z_{th\Delta\omega}^* V_{\omega_1}, \tag{10}$$

where A_T is a constant that depends on the device geometry and material parameters, Q_L is the loaded quality factor, $Z_{th,\Delta\omega}$ is the thermal impedance (7) evaluated at $\Delta\omega$, and $P_{d,\Delta\omega}$ is the $\Delta\omega$ frequency component of the dissipated power in (6). Equation (10) describes the 3IMD signal due to self-heating effects, inside the acoustic transmission line, in terms of the dissipated power. As detailed in the following sub-sections, the dissipated power is due to both electric and acoustic loss, thus both effects contribute to the 3IMD in (10).

2.2.1 3IMD due to viscous losses

Viscosity is introduced in the model as a complex elasticity (Auld, 1990), which translates into a shunt resistance $R_{d,\eta}$ in series with the shunt capacitance C_d in a transmission line implementation. Appendix II details a model transformation to go from the original $R_{d,\eta}$ to an equivalent model in which the viscosity is implemented as a conductance G_d in parallel with the capacitance C_d. The equivalent model allows for an easier extraction of the closed-form expressions.

The instantaneous dissipated power due to viscous damping at each position z along the transmission line of length l (thickness of the piezoelectric layer) is

$$\frac{\partial P_{d,\Delta\omega}(z)}{\partial z} = G_d V_{\omega_1} V_{\omega_2} \cos^2\left(\frac{\pi z}{l}\right), \tag{11}$$

which can be integrated along l to obtain the total dissipated power

$$P_{d,\Delta\omega} = \frac{1}{2} l G_d V_{\omega_1} V_{\omega_2}^*. \tag{12}$$

Equation (12) can be combined with (10) to obtain the peak 3IMD voltage ($V_{\eta,\omega_{12}}$) due to the viscous damping

$$V_{\eta,\omega_{12}} = \frac{1}{2} A_T l G_d Q_L \Delta C_K Z_{th\Delta\omega}^* V_{\omega_1}^2 V_{\omega_2}^* \tag{13}$$

2.2.2 3IMD due to loss in the electrodes

There is certain agreement in considering ohmic losses as a significant dissipation mechanism (Thalhammer et al., 2005) in addition to the viscous damping. As it will be discussed in section II.B.3, electrodes losses are introduced in the circuit model as parasitic series resistances at the input and at the output ports, and their values are determined by fitting the model to the measured scattering parameters in the linear regime. Their contribution to the 3IMD can be calculated by the use of (10) and the power dissipated in the parasitic resistances $P_{\rho\Delta\omega}$:

$$V_{\rho,\omega_{12}} = A_T Q_L \Delta C_K P_{\rho\Delta\omega} Z_{th\Delta\omega}^* V_{\omega_1} \qquad (14)$$

Whereas the parasitic resistance and distributed conductance can be obtained from the measured scattering parameters, that is, they produce distinguishable measurable effect, examination of (13) and (14) looks like both self-heating mechanisms produce the same experimental observable so they may not be distinguishable. This is true if a two-tone experiment at a fixed frequency is performed, but the two effects have different frequency dependence that can be distinguished if the central frequency ω_0 of the 2 tones is swept while keeping the tones spacing $\Delta\omega$ constant. This happens because the frequency pattern of the dissipation due to ohmic losses is different than that produced by viscous losses, as shown in Fig 1. This information is extremely useful to validate the model with 3IMD measurements by looking at the frequency dependence of the 3IMD.

Note that (13) and (14) keep the same definition of thermal impedance $Z_{th,\Delta\omega}$. This is because the electrodes and the piezoelectric layer are thin and made of good thermal conductors, so that the thermal impedance between those layers is negligible, as will be verified with the temperature simulations shown in Section III.B.2.

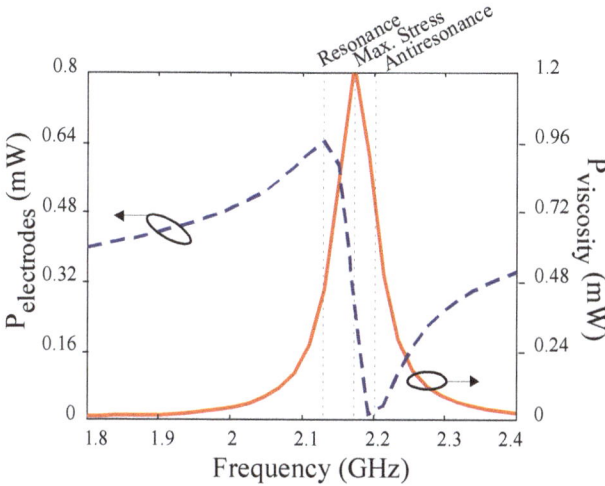

Fig. 1. Simulations of the dissipated power, for an input power of 20 dBm, due to acoustic viscous damping (solid line) and electrode electric losses (dashed line)

2.2.3 Circuit model with self-heating effects

A circuit model implementation to reproduce thermal effects should be capable of predicting dynamic temperature variations. To achieve this, we extend the nonlinear KLM model (Collado et al., 2009) to include the thermal domain (Rocas et al., 2009).

The procedure starts with the one dimensional heat equation along the z direction:

$$\frac{\partial^2 K}{\partial z^2} = \frac{\rho C_p}{k_{th}} \frac{\partial K}{\partial t} - \frac{P_d}{k_{th}}, \qquad (15)$$

where the equivalent distributed parameters can be identified as the volumetric heat capacitance

$$C_{d,th} = \rho C_p \qquad (16)$$

and the thermal resistance

$$R_{d,th} = \frac{1}{k_{th}} \qquad (17)$$

with C_p and k_{th} being the material-specific heat capacity and thermal conductivity, respectively.

With the above-mentioned distributed parameters, a thermal distributed model can be constructed as a cascade of sections of series resistances and shunt capacitance, where each section corresponds to a specific thickness and area. Figure 2 shows a segment with $R_{th}= R_{d,th} \cdot \Delta z/A$ and $C_{th}= C_{d,th} \cdot A \cdot \Delta z$, where A is the area of the cross-section perpendicular to the z direction. In such a thermal equivalent circuit the equivalents of voltage and current are the temperature and heat respectively.

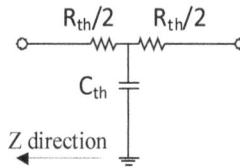

Fig. 2. Implementation of a Δz section of thermal equivalent circuit

The thermal model of a multilayer SMR can be implemented as a cascade of the previously described sections for each material, as shown in Fig. 3. The boundary conditions are the ambient temperature, modeled as a voltage source under the substrate, and the parallel combination of the radiation and convection resistances, terminated with a voltage source at ambient temperature on the upper side of the device (Larson et al., 2002).

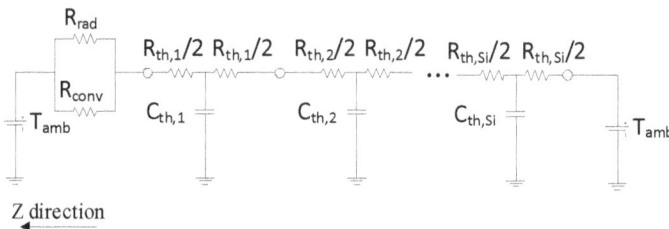

Fig. 3. Thermal model of the upper and lower materials' stacks with boundary conditions

As it can be seen from Fig. 3, the thermal impedance seen from any point along the line has a low-pass filter behavior, which means that for faster variations of the heat source, smaller temperature variations are produced.

The piezoelectric layer is implemented as a cascade of cells, in which the dissipated power due to viscous damping is directly coupled to its correspondent thermal cell. A current source is used because current is the analogue of heat in the thermal domain. The temperature rise is used to modify the distributed acoustic capacitance $C_{NL}(T,K)$, as shown in Fig. 4.

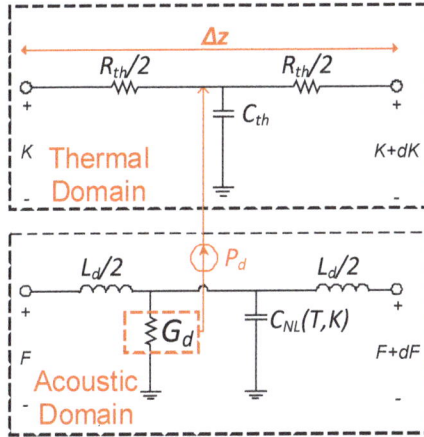

Fig. 4. Implementation of a section of the piezoelectric layer with the acoustic and thermal domains coupled by the generated heat at G_d and the temperature K. L_d is the acoustic distributed inductance $L_d = \rho \cdot A \cdot \Delta z$.

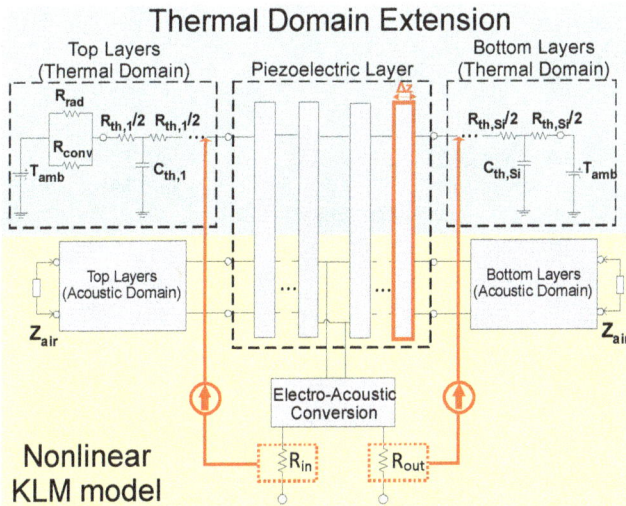

Fig. 5. Complete circuit model with thermo-acoustic model of the piezoelectric layer, top and bottom layers, and lossy electrodes. Electric losses, in the electrodes, and viscous losses, in the piezoelectric layer, produce dissipation that is coupled to the thermal domain to reproduce temperature rise. The temperature rise is used to change the material properties

On the other hand, the parasitic electrodes losses are implemented by use of a lumped resistor at the input and output of the modeled device as shown in Fig. 5. As done for the viscosity, the dissipation in each resistor is coupled to the thermal model as a heat source. In fact, dissipation in the input and output resistors is coupled to the correspondent top and bottom thermal sections that model the electrodes. The complete model can be seen in Fig. 5, where a cell of the piezoelectric layer like that in Fig. 4, is highlighted in red.

In the figure above, the electric-acoustic conversion box includes those elements of the KLM model whose purpose is the electro-acoustic signal conversion (Krimholtz et al., 1970). Additionally, the material layers above and below the piezoelectric are shown as simplified blocks for clarity.

2.2.4 Comparison of formulation and nonlinear simulations

We use the circuit model of Fig. 5, with only a piezoelectric layer, to check the accuracy of the formulation described in the previous section. The circuit model has been simulated, reproducing a two-tone experiment, with Harmonic Balance techniques by use of a commercial CAD software. A simple model is implemented making use of 100 cells to reproduce a 1.25 μm thick and $2.33 \cdot 10^{-8}$ m^2 piezoelectric layer with a quality factor of 1800. The electrodes losses and viscous losses are coupled to a low-pass thermal impedance.

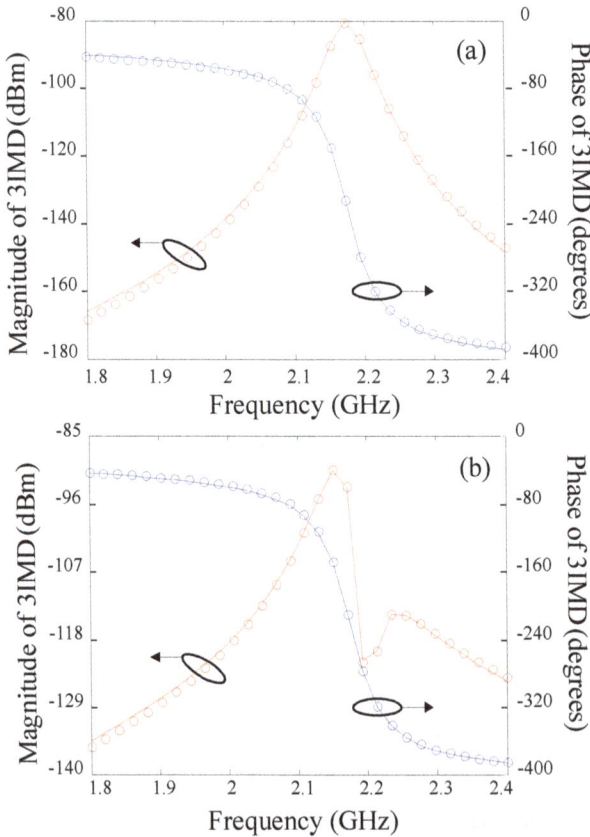

Fig. 6. Comparison of the magnitude and phase of $2\omega_1$-ω_2 calculated with equation (13) (circles) in Fig.6a (viscous losses, no electrode losses) and equation (14) (circles) in Fig.6b (electrode losses, no viscous losses), vs. simulation with the circuit model (solid lines)

In the first set of simulations we keep the tones spacing constant at $\Delta\omega/2\pi$ = 220 Hz and sweep the central frequency ω_0 in a 600 MHz range around the resonance frequency, which is 2.18 GHz. By doing this, we can distinguish the 3IMD produced by viscous self-heating from that produced by electric self-heating by analyzing the resulting frequency dependence. In the former case, we do not connect the dissipation in the electrodes to the thermal domain (Fig.6a), whereas in the latter case we do not connect the dissipation in the piezoelectric layer to the thermal domain (Fig.6b). The 3IMD frequency dependences are a direct consequence of the frequency dependences of the dissipated power. More specifically, a minimum at the anti-resonance frequency appears in Fig. 6.b because there is minimum current flowing through the electrodes at anti-resonance, which can be used in experimental measurements to identify different sources of self-heating effects.

In the second set of simulations we keep the central frequency constant at 2.18 GHz and we change the separation between tones from 100 Hz to 1 MHz. This allows us to reproduce the low-pass filter behavior of the thermal impedance. Figure 7 shows the results of the second set of simulations for a wide range of separation between tones when the self-heating effects are due to viscous losses, where it is clear the low-pass filter behavior of the temperature induced effects. A very similar plot was obtained for electrode losses, which is not shown for simplicity.

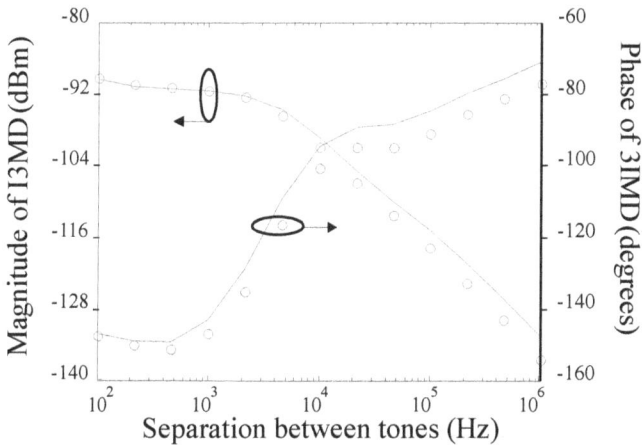

Fig. 7. Magnitude and phase of equation (13) (circles) and simulations with the circuit model (traces) for a wide range of separation between tones

Figures 6 and 7, in addition to giving useful qualitative information about the 3IMD generation due to the self-heating mechanism, show that the formulation of equations (13) and (14) is in very good agreement with the simulations, so that these expressions can be used for a better understanding of the temperature-induced 3IMD in BAW resonators.

3. Experimental results

Four state-of-the-art rectangular Solidly-Mounted Resonators (SMR) from a commercial manufacturer, with different areas summarized in Table 1, have been measured. The resonators have a 1.25 μm thick aluminum nitride layer and a W - SiO_2 Bragg mirror (alternating layers of W and SiO_2), and show quality factors around 1800.

From (8) it is clear that several sources, characterized by $\Delta c^D{}_1$, $\Delta c^D{}_2$ and $\Delta c^D{}_K$, can generate 3IMD. Therefore, we follow a step-by-step procedure that includes several experiments to determine which nonlinear source is designed responsible for each observable at each experiment by use of the circuit model:

- We first adjust the linear model to the measured S-parameters of the devices, so that the electric and viscous losses can be quantified. The procedure consists of a fine tuning of the material properties.
- Second harmonic measurements are performed along the frequency range of interest to extract the intrinsic nonlinear parameter $\Delta c^D{}_1$.
- The term $\Delta c^D{}_2$ also contributes to the 3IMD generation. We use the literature value in (Łepkowski et al., 2005) because this contribution cannot be independently extracted from measurements.
- Third-order intermodulation distortion measurements with sweeping the tone spacing are conducted to quantify the frequency dependence of the thermal impedance, and set the temperature coefficient of stiffened elasticity $\Delta c_K{}^D$ accordingly.

3.1 Linear modeling of the devices under test

Broadband one-port S-parameters measurements of the devices have been performed after an on-wafer OSL calibration. The measurements have been done at a power level of -10 dBm to ensure the linear regime and are used to fit the linear parameters of the circuit model. The only differences between the devices are the resonator area and electrode losses.

Electric losses due to the resistivity of the electrodes are modeled as lumped parasitic resistances, so their values are dependent on the resonator area and have a broadband effect on the linear device response. Table 1 summarizes the device areas and electric resistances.

On the other hand, acoustic losses due to viscosity only have an observable effect at those frequencies where there is substantial electro-acoustic coupling, that is around resonance and anti-resonance. By using the model transformation in Appendix II, the acoustic losses that fit all devices can be described with the same material viscosity value $\eta = 0.033$ N·s·m^{-2}, what verifies the validity of the linear model.

Resonator	Area (m^2)	Electric Resistance R$_s$ (Ω)
A1	6.41e-8	0.28
A2	4.88e-8	0.37
A3	2.33e-8	0.42
A4	1.25e-8	1

Table 1. Tested devices

3.2 Nonlinear characterization

The nonlinear behavior of a BAW resonator arises from different contributions due to intrinsic nonlinear material properties and self-heating mechanisms.

3.2.1 Intrinsic nonlinearities

Intrinsic nonlinearities due to the stiffened elasticity, as shown in (1), predominate above other intrinsic nonlinear material contributions. As a consequence, the second harmonic has the same frequency dependence as the mechanical stress in the piezoelectric layer (Collado et al., 2010). In fact, as done in (Collado et al., 2010), the second harmonic is used to extract a unique unitless value $\Delta c^D{}_1 = 10.5$ of the stiffened elasticity that fits the second harmonic for all the devices.

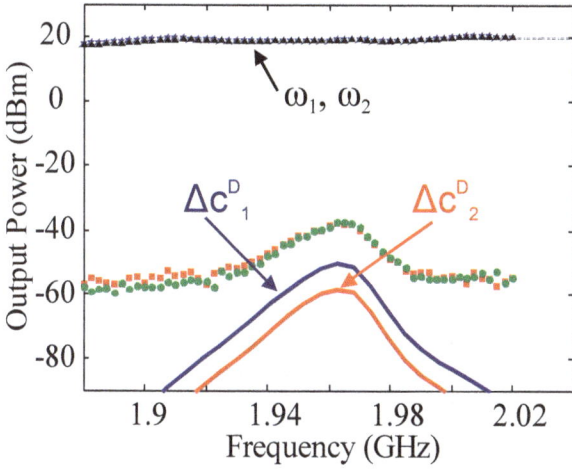

Fig. 8. 3IMD measurements (squares are $2\omega_1-\omega_2$, circles are $2\omega_2-\omega_1$) and circuit simulations of the $\Delta c^D{}_1$ and $\Delta c^D{}_2$ contributions to the 3IMD ($2\omega_1-\omega_2$ and $2\omega_2-\omega_1$ overlap). Both measurements and simulations are done for $\Delta f = 220$ Hz. The intrinsic nonlinearities are not sufficient to explain the measurements. Measurements (squares and triangles) and simulations (dashed lines) of ω_1 and ω_2 are also presented

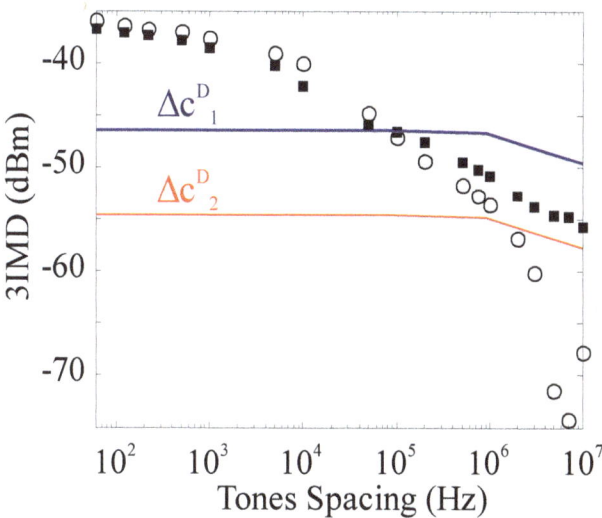

Fig. 9. 3IMD measurements (squares are $2\omega_1-\omega_2$, circles are $2\omega_2-\omega_1$) and circuit simulations of the $\Delta c^D{}_1$ and $\Delta c^D{}_2$ contributions to the 3IMD ($2\omega_1-\omega_2$ and $2\omega_2-\omega_1$ overlap), for several separations between tones. The intrinsic contributions cannot reproduce the envelope frequency-dependent 3IMD level

The parameter Δc^{D_1} is responsible for second harmonic generation, which in turn mixes with the fundamental frequencies ω_1 and ω_2, and gives rise to 3IMD. On the other hand, Δc^{D_2} directly generates a certain level of 3IMD distortion. We use the literature value of Δc^{D_2} = -1 ·10-10 N-1m2 (Łepkowski et al., 2005).

With the above-mentioned values of intrinsic nonlinearity, simulations of 3IMD are performed obtaining values which are below the measured levels, as shown in Fig. 8. This shows that other contributions exist. Figure 9 shows 3IMD measurements at different envelope frequencies, centered at the frequency where the 3IMD is maxima. The measurements reveal a strong dependence with the envelope frequency that cannot be accounted for intrinsic nonlinearities. The dependence of the 3IMD level on the envelope frequency suggests that the 3IMD is dominated by a thermal effect.

3.2.2 Self-heating

The thermal model, as presented in section II.B.3, is implemented by using the literature values of thermal conductivity and specific heat for each layer. The materials stack is composed of more than ten layers. Dissipation on the electrodes and viscous losses are coupled to the thermal domain by means of current sources. By using the model we can determine the temperature distribution along the materials stack, even at different envelope frequencies as shown in Fig. 10. Note that the temperature is almost the same in the piezoelectric layer and the electrodes, which validates the hypothesis in Section II.B.2, of negligible thermal impedance between the electrodes and the piezoelectric layer, to obtain the closed-form expressions.

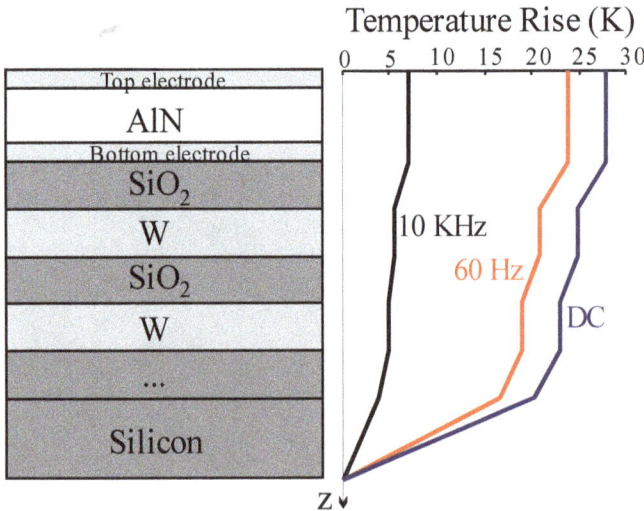

Fig. 10. Simulations of the distribution of the z temperature variation, inside the resonator, calculated with the circuit model. Simulations are done for resonator A1 at P_{in} = 30 dBm, for different separation between tones

A value of $\Delta c^{D}_K/c^{D}_0$ = -15 ppm/K is found to fit the 3IMD best for all resonators. Figure 11 shows the intrinsic and self-heating 3IMD contributions independently as well as the sum of all contributions, which match the measurements for all the tone separations. The model reproduces very well the measured 3IMD values except the asymmetry between lower and higher 3IMD that appears for envelope frequencies around 6 MHz. This asymmetry is considered to be a consequence of a cancellation between different 3IMD contributions and is currently under investigation.

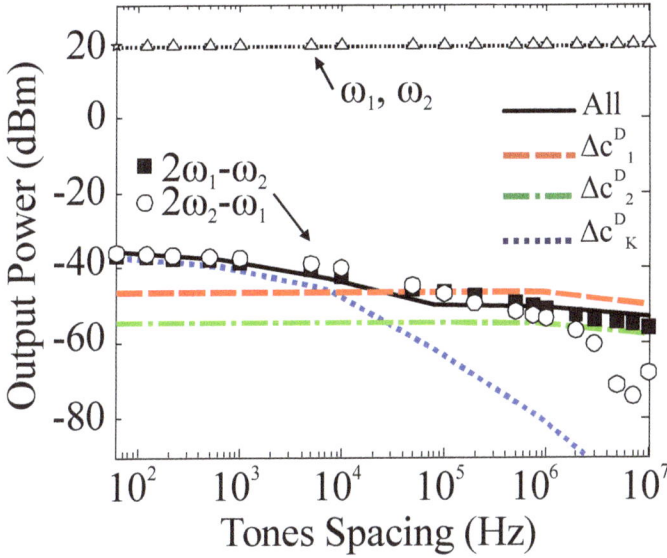

Fig. 11. Measurements (stars, triangles, squares and circles are ω_1, ω_2, $2\omega_1$-ω_2 and $2\omega_2$-ω_1 respectively) and simulations (thin dotted line is ω_1 and ω_2, solid line is $2\omega_1$-ω_2 and $2\omega_2$-ω_1 overlapped) of the fundamentals and the 3IMD for different tones spacing for A1. The dotted line, the dash-dot line and the dashed line are simulations of the 3IMD contribution due to Δc^D_K, Δc^D_2 and Δc^D_1 respectively. The solid line is the 3IMD simulation with all the nonlinear contributions

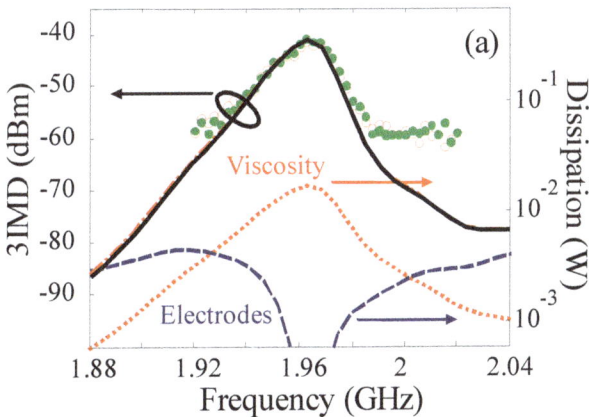

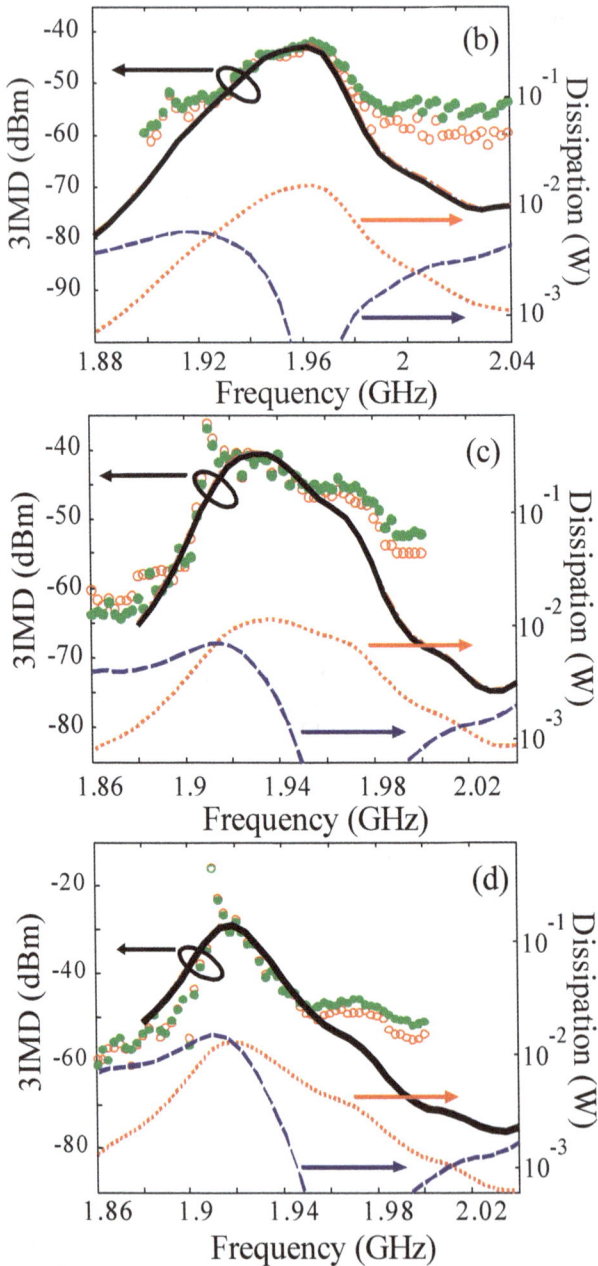

Fig. 12. Measurements (filled circles are $2\omega_1-\omega_2$, empty circles are $2\omega_2-\omega_1$) and simulations (solid line is $2\omega_1-\omega_2$, dash-dot line is $2\omega_2-\omega_1$) of the 3IMD for resonators A1, A2, A3 and A4 in Fig.12.a, Fig.12.b, Fig.12.c and Fig.12.d respectively. The figures also show the dissipation in the electrodes (dashed line) and in the piezoelectric layer due to viscous losses (dotted line)

Figure 12 shows measurements and simulations of the 3IMD about the frequency range of interest, for each resonator, by use of the circuit model. In this experiment, the two tones are swept around the resonating frequencies keeping the separation between tones constant at 100 Hz. The results show good agreement between simulations and measurements above the nonlinear system baseline level, which is around -60 dBm.

The dashed and dotted lines in Fig. 12 show the simulated dissipated power due to electrodes losses and viscosity respectively, which have different frequency dependences according to the maximum electric current and mechanical stress, respectively. 3IMD measurements for resonators A3 and A4 in Fig. 12 show a peak at the antiresonant frequency that is underestimated by the simulations. The frequency dependence in that range points to a possible electric-field contribution to the 3IMD. This contribution is below the system nonlinear baseline level for resonators A1 and A2, and the area scaling has not been successfully reproduced by use of an electric-field dependent permittivity or stiffened elasticity in the acoustic transmission line, so further research is needed.

4. Conclusion

The role of self-heating and material nonlinearities in the generation of 3IMD in bulk acoustic wave devices has been evaluated through measurements, models and equations. Self-heating is found to have a very significant contribution to 3IMD and thus thermal considerations are critical in the device design. The presented circuit model implementation offers the possibility to predict 3IMD in BAW resonators, given their materials stack and geometry. With such information one can use the resonator model to accurately predict 3IMD in filters. Further research will be performed to investigate the relation between the electric-field contribution to 3IMD and the cancellation shown in the measurements. The development of a 3D equivalent thermal model, to take into account complex heat dissipation through the substrate, will also be investigated.

5. Appendix I – 3IMD equations

At each elemental section, and following a similar process than that described in (Collado et al., 2009), the nonlinear capacitance acts as an infinitesimal nonlinear current generator at $2\omega_1 - \omega_2$ (and $2\omega_2 - \omega_1$), when ω_1 and ω_2 are at resonance:

$$\frac{\partial I_{nl,\omega_{12}}(z)}{\partial z} = \frac{1}{2} j\omega_{12} \Delta C_K K_{\Delta\omega}^* V_{\omega_1} \cos\left(\frac{\pi z}{l}\right) \qquad (18)$$

where $K_{\Delta\omega} = Z_{th}(\Delta\omega) P_d(\Delta\omega)$.

Therefore the broadband energy balance all over the acoustic transmission line leads to

$$V_{\omega_{12}} = -j\frac{1}{2} Q_L \frac{\Delta C_K}{C_{d,0}} P_{d\Delta\omega} Z_{th\Delta\omega}^* V_{\omega_1} \left(\frac{1}{j\frac{C_{eq}Q_L}{2W_{0,v}} - 1} \right) \qquad (19)$$

6. Appendix II - Model transformation

Losses are introduced as a complex elasticity by means of the viscous damping coefficient η:

$$c \to c + \eta \frac{\partial}{\partial t} \tag{20}$$

The inverse damping coefficient can also be understood as the conductance per unit length $G_d = \eta^{-1}$. With that, the acoustic telegrapher equations, making use of the analogy between the acoustic and electric domains, can be written as:

$$\frac{\partial V}{\partial z} = -L_d j \omega I \tag{21}$$

and

$$\frac{\partial I}{\partial z} = -\frac{1}{Ac^D + j\omega A \eta} j \omega V. \tag{22}$$

The shunt admittance of the acoustic transmission line implementation, given by (22) and in which $A \cdot c^D = C_d^{-1}$, is a shunt capacitance in series with a resistance. To transform this to be a capacitance in parallel with the loss term, we introduce eq. 8 in eq. 22 and expand the shunt admittance in as a Taylor series. The result is a conductance value in parallel with a nonlinear capacitance of the form:

$$C_d(v,K) = C_{d,0} + \Delta C_1 v + \Delta C_2 v^2 + \Delta C_K K \tag{23}$$

whose terms are related with the material linear and nonlinear properties as follows:

$$G = \omega^2 C_{d,0}{}^2 A \eta \tag{24}$$

$$\Delta C_1 = \frac{c_1^D}{(Ac_0^D)^2} \tag{25}$$

$$\Delta C_2 = -\frac{c_2^D}{A^3 (c_0^D)^2} \tag{26}$$

$$\Delta C_1 = -\frac{c_K^D}{A(c_0^D)^2} \tag{27}$$

7. Appendix III - Broadband loaded quality factor

The loaded quality factor can be defined as (Russer, 2006)

$$Q_L = \frac{Q_0}{1+\beta} \tag{28}$$

where β relates the dissipated power in the acoustic resonator P_{res}, that is the acoustic transmission line, and the externally dissipated power P_{ext} as follows:

$$\beta = \frac{P_{ext}}{P_{res}}. \tag{29}$$

By circuit analysis of the KLM circuit model, it can be found that β is

$$\beta = \frac{\mathrm{Re}\left(\dfrac{1}{\left(50 + jX_m + \dfrac{1}{j\omega C_0}\right)^*}\right)}{\mathrm{Re}\left(\dfrac{1}{\left(Z_{in} - jX_m - \dfrac{1}{j\omega C_0}\right)^*}\right)} \tag{30}$$

where Z_{in} is the input impedance of the device and X_m is the series reactive term of the KLM model (Krimholtz et al., 1970). Q_0 in (28) represents the unloaded quality factor, that is obtained from S-parameters using (Feld et al., 2008)

$$Q_0 = \omega \tau \frac{|S_{11}|}{1 - |S_{11}|^2} \tag{31}$$

8. References

Auld B. A., Acoustic Fields and Waves in Solids (Krieger, Malabar, Florida), Vol. I, 1990

Camarchia V., Cappelluti F., Pirola M., Guerrieri S. D., Ghione G. 2007. Self-Consistent Electrothermal Modeling of Class A, AB, and B Power GaN HEMTs Under Modulated RF Excitation. *IEEE Transactions on Microwave Theory and Techniques*, vol. 55, no. 9, Sept. 2007, pp. 1824-1831.

Cho Y., Wakita J. 1993. Nonlinear equivalent circuits of acoustic devices. *Ultrasonics Symposium, 1993. Proceedings, IEEE 1993*, pp. 867-872 vol. 2, 31 Oct-3 Nov 1993

Constantinescu F., Nitescu M., Gheorghe A. G. 2008. New Nonlinear Circuit Models for Power BAW Resonators. *ICCSC 2008. 4th IEEE International Conference on Circuits and Systems for Communications*, pp. 599-603, 26-28 May 2008

Collado C., Rocas E., Padilla A., Mateu J., O'Callaghan J. M., Orloff N. D., Booth J. C., Iborra E., Aigner R. 2010. First-order nonlinearities of bulk acoustic wave resonators. *IEEE Transactions on Microwave Theory and Techniques*, submitted.

Collado C., Rocas E., Mateu J., Padilla A., O'Callaghan J. M. 2009. Nonlinear Distributed Model for Bulk Acoustic Wave Resonators. *IEEE Transactions on Microwave Theory and Techniques*, vol. 57, no. 12, Dec. 2009,pp. 3019-3029

Collado C., Mateu J. and O'Callaghan J. M. 2005. Analysis and Simulation of the Effects of Distributed Nonlinearities in Microwave Superconducting Devices. *IEEE Trans. Appl. Supercond.* vol. 15, No. 1, March 2005, pp. 26-39.

Feld D. A. 2009. One-parameter nonlinear mason model for predicting 2nd & 3rd order nonlinearities in BAW devices. *2009 IEEE International Ultrasonics Symposium (IUS)*, pp. 1082-1087, 20-23 Sept. 2009

Feld D. A., Parker R., Ruby R., Bradley P., Shim D. 2008. After 60 years: A new formula for computing quality factor is warranted. *2008 IEEE International Ultrasonics Symposium*, pp. 431-436, 2-5 Nov. 2008

Ivira B., Benech P., Fillit R., Ndagijimana F., Ancey P., Parat G. 2008. Modeling for temperature compensation and temperature characterizations of BAW resonators at GHz frequencies. *IEEE Transactions on Ultrasonics, Ferroelectrics and Frequency Control*, vol. 55, no. 2, February 2008, pp. 421-430.

Krimholtz R., Leedom D.A., Matthaei G.L. 1970. New equivalent circuits for elementary piezoelectric transducers. *Electronics Letters*, vol. 6, no. 13, pp. 398-399, June 1970

Lakin K. M., McCarron K. T., McDonald J. F. 2000. Temperature compensated bulk acoustic thin film resonators. *2000 IEEE International Ultrasonics Symposium*, pp. 855-858 vol. 1, Oct 2000

Larson III J. D., Oshrnyansky Y. 2002. Measurement of effective kt2, q, Rp, Rs vs. Temperature for Mo/AlN FBAR resonators. *2002 IEEE International Ultrasonics Symposium*, pp. 939- 943 vol. 1, 8-11 Oct. 2002

Łepkowski S. P., Jurczak G. 2005. Nonlinear elasticity in III-N compounds: Ab initio calculations. *Physical Review B*, vol. 72, 245201, 2005, pp 1-12.

Nakamura H., Hashimoto K.-y., Ueda M. 2010. Nonlinear effects in SAW and BAW components. *2010 IEEE International Ultrasonics Symposium (IUS)*, Short Course, 11-14 Oct. 2010

Parker A. E., Rathmell J. G. 2004. Self-heating process in microwave transistors. *Workshop on Applications in Radio Science*, Hobart, TAS, Australia, 18-20 Feb. 2004.

Petit D., Abele N., Volatier A., Lefevre A., Ancey P., Carpentier J.-F. 2007. Temperature Compensated Bulk Acoustic Wave Resonator and its Predictive 1D Acoustic Tool for RF Filtering. *2007 IEEE International Ultrasonics Symposium*, pp. 1243-1246, 28-31 Oct. 2007

Rocas E., Collado C., Orloff N. D., Mateu J., Padilla A., O'Callaghan J. M., Booth J. C. 2010. Passive intermodulation due to self-heating in printed transmission lines. *IEEE Transactions on Microwave Theory and Techniques*, accepted for publication.

Rocas E., Collado C., Booth J. C., Iborra E., Aigner R. 2009. Unified model for Bulk Acoustic Wave resonators' nonlinear effects. *2009 IEEE International Ultrasonics Symposium (IUS)*, pp. 880-884, 20-23 Sept. 2009

Rocas E., Collado C., Mateu J., Campanella H., O'Callaghan J. M. 2008. Third order Intermodulation Distortion in Film Bulk Acoustic Resonators at Resonance and Antiresonance. *IEEE MTT-S International Microwave Symposium Digest*, pp.1259-1262, June 2008.

Russer P. 2006. Electromagnetics, microwave circuit, and antenna design for communications engineering (Artech House), 2006

Thalhammer R., Aigner R. 2005. Energy loss mechanisms in SMR-type BAW devices. *2005 IEEE MTT-S International Microwave Symposium Digest*, pp. 4, 12-17 June 2005

Ueda M., Iwaki M., Nishihara T., Satoh Y., Hashimoto K.-Y. 2008. A circuit model for nonlinear simulation of radio-frequency filters using bulk acoustic wave resonators. *IEEE Transactions on Ultrasonics, Ferroelectrics and Frequency Control*, vol. 55, no. 4, pp.849-856, April 2008

Vuolevi J. H. K., Rahkonen T., Manninen J. P. A. 2001. Measurement technique for characterizing memory effects in RF power amplifiers. *IEEE Transactions on Microwave Theory and Techniques*, vol. 49, no. 8, Aug. 2001, pp. 1383-1389.

9

Ultrananocrystalline Diamond as Material for Surface Acoustic Wave Devices

Nicolas Woehrl and Volker Buck
University Duisburg-Essen and CeNIDE, Duisburg
Germany

1. Introduction

Diamond is one of the most promising materials for SAW applications due to the highest sound velocity and thermal conductivity. Unfortunately single crystals and epitaxial CVD-films are expensive and beyond that conventional CVD grown microcrystalline diamond films feature large facet structures with high roughness inapplicable for this application. Ultra-Nanocrystalline diamond (UNCD) films grown in a microwave plasma enhanced chemical vapor deposition (MPECVD) system on Si substrates possess a smooth surface making it an ideal material for SAW applications. Moreover, due to its nanocrystalline structure, the film properties of the UNCD material can be tailored in a wide range to adjust them to the specific needs of a SAW filter. For this task a profound understanding of the growth process of UNCD and the dependency of the film performance from the film properties is needed. In addition, a simple and quick method to characterize the properties of the UNCD films is necessary. Laser-induced SAW pulse method, which is fast and accurate, is demonstrated to measure the mechanical and structural properties of the UNCD films. AFM measurements were done to correlate the SAW pulse method results with the surface roughness of the deposited films.

Another advantage of the UNCD films is, that highly C-axis textured aluminum nitride (AlN) films can be grown directly on UNCD films by DC-sputtering. Using this technique a feasibility study for SAW devices has been successfully performed.

2. Saw filters

SAW devices are electromechanical products commonly used in high frequency applications such as filters, oscillators and transformers and are based on the transduction of acoustic waves. SAW filters are now widely used in mobile telephones applications for filtering and provide significant advantages in performance, cost, and size over other filter technologies such as quartz crystals (based on bulk waves), LC filters, and waveguide filters by offering a high degree of frequency selectivity with low insertion loss in compact size (Campbell, 1989). In SAW filters an Interdigital Transducer (IDT) that is attached to a piezoelectric material converts electrical signals to a mechanical wave. The piezoelectric effect and the electric field generated by the IDT are distorting the crystal close to its surface.

The oscillations of the crystal lattice can than add up by constructive wave interference and superimpose to a surface wave before being converted back to an electrical signal by further electrodes.

The passing frequency of a SAW filter can be calculated by

$$f_0 = \frac{v_0}{\lambda}$$
(1)

where λ represents the wavelength of the acoustic surface wave (corresponding to twice the distance of the fingers of the IDT) and v_0 is the crystal-dependent velocity oft the surface wave. The operation frequency of a SAW device is closely related to the spacing of the interdigital transducer (IDT) that is significantly limited by the photolithograph capability (Springer et al., 1999). Thus one way to achieve higher passing frequencies is to use crystals with a higher speed of sound, such as sapphire (Caliendo, 2003), SiC (Takagaki, 2002) or diamond (Yamanouchi et al., 1989)(Nakahata et al., 1992).

3. Diamond as SAW material

The material with the highest speed of sound is diamond with 18000 m/s. Besides the high speed of sound, diamond features other remarkable properties such as high thermal conductivity and high hardness to name only a few. Due to its extraordinary properties natural and HPHT diamond is used for a long time as a material for tools, especially for grinding or sawing of rocks. Since the 1980s the microcrystalline diamond deposited by thin film technology is increasingly used. One major problem with the microcrystalline diamond films deposited in CVD processes – especially for microelectronics and micromechanical applications with their decreasing structural sizes - is the high surface roughness (Malshe et. al., 1999). Moreover, high surface roughness results in large propagation loss, reducing the applicability of the material. Although Sumitomo Electrics developed SAW filters and resonators with various bandwidth in the 2-5 GHz range it turned out that the polishing of the rather rough CVD diamond surface was too expensive and time consuming due to the chemical inertness and highest hardness of diamond and the SAW filters were never produced in an industrial scale (Fujimori, 1996). Even if one solution to this problem was demonstrated by using the unpolished nucleation side of freestanding CVD diamond (Lamara et. al., 2004) this idea never went into production.

Another drawback of the microcrystalline diamond films is that the homogeneous deposition of such films on substrates with a high aspect ratio is difficult because the films consist of relatively large crystals.

4. Nanocrystalline diamond as SAW material

The growing interest in nanotechnology and nanostructured materials has encouraged the research of diamond films with reduced grain size. By reducing the grain size those films feature rather unique combinations of properties making them potential materials for emerging technological developments such as Nano/Micro- Electro-mechanical Systems (N/MEMS) (Auciello et. al., 2004) (Hernandez Guillen, 2004), optical coatings, bioelectronics (Yang et. al., 2002), tribological applications (Erdemir et. al., 1999) and also surface acoustic wave (SAW) filters (Bi et. al., 2002).

The nanostructured films differ from the microcrystalline films in grain size and in roughness of the surface as shown in Fig. 1.

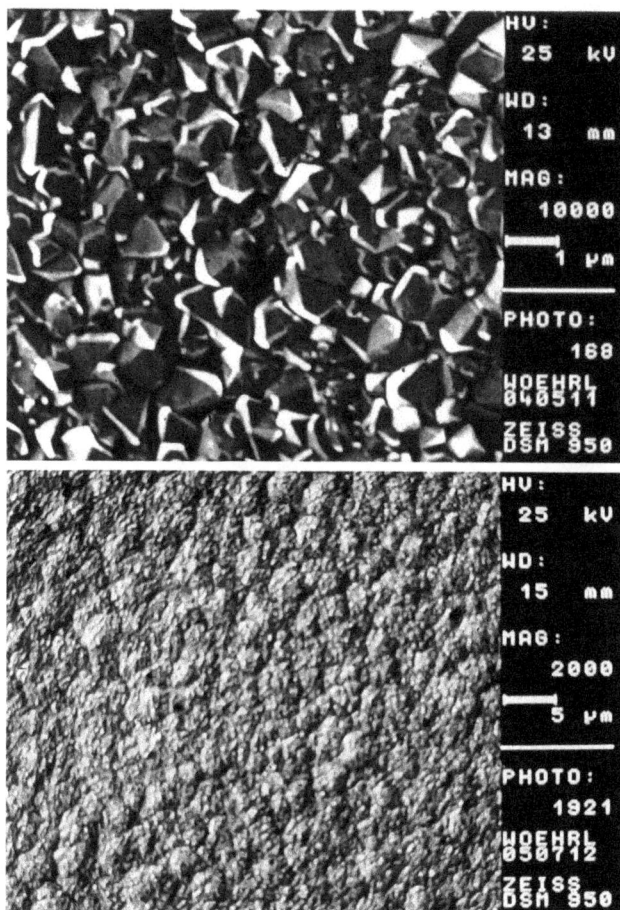

Fig. 1. Morphological comparison of microcrystalline diamond film (upper picture) and UNCD film (lower picture). The scale bar in the upper picture corresponds to 1 μm while the scale bar in the lower picture corresponds to 5 μm

The terms nanocrystalline (NCD) and ultrananocrystalline diamond (UNCD) were coined by the Argonne National Laboratory group that performed the pioneering work in this field. These terms were introduced to establish a differentiation to the microcrystalline diamond films that differ not only in film properties but also in the way they are deposited. The technology developed at Argonne National Laboratory started from deposition of hydrogen free plasmas using fullerenes in Ar (Ar/C_{60}) and was thereafter extended to hydrogen diluted plasmas using Ar/CH_4 and gas mixtures containing only about 1 % hydrogen (either added intentionally or through the thermal decomposition of CH_4) (Gruen, 1999). UNCD is grown from Argon-rich plasma giving it a very fine and uniform structure with grain sizes between 2 and 15 nm (Auciello et. al., 2004). The grain sizes are independent from film thickness due to the high secondary nucleation of new growth sites during the whole deposition that is not taking place in the standard growth of diamond. This can be shown within the experimental errors when measuring the Young's modulus (GPa) as a function of the deposition time (Shen et. al., 2006).

UNCD consists of pure sp³ crystalline grains that can be separated by atomically abrupt (0.5 nm) grain boundaries or embedded in an amorphous 3D matrix. By reducing the grain size of microcrystalline diamond films the amount of material between the grains is increased. This matrix in the films can contribute to a large fraction of the overall film, sometimes exceeding 10 % of the total volume, giving those films a great proportion of non-diamond or disordered carbon (Auciello et. al., 2004]. But also values down to 5 % sp²-bonded carbon have been reported and determined by UV Raman spectroscopy and synchrotron based near-edge X-ray absorption fine structure measurements (NEXAFS) (Gruen, 1998).

In fact the overall volume and structure of the film matrix significantly determine the properties of nanocrystalline diamond films giving another degree of freedom for the material. The well-aimed use of an amorphous matrix for nanocrystalline diamond grains leads to an enormous field of new materials, because a whole class of carbon based materials (diamondlike carbon, DLC) can be used as matrix that may contain carbon solely (a-C) or carbon and hydrogen (a-C:H) as well as other components like metals (Me-C:H); additionally other dopants like silicon, oxygen, halogens or nitrogen may be added with considerable effect on the film properties. By combining soft matrix properties with the hard diamond crystals on the nanoscale it is possible to combine hard with elastic properties and get a material that is hard and tough at the same time. With tailoring the mechanical stress in the films or the coefficient of thermal expansion it was possible to tailor yet other very important mechanical properties for the application of UNCD films by adjusting the overall matrix fraction to the film volume (in the case of a 3D matrix surrounding the nanocrystals) (Woehrl & Buck, 2007) (Woehrl et. al., 2009).

Thus, when comprehensively characterizing UNCD films, one also has to analyze the matrix properties. Since the carbon atoms in the matrix have no crystalline configuration and are indeed amorphous, conventional techniques known from the analysis of amorphous carbon films can be used.

5. Deposition of UNCD films

It is well accepted that the initial nucleation is one decisive factor for the subsequent CVD diamond film growth. While a low nucleation density can lead to van-der-Drift growth – known as the "survival of the largest" – high initial nucleation leads to shorter coalescence time and lower surface roughness. Due to the fact that substrate pre-treatment can significantly increase initial nucleation, the pre-treatment is an important process step already predetermining the film properties (Liu & Dandy, 1995). Three effective seeding methods are known: Mechanical scratching of the substrate surface (see e.g. (Buck & Deuerler, 1998)), enhancing the nucleation by applying a bias voltage to the substrate in the early stages of deposition (Yugo, 1991), and ultrasonically activating the substrate in a suspension containing diamond powder (Lin et. al., 2006)(Sharma et. al., 2010). Nucleation densities of 10^{10} cm⁻² or more were achieved with either of these methods. The latter method was used for the substrates in this work mainly because of the good reproducibility and uniformity even with larger substrates. Details on the pre-treatment and the deposition parameters used for the UNCD films deposited in this work are given below.

Ultra-Nanocrystalline diamond (UNCD) films were synthesized by microwave plasma enhanced chemical vapour deposition technique using a 2.45 GHz IPLAS CYRANNUS® I-6" plasma source. The nanocrystalline films were deposited from an $Ar/H_2/CH_4$ plasma.

As standard substrates in this work (100) oriented double side polished silicon wafers with a thickness of 425 μm were used. The substrates were usually cut from a wafer to a size of about 20 x 20 mm.

To enhance the nucleation of diamond the substrates were ultrasonically scratched for 30 min with a scratching solution consisting of diamond powder (~ 20 nm grain size), Ti powder (~ 5 nm particle size) and Ethanol in a weight percent ratio of 1:1:100 (wt%). Afterwards the substrates were ultrasonicated for 15 min in Acetone to clean the surface from any residues of the scratching solution (Lin et. al., 2006) (Buck, 2008). After the substrate pre-treatment they were immediately installed into the vacuum chamber placed on top of the molybdenum substrate holder and the recipient was pumped down to high-vacuum.

The plasma is ignited at ca. 1 mbar pressure with a process gas mixture of hydrogen (≈ 3 %) in argon (≈ 97 %) and a MW-power of 1 kW. After the ignition the pressure was slowly increased up to the deposition pressure (typically 200 mbar) during a 30 min period. This 30 min step is due to two reasons: Firstly the substrate surface is cleaned by the etching effect of the plasma. Secondly the temperature of the substrate is slowly increased in the process of the rising pressure. In doing so the substrate is already close to the targeted deposition temperature before switching to the deposition parameters and introducing the carbon carrier gas into the chamber. During the whole process of increasing the pressure the MW power coupling into the plasma is adjusted to the changing conditions. After reaching the desired deposition pressure the carbon carrier gas was introduced therewith starting the deposition process.

The nanocrystalline films shown here were deposited at a pressure of 200 mbar from an $Ar/H_2/CH_4$ plasma. To investigate the influence of the hydrogen admixture on the properties of the deposited films, the percentage of hydrogen in the process gas was varied between 2 % and 8 % as shown in Table 1.

The MW-power was kept constant at 1 kW and the films were deposited for 5 h.

Pressure	200 mbar
Gasflow	400 sccm
H_2 fraction	2 % - 8 %
CH_4 fraction	0,8 %
Ar fraction	91,2 % - 97,2 %
MW-power	1 kW
Deposition time	5 h
Substrate	Pretreated Silicon (100)

Table 1. CVD Deposition Parameters

6. Morphology of the films

The deposition parameters were systematically varied to investigate the influence on film structure and film properties with special attention to the speed of sound and the roughness of the films as most important properties for the application as SAW filters. Because of that the main focus was on deposition parameters that influence the diamond grain size and matrix. It is expected that both are directly influencing the elastic modulus of the films and thus the speed of sound. One important parameter that is influencing the crystal size is the admixture of hydrogen in the process gas. The higher the hydrogen fraction the bigger the crystals grow. (Woehrl & Buck, 2007)

In previous publications it was suggested that different species for the nucleation on the one hand and the growth of diamond grains on the other hand exist. The ratios of these species determine the macroscopic structure of the growing films by influencing the rate of secondary nucleation and therefore the matrix density and the grain size of the growing crystals. A higher amount of the nucleation species leads to smaller crystals and more material between the grains. A higher amount of growth species allows the grains to grow faster (thus a higher growth rate) suppressing the secondary nucleation. In the literature, C_2 was suggested to be the nucleation species (Gruen, 1999) as strong emission of the C_2 dimer could be found in the plasmas used for the deposition of fine-grained UNCD films. On the other hand the CH_3 radical is generally believed to be the growth species of diamond films (May & Mankelevich, 2008). Without taking part in the discussion concerning specific details of growth and nucleation species, previously published data can be interpreted in a way that these two competitive processes determine the structure of the deposited films.

Fig. 2. Morphology of UNCD films deposited with different hydrogen admixtures. The scale bars in all three pictures correspond to 2 μm

The atomic force microscope (AFM) is a scanning probe type microscope that offers a resolution of less than a nanometer that is by a factor of 1000 better than the optical diffraction limit. The AFM consists of a cantilever with a sharp tip with a radius of curvature in the order of nanometers at its end that is used to scan the sample surface. When the tip is brought close to the surface, atomic forces between the tip and the sample lead to a deflection of the cantilever. The deflection of the cantilever is then measured by a laser that is reflected from the cantilever onto an array of photodiodes. In comparison to the scanning electron microscope (SEM) that is measuring a two- dimensional image of a sample not necessarily corresponding to the morphological features, the AFM provides a true three-dimensional topographical image of the surface giving information about the roughness of the investigated surface. While specimens measured in SEM needs to be conducting and are therefore often coated with a thin metal film (e.g. gold) irreversibly alter the film properties, AFM measurements do not require such special treatments. While the SEM can easily measure an area in the order of square millimeters with a depth of field on the order of millimeters the AFM is usually restricted to a maximum scanning area of around 150 μm^2 with a depth of field in the order of micrometers. Another characteristic that has to be considered for high resolution AFM is the fact that the quality of an image is limited by the radius of curvature of the probe tip and can lead to image artifacts. (Sarid, 1991)

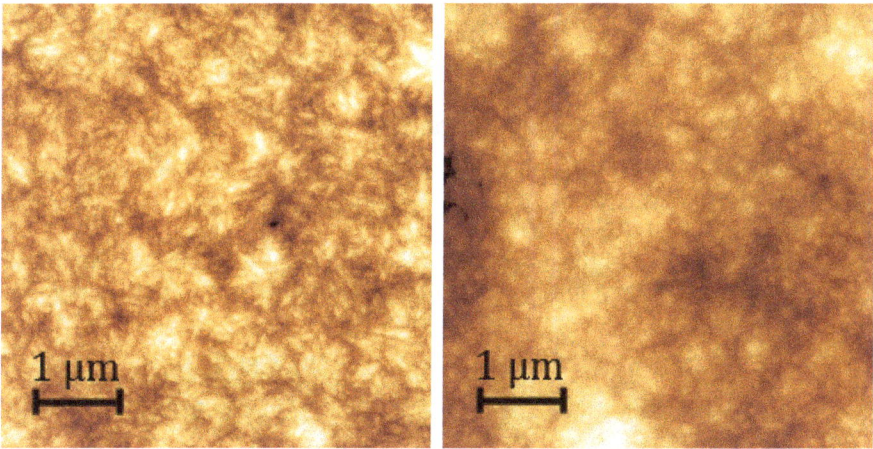

Fig. 3. AFM measurements of UNCD samples deposited with 2,5 % H_2 (left) and 6 % H_2 (right). Both images cover a 5 x 5 μm area

Fig. 2 shows UNCD films deposited with different admixtures of hydrogen to the process gas. It is clearly seen that the hydrogen is influencing the morphology of the deposited films. In fact the crystals are larger and the surface is rougher at hydrogen admixtures of 7% compared to the films deposited at lower admixtures.

Fig. 3 shows an AFM measurement of 5 μm thick UNCD films on a Si substrates. The measured area on the sample was 25 μm^2. The RMS-roughness (root-mean-squared roughness) of the surface is measured to be R_q= 21.1 nm for the sample deposited at 2.5 % H_2 (left picture) and R_q= 51.3 nm for the sample deposited at 6 % H_2 (right picture).

SEM as well as AFM measurements show that higher hydrogen admixture in the process gas lead to larger diamond crystals and rougher surfaces.

The RMS-roughness measurements as a function of the hydrogen admixture are shown in Fig. 4.

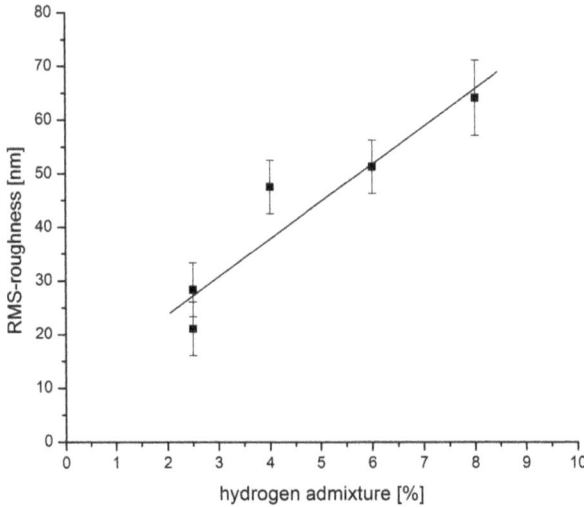

Fig. 4. RMS-roughness measurements as a function of hydrogen admixture in process gas

7. Influence of nitrogen admixture on morphology

An especially appealing field of application for UNCD is nitrogen doped semiconducting films. UNCD films are usually insulating, but n-doping is easily possible by admixture of nitrogen to the process gas. (Gruen, 2004)
To investigate the influence of the nitrogen admixture in the plasma on the film properties, more films were deposited at a pressure of 200 mbar with admixtures of nitrogen from 0 % to 7.5 %.

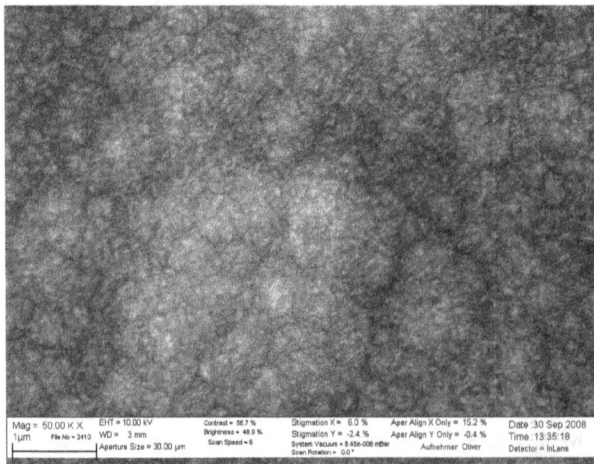

Fig. 5. High resolution SEM measurement of a UNCD film deposited with 2.5 % hydrogen and 2.5 % nitrogen admixture. The scale bar shown corresponds to 1 μm

High-resolution SEM pictures were taken to investigate the influence of hydrogen and nitrogen admixture on the morphology of the films. Fig. 5 shows a film deposited with 2.5 % hydrogen and 2.5 % nitrogen in the plasma. The diamond grains appear to be very fine. Increasing the nitrogen admixture to 7.5 % and keeping the hydrogen admixture at 2.5 % changes the shape of the diamond grains. They appear to be needle-shaped as shown in Fig. 6. These measurements show that the nitrogen admixture can influence the shape of the diamond grains.

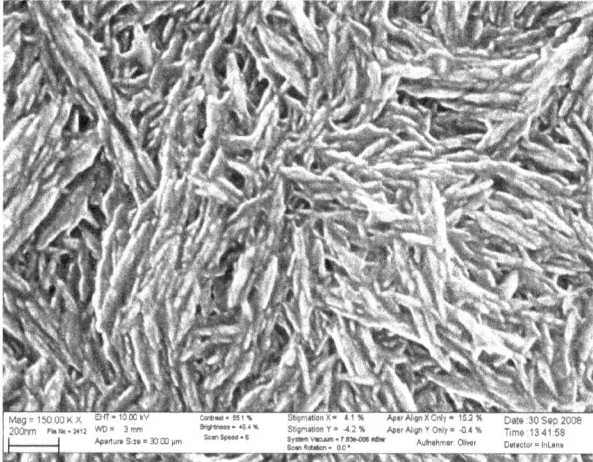

Fig. 6. High resolution SEM measurement of a UNCD film deposited with 2.5 % hydrogen and 7.5 % nitrogen admixture. The scale bar shown corresponds to 200 nm

It is expected that the change in the crystal shape will have a strong influence on the propagation speed of sound in the material giving yet another degree of freedom when designing the material for specific applications.

8. SAW pulse technique

The low surface roughness of UNCD films on the one hand and the high speed of sound in single crystalline diamond on the other hand are making UNCD a very promising material for SAW application. Yet the decisive question is whether the abundance of grain boundaries in the films or the amorphous matrix surrounding the grains will change this picture by e. g. damping the excellent propagation characteristics of surface acoustics waves. The laser-induced SAW pulse method is capable of measuring the SAW-related (i.e. mechanical and structural) properties of thin films (Weihnacht et. al, 1997) (Schenk et. al., 2001) and was used in this work. The applicability of this method for investigating the film properties of polycrystalline diamond films was demonstrated in previous publications (Lehmann et. al., 2001). This method allows measuring all necessary material constants and the wave excitation and propagation parameters decisive for the performance of the SAW material. The biggest advantage of this method is, that it is not necessary to prepare a piezoelectric layer or patterning an interdigital transducer (IDT) structure on the surface, and that rather thin films can also be measured without being disturbed by effects from the Si substrate. The method is a fast and accurate way to measure acoustic wave propagation effects in thin film systems (Schneider et. al. 1997). Pioneering work on utilizing surface acoustic waves as a tool in material science has been done by P. Hess, a general overview can be found in (Hess, 2002).

A somewhat different setup has been used in this work and is schematically shown in Fig. 7. This setup is commercially available at Fraunhofer IWS Dresden[1].

A pulsed laser beam (N_2-laser at 337.1 nm, 0.5 ns pulse duration) is focused on the substrate by a cylindrical lens to excite a line-shaped broadband SAW pulse via a thermo-elastic mechanism. A piezoelectric PVDF polymer foil, pressed onto the sample surface by a sharp steel wedge (width around 5 µm), is used as a broadband sensor for detecting the SAW pulse propagated along the surface of the thin film system. SAW propagation measurements are performed for different propagation lengths between a few mm and some cm. The signal will then be amplified, digitized by an oscilloscope and converted to complex valued spectra (i.e. amplitude and phase spectra) by a fast Fourier transform algorithm. By doing so for different well-known propagation lengths on the one hand the SAW phase velocity dispersion can be determined accurately from the accompanying phase spectra. Knowledge of the velocity dispersion of a film system is decisive, because it gives the possibility to recover the materials parameters (e.g. elastic constants, mass density and film thickness). To derive the elastic properties, a theoretical approach, modeling the films as an isotropic layer but taking into account the anisotropy of the silicon substrate, was fitted to the measured dispersion data. The fact that we have a specimen that consists of a film on top of a substrate introduces a length scale, and thus generates the observed dispersion effect from that the elastic and mechanical properties can be derived.

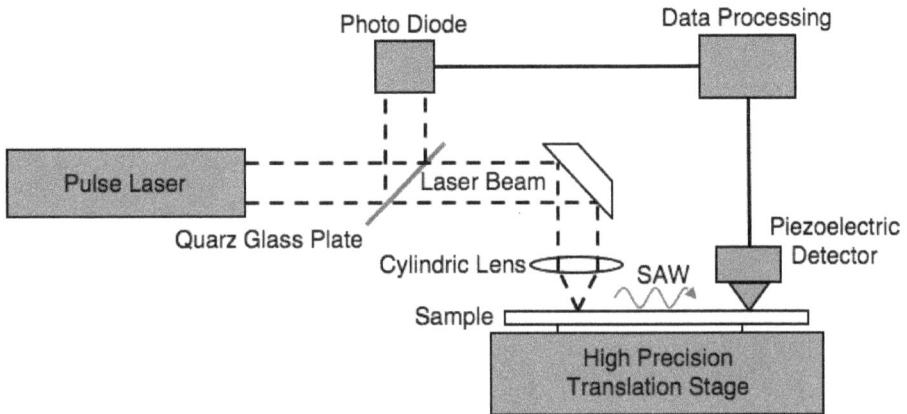

Fig. 7. Principle of SAW pulse technique

A measurement of the SAW phase velocity as a function of frequency as well as the fitted data is shown in Fig. 8. The phase velocity increases with frequency in the case of diamond on silicon substrate ('anomalous dispersion' or 'stiffening case'), because the smaller wavelengths, propagating predominantly in the film, have higher phase velocity.

---

[1] LAWave® (http://www.iws.fhg.de/projekte/062/e_pro062.html)

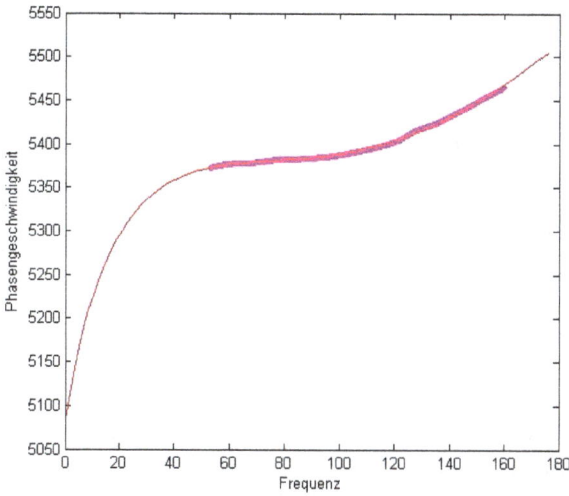

Fig. 8. Measured velocity dispersion and fitted data

Beyond that the damping of the amplitude spectra with increasing propagation length can deliver an estimation of SAW propagation losses due to scattering at thin film imperfections.

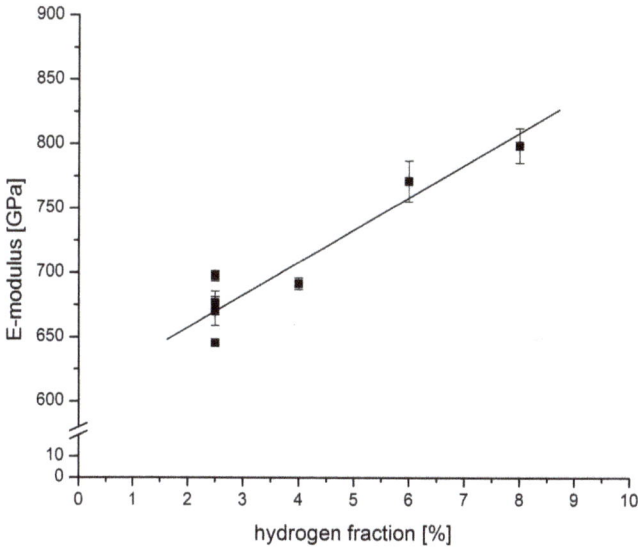

Fig. 9. E-modulus as a function of hydrogen admixture

As expected the elastic modulus is higher (the material is stiffer) for higher admixtures of hydrogen (Fig. 9). This can be explained by the larger diamond crystals and a smaller contribution of amorphous matrix and the fact that the elastic modulus of the amorphous matrix is significantly lower than the modulus of the diamond grains. While the elastic modulus for diamond is around 1220 GPa the elastic modulus of the deposited UNCD films can reach ca. 65 % of this value.

9. Influence of nitrogen and oxygen on mechanical properties

The influence of the nitrogen admixture on the elastic modulus of the deposited films was measured by nanoindentation.
The films that were deposited with additional nitrogen are less stiff compared to films where no additional nitrogen was used. The elastic modulus of the UNCD films deposited with 2.5 % nitrogen in the plasma was measured to be around 370 GPa and increasing the nitrogen admixture even higher to 7.5 % in the plasma resulted in UNCD films with values for the elastic modulus as low as 100 GPa. Thus it was shown that UNCD films deposited with additional nitrogen are unsuitable for the application as SAW device.
An opposite trend can be found when oxygen is used as admixture to the process gas. It was shown that the Young's modulus can be increased up to 950 GPa (ca. 75 % of single crystalline diamond). The reason can be found in the effective etching of sp^2-bonded carbon by the oxygen in the plasma and thus bigger diamond crystals (Shen et. al., 2006).

10. Feasibility study

As a feasibility study SAW resonators with sputtered AlN film as piezoelectric transducer have been produced. Fig. 10 shows the concept of the fabricated AlN-UNCD layered SAW resonator.

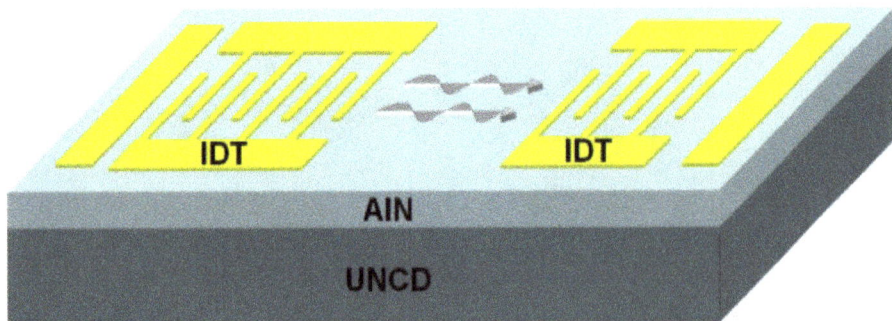

Fig. 10. Schematic Structure of AlN-UNCD layered SAW resonator with golden IDT patterns shaped by photolithography

In the previous chapters it was shown that UNCD films are very suitable as basic material for SAW applications. It was shown that the addition of hydrogen on the one hand improves the elastic constants (towards the value of diamond single crystals), and on the other hand increases the roughness (to values of microcrystalline diamond films), which leads to large propagation loss. Thus a compromise has to be made. The process parameters used for this feasibility study are given in table 2.

Pressure	240 mbar
Gasflow	400 sccm
H_2 fraction	2 %
CH_4 fraction	0,8 %
Ar fraction	97,2 %
MW-power	1 kW

Table 2. Deposition conditions

In order to induce a surface acoustic wave in the UNCD material, a piezoelectric layer is necessary. AlN was chosen for this feasibility study due to being the material with the highest phase velocity (6700 m/s) among piezoelectric materials (Ishihara et. al., 2002). The applicability of AlN thin films on various CVD diamond substrates was demonstrated before (Chalker et. al., 1999).

AlN is an intrinsic piezoelectric material; the wurtzite structure is thermodynamically stable. Several methods for deposition of AlN-films have been reported e.g. MOCVD (Tsubouchi & Mikoshiba, 1985), MBE (Weaver et. al., 1990) and reactive DC or RF sputtering (Akiyama et. al., 1998)(Karmann et. al., 1997). Reactive sputtering processes have the advantage of low substrate temperatures (Dubois & Muralt, 2001)(Naik et. al., 1999)(Tait & Mirfazli, 2001)(Assouar et. al., 2004). Here, magnetron sputtering processes was chosen, for being a common and reliable industrial process.

However, highly (002) oriented films with smooth surfaces are required. Thus deposition parameters (power, pressure, N_2 ratio and substrate temperature) have to be systematically optimized to reach this goal. The influence of oxygen on the film structure was demonstrated before (Vergara et. al., 2004) showing that a low residual gas pressure is crucial for the desired film properties. Therefore a vacuum chamber with turbo molecular pump and a load lock system was used in this work to assure clean conditions. By that, highly oriented AlN films with very smooth surface were deposited on UNCD films that turned out to possess good piezoelectric properties. (Lee et. al., 2007). DC power was 300 W at a pressure of 0.4 Pa and 50 sccm N_2 gas flow at 300°C. The film thickness of the AlN films was ca. 3.5 µm and structure, morphology and bonding structure were characterized by X-Ray diffractometry (XRD), scanning electron microscopy (SEM), atomic force microscopy (AFM), Raman spectroscopy (Renishaw, RA100) and NEXAFS in synchrotron technique.

On top of the AlN film a gold film was deposited by sputtering which was shaped by conventional photolithography. The resonator consists of a central IDT with reflectors at each side (Fig. 11).

The produced SAW Resonators were analyzed due to their performance. Thickness of UNCD as well as AlN have been systematically varied (2 µm to 6.2 µm for UNCD, 1.4 µm to 3.5 µm for AlN). It was measured that with increasing thickness of AlN and UNCD films the resonance frequency increases as well and the resonance peak become clearer. The increase of resonance frequency and thus of SAW velocity is due to reduced influence of the low SAW velocity of the Si substrate. The clearer resonance peak means larger coupling coefficient, which is due to the relative thickness of AlN piezoelectric layer increasing.

Furthermore the influence of the IDT pair number on the SAW resonator performance was investigated (100 Pairs to 200 Pairs). It was measured that the resonance frequency and the resonance strength kept almost the same while doubling the IDT pair numbers.

This feasibility study indicates that the SAW velocity and coupling coefficient only depend on the relative thickness of ALN and UNCD films, but are not affected by IDT pattern.

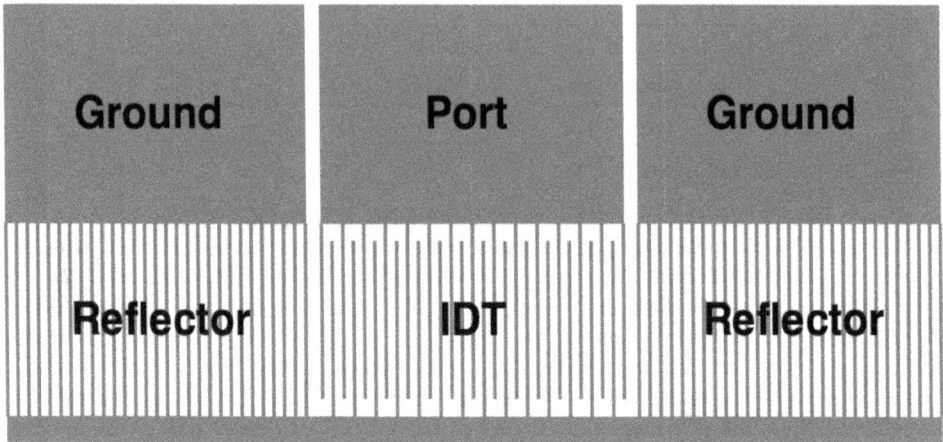

Fig. 11. Schematic Pattern design of SAW Resonator. The actual device consists of significant more lines

11. Acknowledgment

The authors like to thank Dr. Dieter Schneider at Fraunhofer IWS Dresden for the E-modulus measurements of the UNCD films and Hanna Bukowska, University Duisburg-Essen for the AFM measurements.

12. References

Akiyama M., Xu C.-N., Nonaka K., Shobu K., Watanabe T., *Thin Solid Films*, 315 (1998) 62

Assouar M. B., El Hakiki M., Elmazria O., Alnot P., Tiusan C., *Diamond Relat. Mater.* 13 (2004) 1111

Auciello O., Birrell J.,. Carlisle J. A, Gerbi J. E., Xiao X., Peng B., Espinosa H. D., *J. Phys.: Condens. Matter* 16 (2004) R539

Bi B., Huang W. -S., Asmussen J., Golding B., *Diamond Relat. Mater.* Vol. 11, Issues 3-6 (2002) 677-680

Buck V., Deuerler F., *Diamond Relat. Mater.* 7 (1998) 1544

Buck V., *J. Opto. Adv. Mater.* 10 (2008) 85

Caliendo C., *Appl. Phys. Lett.* 83 (2003) 4851

Campbell C., Academic Press (1989)

Chalker P. R., Joyce T. B., Johnston C., Crossley J. A. A., Huddlestone J., Whitfield M. D., Jackman R. B., *Diamond Relat. Mater.* 8 (1999) 309

Dubois M.-A., Muralt P., *J. Appl. Phys.*, 89 (2001) 6389

Erdemir A., Fenske G. R., Krauss A. R., Gruen D. M., McCauley T., Csencsits R. T., *Surf. Coat. Techn.*, 120-121 (1999) 565- 572

Fujimori N., Recent Progresses in Electrical Application of Diamond, *7th European Conference on Diamond, Diamond-like and Related Materials*, 8-13 September 1996

Gruen D. M., *MRS Bulletin*, Vol. 23, No. 9 (1998) 32

Gruen D. M., *Annu. Rev. Mater. Sci.* 29 (1999) 211

Gruen D. M., Krauss A. R., Auciello O. H., Carlisle J. A., (2004) US Patent No. 6793849

Hernandez Guillen F. J., Janischowsky K., Ebert W., Kohn E., *Phys. Stat. Sol.*, A, Appl. res., vol. 201, no11 (2004) 2553-2557

Hess P., *Physics Today*, March issue (2002) 42

Ishihara M., Nakamura T., Kokai F., Koga Y., *Diamond Relat. Mater.* 11 (2002) 408

Karmann S., Schenk H. P. D., Kaiser U., Fissel A., Richter W., *Mater. Sci. Eng.*, B50 (1997) 228

Lamara T., Belmahi M., Elmazria O., Le Brizoual L., Bougdira J., Rémy M., Alnot P., *Diamond Relat. Mater.* 13 (2004) 581

Lee Y. C., Lin S. J., Buck V., Kunze R., Schmidt H., Lin C. Y., Fang W. L., Lin I. N., *Diamond Relat. Mater.* 17 (2008) 446

Lehmann G., Schreck M., Hou L., Lambers J., Hess P., *Diamond Relat. Mater.* 10 (2001) 686

Lin I. N., Lee Y. C., Lin S. J., Lin C. Y., Yip M. C., Fang W., *Diamond Relat. Mater.*, Vol 15 No 11-12 (2006) 2046

Liu H., Dandy D. S., Diamond Chemical Vapor Deposition, Nucleation and early growth stages, Noyes Publications (1995)

Malshe A. P., Park B. S., Brown W. D., Naseem H. A., *Diamond Relat. Mater.* 8 (1999) 1198

May P. W., Mankelevich Yu. A., *J. Phys. Chem.* (2008)

Naik R. S., Reif R., Lutsky J. J., Sodini C.G., *J. Electrochem. Soc.*, 146 (1999) 691

Nakahata H., Hachigo A., Shikata S., Fujimori N., *IEEE Ultrason. Symp. Proc.* (1992) 377

Sarid D., Scanning Force Microscopy, Oxford Series in Optical and Imaging Sciences, Oxford University Press, New York (1991)

Schenk H. P. D., Feltin E., Vaille M., Gibart P., Kunze R., Schmidt H., M. Weihnacht, E. Dogheche, *Phys. Stat. Sol.*, A Appl. Res. 188 (2) (2001) 537

Schneider D., Schwarz T., Scheibe H.-J., Panzner M., *Thin Solid Films* 295 (1997) 107

Shen Z. H., Hess P., Huang J. P., Lin Y. C., Chen K. H., Chen L. C., Lin S. T., *J. Appl. Phys.*, 99 (2006) 124302

Sharma R., Woehrl N., Barhai P. K., Buck V., *J. Opt. Adv. Mater.*, Vol.12 Iss. 9 (2010) 1915

Springer A., Hollerweger F., Weigel R., Berek S., Thomas R., Ruile W., Ruppel C., Guglielmi M., *IEEE Trans. Microwave Theor. Tech.* 47 (1999) 2312

Tait R. N., Mirgazli A., J. *Vac. Sci. Technol.* A, 19 (2001) 1586

Takagaki Y., Santos P.V., Wiebicke E., Brandt O., Schoenherr H.P., Pliig K., *Appl. Phys. Lett.* 81 (2002) 2538

Tsubouchi K., Mikoshiba N., *IEEE Trans. Sonic Ultrason.*, SU-32 (1985) 634

Vergara L., Clement M., Iborra E., Sanz-Hervás A., García López J., Morilla Y., Sangrador J., Respaldiza M. A.

Weaver W., Timoshenko S. P.,. Young D. H, *Vibration Problems in Engineering*, John Wiley & Sons., (1990)

Weihnacht M., Franke K., Kaemmer K., Kunze R., Schmidt H., *IEEE Ultrason. Symp. Proc.* (1997) 217

Woehrl N., Buck V., *Diamond Relat. Mater.* 16 (4-7) (2007) 748

Woehrl N., Hirte T., Posth O., Buck V., *Diamond Relat. Mater.* 18 (2009) 224

Yamanouchi K., Sakurai N., Satoh T., *IEEE Ultrason. Symp. Proc* (1989) 351

Yang W. et al., *Nature Materials* 1, (2002) 253257
Yugo S., Kanai T., Kimura T., Muto T., *Appl. Phys. Lett.* 58, (1991) 1036

10

Aluminum Nitride (AlN)
Film Based Acoustic Devices:
Material Synthesis and Device Fabrication

Jyoti Prakash Kar[1] and Gouranga Bose[2]

[1]Department of Electronics Engineering, University of Tor Vergata, Rome
[2]Department of Applied Electronics and Instrumentation Engineering,
Institute of Technical Education and Research, Bhubaneswar, Orissa
[1]Italy
[2]India

1. Introduction

Enormous growth has taken place in electronics, especially in the field of RF communications towards the beginning of 21st century and continuously striving for better communication performance. Presently, the key concerns of RF communications is bandwidth, in the range of low/medium GHz range, to avoid frequency crowding, especially for wireless communication mobile handsets and base stations (Kim et al., 2004). In addition, reduction in signal loss, low power consumption, scaling down device size, reduction in materials and fabrication costs, and packaging of the device are main issues today. Some of these issues can be resolved, if the new generation of electroacoustic devices can be monolithically integrated with integrated circuit (IC). Conventional electroacoustic devices, used in the communication e.g. Surface Acoustic Wave (SAW) and Bulk Acoustic Wave (BAW) based systems, are widely used for today's wireless communication. These devices are typically made on a single crystal piezoelectric substrate such as quartz, lithium niobate, and lithium tantalate (Assouar et al., 2004). Unfortunately, these substrates based electroacoustic devices are made separately and then it is wired with the signal processing chip, which has several limitations, in particular low acoustic wave velocity and high frequency device fabrication. To resolve these two core issues, thin film materials based electroacoustic devices are actively under consideration [Bender et al., 2003]. Where, a crystalline film is grown on a particular substrate, especially silicon wafer and electroacoustic device is made out of crystalline film. Thus, the electroacoustic device can be integrated with the signal processing circuit. Apart from the silicon wafer as a base material for crystalline film deposition, a variety of other substrates are also explored for academic and technology interests. Furthermore, to get electroacoustic devices of better quality in terms of high frequency and high quality factor (Q), the piezoelectric property of the film is also exploited with different type of device concept called "Micro-Electro-Mechanical Systems" (MEMS). Thin film bulk acoustic resonators (TFBAR) comes under this MEMS devices, where the crystalline film is made to resonate at RF frequency. These MEMS devices have smaller size, lower insertion loss and higher-power handling capabilities than conventional SAW devices (Lee et al., 2004).

Generally, thin piezoelectric films, such as aluminum nitride (AlN), zinc oxide (ZnO) and lead zirconium titanate (PZT) are used for high frequency acoustic devices (Loebl et al., 2003; Yamada et al., 2004; Schreiter et al. 2004). AlN has higher SAW velocity, lower propagation loss, and higher thermal stability in comparison to ZnO; whereas, PZT thin films need selective substrates for deposition and thereafter, needs post-deposition poling to get specific cystal orientation. Thus, AlN seems to have edge over the ZnO and PZT films for electroacoustic devices. The critical factor of piezoelectric AlN thin film is its crystal orientation and morphology. Furthermore, to integrate with the signal processing chip, it is also essential that AlN film should be compatible to the complementary metal oxide semiconductor (CMOS) fabrication processes. In addition, AlN being a dielectric material, it can be used as an insulating material in integrated circuits as well as a piezoelectric material in electroacoustic device. Thus, it is imperative to study the presence of electrical charges and the nature of generation of defects in the AlN film along with its morphology. Usually, there are four types of electric charges present in the insulating film; namely, bulk charges (Q_{in}) and interface (D_{it}) charges, fixed charges (Q) and mobile charges (Q_m). In present IC processing, the presence of fixed charges (Q) and mobile charges (Q_m) are eliminated upto a large extent. Furthermore, the bulk charges (Q_{in}) and interface (D_{it}) charges are reduced further by the optimization of growth parameter and the post-deposition treatments. Reduction in the bulk charge (Q_{in}) and interface charge (D_{it}) density is most essential in cantilever beam based MEMS resonator, otherwise the electrostatic force produced by the these charges may stuck cantilever beam on the substrate (Luo et al., 2006). Most of the MEMS are made out of single crystal silicon substrate utilizing well-matured IC fabrication technology. This poses a challenge to be compatible with a new generation of functional materials. Apart from the electrical charges, the selective etching of piezoelectric materials and silicon for electroacoustic device fabrication is a key technology.

2. Properties of AlN film

AlN is a III-V family compound having hexagonal wurtzite crystal structure with lattice constants a = 3.112 Å and c = 4.982 Å (Yim et al., 1973). In this structure, each Al atom is surrounded by four N atoms, forming a distorted tetrahedron with three Al---$N_{(i)}$ (i = 1, 2,3) bonds named B_1 and one Al---N_0 bond in the direction of the c-axis, named B_2. The bond lengths of B_1 and B_2 are 1.885 Å and 1.917 Å, respectively. The bond angle N_0---Al---N_i is 107.7° and that for N_1---Al---N_2 is 110.5 ° (Xu et al., 2001).

AlN has gained ground in semiconductor industry because of its unique electrical, mechanical, piezoelectric and other properties (Table 1). Some of these noteworthy properties are wide bandgap, high thermal conductivity, high SAW velocity, moderately high electromechanical coupling coefficient, high temperature stability, chemical stability to atmospheric gases below 700 °C, high resistivity, low coefficient of thermal expansion (close to Si), high dielectric constant and mechanical hardness (Xu et al., 2001; Strite et al., 1992; Wang et al., 1994). Its high thermal conductivity (about 100 times that of SiO_2 and roughly equal to that of silicon) and electrical insulating property can prove to be a good dielectric layer for a new generation of integrated circuit devices, particularly in metal insulator semiconductor (MIS) devices. High heat dissipation of AlN can significantly enhance device lifetime and efficiency. AlN film with (002) preferred orientation (c-axis) has maximum piezoelectricity among all other orientations of its crystal structure (Naik et al., 1999). Furthermore, its lattice matching is near to that of silicon and thus less stress is expected to be generated at the AlN/silicon interface. Owing to these properties, AlN films have received great interest as an electronic material for thermal dissipation, dielectric and passivation layers for ICs, acoustic devices, resonators and optoelectronic devices.

Bandgap	6.2 eV, direct
Thermal conductivity	2.85 $Wcm^{-1}K^{-1}$
Coefficient of thermal expansion	$4-5\times10^{-6}$ K^{-1}
Refractive index	1.8-2.2
Dielectric constant	8.5
Electrical resistivity	$10^{11}-10^{13}$ Ω.cm
SAW velocity	6000 m/sec
Melting point	2490 $^{\circ}C$
Hardness	9 Mhos

Table 1. Properties of AlN

3. Synthesis of AlN film

Depending on the intended application, various techniques have been implemented for synthesizing AlN films; namely, molecular beam epitaxy (MBE), reactive evaporation, pulsed laser deposition (PLD), chemical vapour deposition (CVD) and sputtering. Among these techniques, sputtering has the advantage of low-temperature deposition, ease of synthesis, less expensive, non-toxic, good quality films with a fairly smooth surface [Kar et al., 2006; Kar et al., 2007]. In addition, sputtering technique has also CMOS process compatibility. In sputtering technique, plasma is created between the two electrodes by applying high voltage in low pressure. The plasma region contains, positive ions, electrons and neutral sputtering gas, thus the plasma behaves like a conducting medium. Usually, argon gas is used as a sputtering gas. The material that is to be sputtered is called target and it is fixed to the negatively charged electrode. The other electrode is called anode, which is grounded so that the ratio of the target to anode area is significantly reduced. This electric configuration of the sputtering system makes high electric field at the target and that enhances the rate of sputtering. During sputtering process, the energetic ions strike the target and dislodge (sputter) the target atoms. These dislodged atoms travel through the plasma in a vapour state and stick to the surface of wafers, where they condense and form the film. AlN film can be deposited either by directly using the AlN target or by sputtering of aluminum metal in presence of argon and nitrogen gas. The sputtered aluminum atoms react with the nitrogen gas and form AlN film. This process of film deposition is called "reactive sputtering deposition". The sputtering parameters are required to be optimized for desired morphological and electrical properties. These deposition parameters are mainly sputtering pressure, wafer to target distance, sputtering power and wafer temperature. AlN film deposition by reactive sputter deposition technique requires nitrogen as a reactive gas, where it is introduced into the sputtering chamber along with inert argon gas. Argon ions produced in the plasma due to sputtering power and thereafter they strike to the aluminum target and sputter aluminum atoms. These aluminum atoms react with nitrogen and form AlN compound and that deposit on the wafer. Hence, the gas flow ratios need to be optimized. To increase sputtering rate, magnets are placed under the aluminum target, so that magnetic field and the electric field are perpendicular to each other. This configuration of sputtering system is called "magnetron sputtering technique". In the magnetron sputtering, electrons travel in spiral motion in the plasma region. This increases the collision of electrons to neutral argon atoms significantly and that increases argon ions in many folds, thus sputtering rate becomes high.

AlN film can be deposited by DC (direct current) and RF (radio frequency, 13.56 MHz) magnetron sputtering modes. In the DC mode of sputter deposition, the target material must be conductive, so that plasma can sustain. If trace of impurity is present in the system, the surface of the aluminum target becomes contaminated and target poisoning takes place. On the other hand, RF sputtering has the major advantages to produce good quality film, high deposition rate and less chance of target poisoning. For these reasons, RF sputtering technique is preferred than the DC sputtering technique. To obtain well oriented crystalline AlN films for SAW and MEMS structures, the RF sputtering parameters need to be optimized. The sputtering parameters are: RF power, substrate temperature, sputtering pressure, nitrogen concentration and target-substrate distances (D_{ts}). AlN films are deposited on CMOS IC compatibility silicon (100) wafer by the RF reactive magnetron sputtering. The change in morphological and electrical properties of the AlN films with the growth parameters are reported in following section.

3.1 RF power

Amorphous AlN film is found at lower RF sputtering power (100 W), but films became (002) oriented at a sputtering power of 200 W. Further increase of RF power to 400 W, a significant increase in (002) orientation has taken place. This is due to the increase of kinetic energy of atoms that leads to atomic movements on the substrate surface as a result of higher RF power. These newly arrived surface atoms are called "ad-atom". Higher sputtering power increases the AlN grain size that leads to increase in surface roughness as shown in scanning electron microscope (SEM) images (Fig. 1) (Kar et al., 2009).

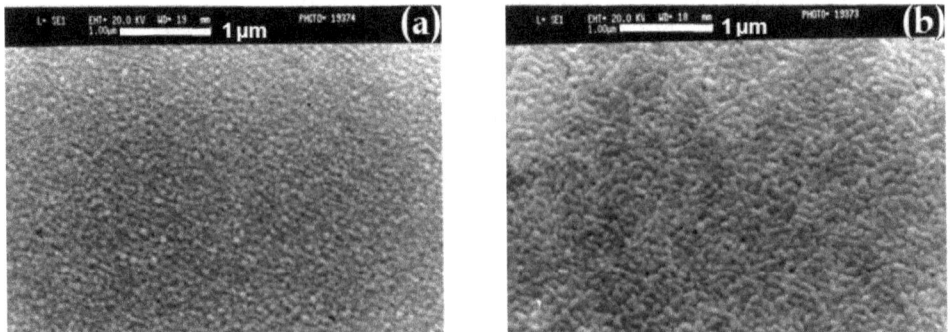

Fig. 1. SEM micrographs of AlN films deposited at (a) 200 W, and (b) 300 W

3.2 Substrate temperature

The structural and morphological properties of the deposited AlN films are strongly dependent on the kinetics of the sputtered atoms arrived at the substrate. The kinetics of sputtered atoms depends on the sputtering parameters. For instance, substrate temperature increases the ad-atom mobility and changes the film morphology significantly. One such illustrations of morphological change with temperature are seen from the X-Ray diffraction (XRD) studies. It is clearly seen from the XRD studies that the c-axis oriented AlN (002) peaks become prominent at moderate temperature range (200–300 °C), but degrades significantly at 400 °C (Kar et al., 2006). This can be attributed to the structural disorderness resulting from the incorporation of impurity atoms at higher temperature (Wang, 2000). The amount of contamination depends on the sputtering deposition system and process related factors, such as base pressure, temperature, gas purity and the partial pressure of moisture, etc. (Naik et al., 1999). Furthermore, smaller grain size with smoother surfaces is observed at lower deposition temperature, and that increases with temperature (Fig. 2). A possible reason may be that the smaller grains grow and merge to form bigger grain, due to the higher thermal energy.

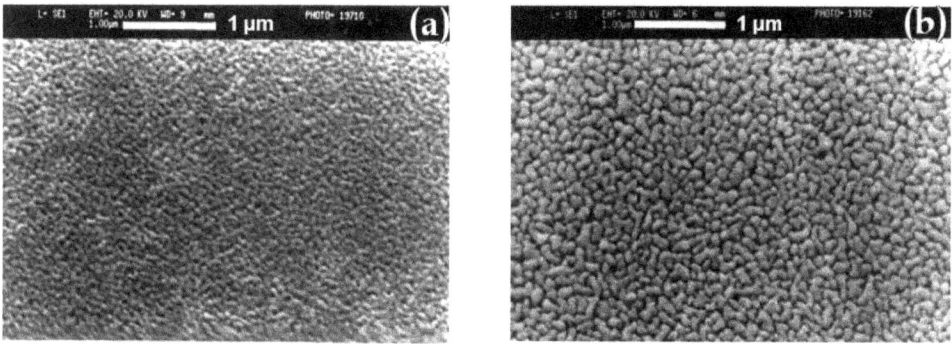

Fig. 2. SEM micrograph of AlN films deposited at (a) 100 °C, and (b) 400 °C

3.3 Sputtering pressure

The variation in crystal orientation with different sputtering pressure are observed from the XRD studies, where the intensity of (002) orientation increases with the deposition pressure and attained a maximum value at 6×10^{-3} mbar. On further increase to a deposition pressure of 8×10^{-3} mbar, the (002) crystal orientation of the AlN film is changed abruptly to the (100) orientation with lesser intensity. The deposited atoms may have altered their direction, energy, momentum and mobility due to the decrease in mean free path of the atoms with sputtering pressure. The hexagonal wurtzite structure of AlN has two kinds of Al–N bond named as B_1 and B_2. These bonds B_1 and B_2 together correspond to (110) and (002) planes, where B_1 corresponds to (100) plane. The formation energy of B_2 is relatively larger that of B_1 (Cheng et al., 2003). Hence, the energy required for sputtering species to orient along c-axis is larger than the other possible planes. At low pressure, sputtering species possess enough energy to form hexagonal wurtzite crystalline structure on substrate surface. It is also reported that the surface roughness of the film increases with the increase in deposition pressure. The grain size is increased till 6×10^{-3} mbar deposition pressure and then it reduced to 80 nm at 8×10^{-3} mbar (Kar et al., 2006). In addition, inhomogeneous patterns on the surface are also observed at this higher pressure (Fig. 3). It is also observed that the AlN film has changed its orientation with less Al-N bond density and reduction of grain size at 8×10^{-3} mbar sputtering pressure. Hence, it is inferred that the structural disorder and/or the change in the Al-N bond density/angles must have taken place at this particular sputtering pressure.

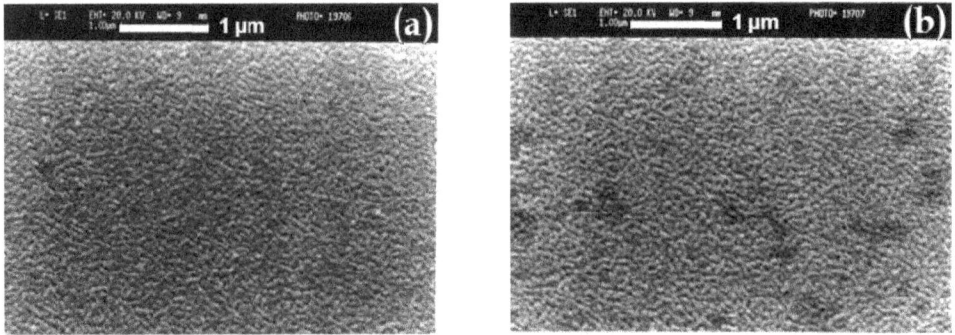

Fig. 3. SEM micrograph of the AlN films deposited at (a) 2×10^{-3} mbar, and (b) 8×10^{-3} mbar

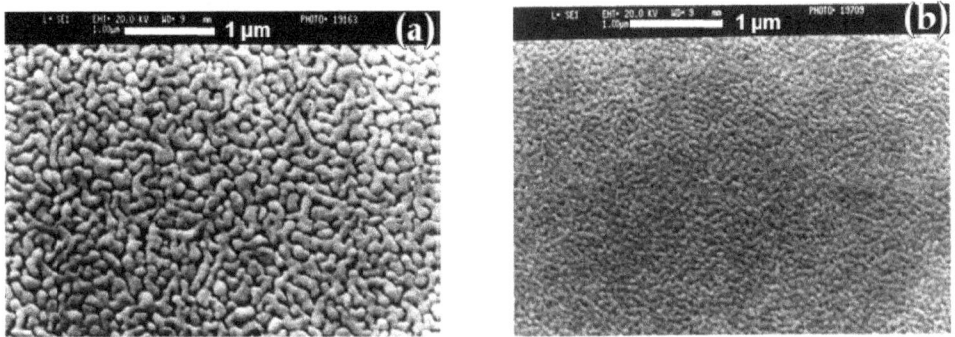

Fig. 4. SEM micrograph of AlN films deposited at (a) 20 % N_2, and (b) SEM image of AlN film for D_{ts} of 5 cm

3.4 Gas flow ratio

At lower nitrogen concentration, the intensity of (100) peak is relatively more prominent than (002), but the trend reverses with higher nitrogen concentration (Kar et al., 2006). At 80% N_2, a highly oriented (002) peak is observed without trace of (100) orientation. Lower argon and higher nitrogen gas concentration results slower aluminum sputtering rate. If the time interval for the arrival of Al species at the wafer surface is slower, the Al atom gets enough time to react with N_2. This increases the probability of Al-N bond formation and bonded Al-N molecules get more time to adjust themselves along (002) orientation on the substrate. On the other hand, at higher argon concentration, Al does not get enough time for complete nitridation due to higher sputtering rate. In addition, faster arrival of the Al at the substrate surface results not only in a poor AlN bond, but also provides less time for the newly formed AlN to arrange itself along c-axis. A surface texture of smaller grain size, smoother, homogeneous and dense granular microstructures has been observed at higher concentrations of nitrogen. This indicates a low surface mobility of the ad-atoms at high nitrogen concentration. In contrast, bigger grain size with increased roughness is observed at lower nitrogen concentration, where the newly formed smaller grain merges together with a previously formed grain and becomes bigger in size. The size and distribution of the micro-grains is quite uniform at 80% nitrogen concentration. At lower nitrogen concentrations, Ar^+ ions transfer more energy to the Al target during bombardment, generating more aluminum atoms that make clusters with incomplete nitridation of aluminum on the wafer surface. This leads to formation of fewer bonds, a poor c-axis orientated and a rough film (Fig 4 (a)).

3.5 Target-substrate distance

The kinetics of the sputtered species arriving at the substrate controls the ad-atom mobility and atomic rearrangement that governs the microstructure of the film. From the XRD studies, it is observed that the intensity of c-axis orientation of the film decreases with increase in target to substrate distance D_{ts} (Kar et al., 2008). At shorter D_{ts}, the Ar ions travel almost normal to the target due to the high electrical field and knock out Al atoms around perpendicular to the target. Because of short deposition path, the probability of collisions of the Al atom with gas atoms is low. Therefore, a good quality film is obtained at lower D_{ts} (5 cm). On the other hand, at larger D_{ts}, the chances of Al collision with gas molecules is increased. In this process Al atoms lose its kinetic energy significantly as well as alter deposition angles. These randomly arriving Al atoms, with lesser energy, cause self-shadowing effects and reduce atomic migration that leads to generation of voids in the film (Lee et al., 2003). The grain size of the AlN film increases with D_{ts}. For lower D_{ts}, smaller grain with minimum surface roughness is observed (Fig. 4(b)) and a coarser grain is found at the highest D_{ts} (8 cm). Surface roughness of the synthesized AlN films are also increases with D_{ts}. The kinetic energy of deposited species is considered to be a major factor for the grain size and the surface roughness of the film.

3.6 Variation of electrical properties with sputtering parameter

The AlN film can be used as a dielectric layer in IC; hence, the electric charges are essential to study with the sputter deposition parameters. Electric charges like Q_{in} and D_{it} are highly governed by the sputter deposition parameters. A decrease in the Q_{in} is observed with sputtering power, where as D_{it} is found to be minimum at moderate RF power. At higher temperature, better electrical properties in the bulk as well as the interface of sputtered AlN films are reported; this is mainly due to the formation of bigger grain size and its associated effects. It is reported that the defects produced by stress, voids and incorporation of gases are main responsible cause for the monotonic increase in Q_{in}. The D_{it} has a minimum value at 6×10^{-3} mbar sputtering pressure. The Q_{in} and D_{it} increases with nitrogen concentration. This will have a deleterious effect for silicon-based devices at higher nitrogen concentration. Rise in the Q_{in} and D_{it} with the increase in D_{ts} is also reported. It is seen that at larger D_{ts}, the morphological as well as the electrical properties of the AlN films deteriorates, whereas, at shorter D_{ts} the quality of the film comes out to be better (Kar et al., 2007). Apart from the electric charges, it is observed that better crystallinity posses AlN films of higher dielectric constant.

4. Post-deposition annealing effect

AlN film may see high temperature, if AlN film is monolithically integrated during IC fabrication. Post-deposition heat treatment significantly affects the morphology and electric charges of AlN film. The post-deposition thermal treatments (annealing) of AlN film are generally carried out by two distinguished modes; namely, Rapid Thermal Annealing (RTA) and conventional furnace annealing.

4.1 Rapid thermal annealing (RTA) process

The XRD studies show that the intensity of c-axis (002) orientation increases upto annealing temperature of 800 °C in nitrogen ambient and then it marginally decreases at 1000 °C (Kar et al., 2005). The shift in XRD diffraction peaks is reported at higher temperatures, which may be due to the generation of stress. Granular worm-like nanostructures are found in as-deposited AlN films (Fig. 5), whereas cracks are observed for annealing at 1000 °C. A short duration of heat pulse by RTA is barely sufficient to modulate the film surface, but not enough to activate the grains of the AlN films to merge themselves to form bigger grains. Appearances of cracks are due to the stress developed in the film. The thermal coefficient mismatch between the AlN and silicon substrate may be generated from the fast ramp up and ramp down annealing heat cycle during rapid thermal annealing. The position and density of cracks depend strongly on the defects, dislocations and the structural relaxation of grain boundaries. The surface roughness is considerably increased for the film annealed at higher temperatures due to the surface modulation.

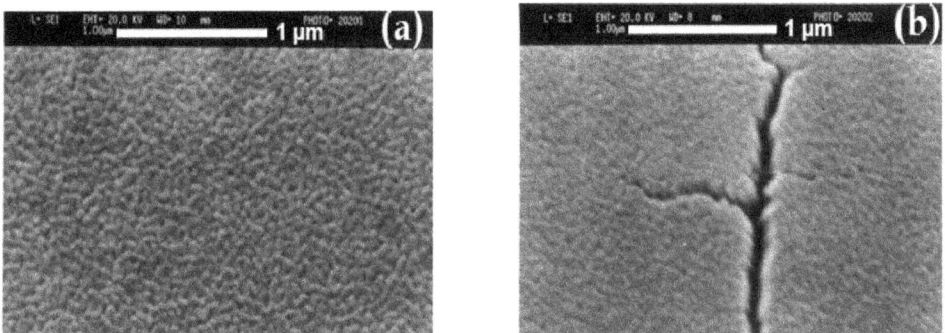

Fig. 5. SEM micrograph of AlN films RTA processed at (a) as-deposited, and (b) 1000 °C

4.2 Furnace annealing

The intensity of the (002) peak increases with furnace annealing temperature, where the atoms acquire adequate activation energy to become (002) oriented. Sometimes, many of the atoms may not be at the crystal lattice site in the as-deposited AlN film, which causes the lattice strain and the formation of microvoids. During conventional furnace annealing, atoms get enough time to acquire sufficient kinetic energy and occupy relative equilibrium positions that minimize the lattice strain and microvoids, which results in a better crystalline film. Furthermore, the furnace annealing process minimizes the dislocations and the other structural defects and forms a better stoichiometric material. From the SEM micrographs, it is observed that the granular worm-like textures grow bigger in size with increased surface roughness as a result of annealing (Fig. 6). The possible reasons for increase in the grain size and the surface roughness may be due to atomic migration in the film towards the lower surface energy with annealing temperature (Kar et al., 2009).

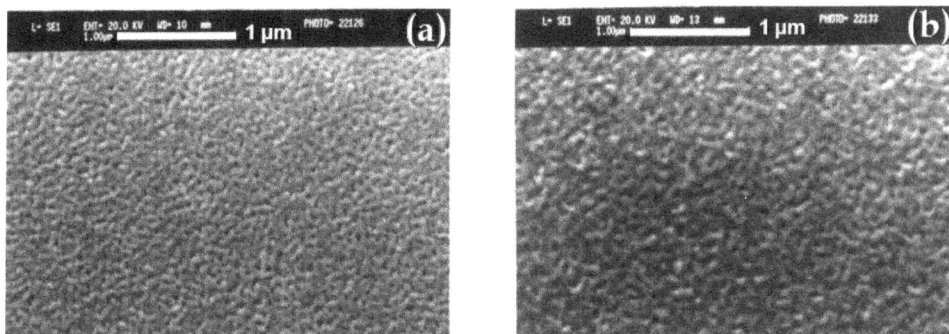

Fig. 6. SEM micrograph of AlN films annealed at (a) as-deposited, and (b) 800 °C

4.3 Effect of annealing on electrical properties

In both types of annealing, the Q_{in} is increased with temperature. The probable reasons for the rise in the Q_{in} may be due to the generation of trap centres with annealing in the nitrogen ambient. In RTA, the D_{it} is found to be strongly dependent on the annealing temperature and it significantly reduced at 600°C. With furnace annealing, the D_{it} decreases with increase in annealing temperature.

5. Growth of AlN films on different substrates

Many a times, AlN films are made on an insulator (SiO_2) for isolation or it is deposited over the metallic electrodes for thin film resonators (TFR). In future, AlN film on high speed semiconductor substrates such as GaAs, InP can be exploited for high speed signal processing and Micro-Opto-Electro-Mechanical Systems (MOEMS) applications. Hence, integration of AlN films on GaAs and InP substrates for a new generation of high-speed devices/subsystems, especially for telecommunications, and radar applications are required. Growth and surface morphology of a deposited film depends not only on the kinetics of the arriving species at the substrate, but also on the nature of the substrates chosen, even if they belong to the same family. In addition, substrate orientation, thermal conductivity and thermal expansion coefficients play vital roles in film growth and its morphology. C-axis oriented AlN films are deposited on Si and SiO_2/Si substrates by RF reactive magnetron sputtering, where the degree of orientation decreases with increase in oxide thickness. The surface roughness of the films deposited on SiO_2/Si is higher. AlN films are also deposited on GaAs and InP substrates by reactive magnetron sputtering technique under identical deposition conditions. c-axis (002) oriented films are observed on GaAs substrates; whereas, AlN (100), (002) and (102) oriented peaks are seen in case of InP substrates. Surface morphology of the films deposited on Si and InP substrates seems to be similar, but the films on InP are little rougher with the development of nano-pores. AlN films, grown on GaAs substrates, forms bump like structures (Kar et al., 2009), which may be due to thermal and/or lattice mismatch. It is important to note that the crystallinity and stochiometry of the initial layer of AlN film also plays a significant role in the creation of defects and mismatches (Ahmed et al., 1992). Crystal orientation of AlN films is also a strong function of the bottom metal electrodes. AlN films deposited on metals (Al, Cu, Cr, Au) are c-axis oriented, whereas the films deposited on Al and Cu are rough with larger grains.

6. AlN film based acoustic device

SAW and BAW are two import types of acoustic wave devices used in RF comunication. These devices can be realized either on a solid substrate/film or through micromachined suspended beam structures. In SAW devices, an elastic wave travels on the surface of a piezoelectric material and displaces the atoms about their equilibrium positions at the interface of piezoelectric film and solid substrate. The neighboring atoms at the interface then produce restoring forces to bring the displaced atoms back to their original positions. SAW can be generated by placing two inter digital transducers (IDT) either sides of the substrate. These IDTs have alternating periodic fingers (Fig. 7). RF signal is applied one of these alternating polarity fingers (IDT) that produces elastic mechanical wave in the substrate. This wave travels along the substrate and also collected by placing another IDT on the piezoelectric material, some distance away from the first IDT. The second IDT collects the RF signal, which can be retransformed into the electric signal. As the elastic-mechanical wave has the speed of acoustic wave, it introduces the delay of signal by 10^3 order. This is the prime use of SAW device. The periodicity p, (centre-to-centre spacing between neighbouring IDT fingers of same polarity) becomes the wavelength of the acoustic wave, and dictates its frequency $f = v/p$, where f and v represent acoustic wave frequency and acoustic propagation velocity, respectively.

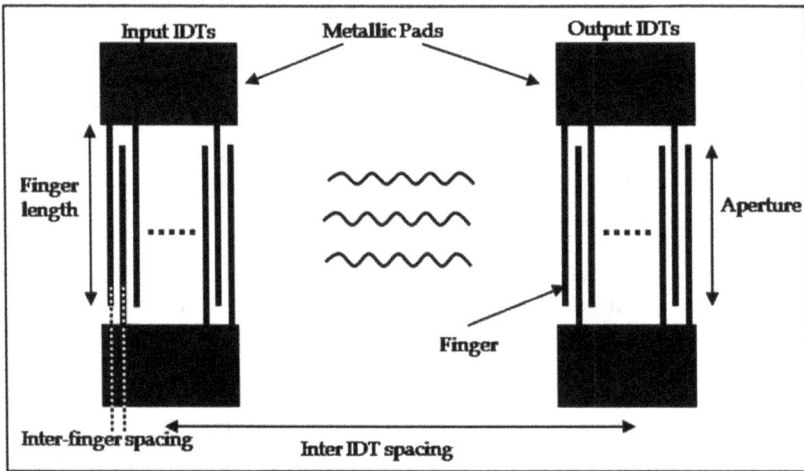

Fig. 7. Schematic diagram of SAW device

Thin Film Bulk Acoustic Resonator (FBAR) device consisting of a piezoelectric material sandwiched between two electrodes and is acoustically de-coupled from the surrounding medium. FBAR devices, using AlN piezoelectric with thickness ranging from tenth of micrometers to several micrometers, resonate in the cellular bands of cell phones and other wireless applications. On applying voltage across the electrodes, the piezoelectric thin film undergoes a shear deformation, and a BAW resonance occurs in the AlN film due to coherent reflection at the top and bottom boundaries of the metal film or plate electrodes. The frequency of resonance is dependent on the physical structures; hence, desired resonant frequency can be obtained by tailoring physical dimension of the structure. For RF frequency, physical dimension of resonators can be realized by using MEMS technology.

MEMS resonators are comprised of a microscale mechanical element, which converts mechanical to electrical signal and vice versa. One of the prominent resonator structures is MEMS cantilever, which is based on thin piezoelectric films. Film resonates when an ac voltage is applied across the film. Resonator can be made without piezoelectric material (electrostatic, capacitive resonator), but it suffers with large resistance, in the range of MΩ, and depends on driving voltage. On the other hand, piezoelectric resonators have smaller resistance of the order of KΩ and are more suitable for UHF device applications. (Lakin, 1999; Quandt et al., 2000; Humad et al., 2003). In addition, the output is easier to sense in a piezoelectric resonator. Furthermore, a piezoelectric resonator has certain advantages over the electrostatic resonator (capacitive resonators), such as low current consumption and lower actuation voltages (Olivares et al.; 2005). But the quality factor (Q) of piezoelectric resonator is smaller than that of a capacitive resonator. The quality factor of any resonator is proportional to the decay time, and is inversely proportional to the bandwidth around resonance. Higher Q represents higher frequency stability and accuracy capability of the resonator (De Los Santos, 1999).

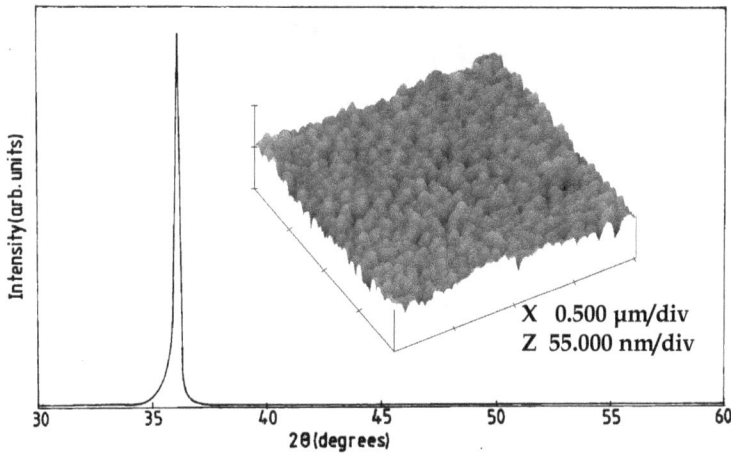

Fig. 8. XRD pattern, and AFM image (inset) of sputtered AlN film for SAW

6.1 Evaluation of AlN films through SAW devices

Higher RF power (400 W) and nitrogen concentration (80%), moderate substrate temperature (200 °C) and sputtering pressure (6×10^{-3} mbar), lower target-substrate distance (5 cm) is suitable for the growth of smooth, highly c-axis oriented AlN film with better electrical properties. A c-axis (002) oriented peak is recorded at 2θ value of 36.1° (Fig. 8). The atomic force micrograph of the film shows dense microstructure with continuous grain growth (inset of Fig. 8). This kind of film is suitable for SAW devices. In a typical case, each IDT consisted of 25 pairs of fingers/electrodes with 30 μm centre-to-centre spacing between the two neighbouring fingers comprising a pair (p/2). The width of each finger/electrode is designed to be 15 μm (p/4) with each of 6.0 mm length and 5.0 mm overlap, producing a SAW filter with an acoustic wavelength of 60 μm (Kar et al., 2009). The SAW device parameters are: AlN film thickness = 0.92 μm, acoustic wavelength = 60 μm, SAW velocity = 5058 m/sec, electromechanical coupling coefficient (K^2) = 0.34%.

Time response, due to acoustic frequency alone, is found after the gating out the response due to electromagnetic feed through. For the centre to centre distance between the two IDT's (d) = 7 mm, the main SAW signal centred at time (delay) = d/ V_{SAW} = 7/5058 = 1.384 µs is obtained. The central acoustic frequency (f_0) response after the gating out is observed at 84.304 MHz (Fig. 9).

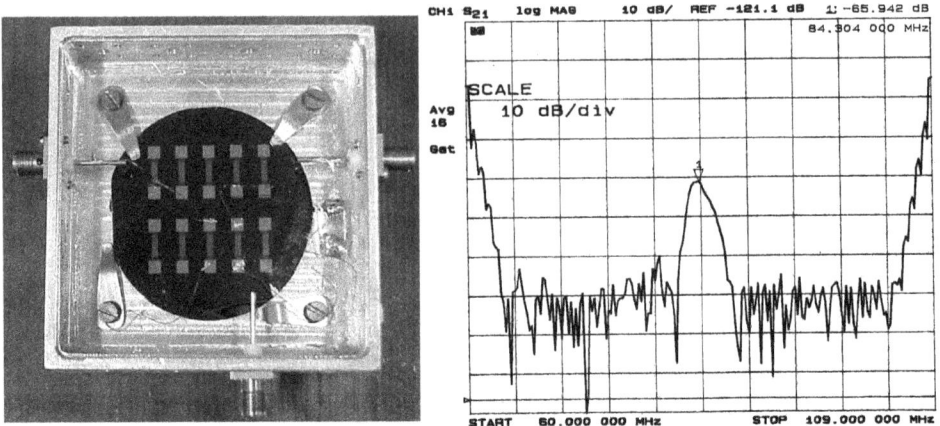

Fig. 9. (a) Optical image of AlN based SAW devices, and (b) Measured response of the SAW device

6.2 Fabrication of AlN film based MEMS

Anisotropic etching of silicon is a key technology for fabrication of various three-dimensional structures such as thin membranes and microbridges for MEMS. Generally, anisotropic silicon etching is done by potassium hydroxide (KOH) or ethylenediamine pyrochatechol (EDP) etchant (French, 2001; Ni et al., 2005). Another technique, which is more versatile, CMOS process compatible and nontoxic, provides better selective etching using doped tetramethylammonium hydroxide (TMAH) etchant (Biswas et al., 2006). The characteristics of these three etchants are listed in Table 2. Many AlN and Al based MEMS structures are also isolated from silicon by silicon dioxide. Hence, their selective etching is very important. Diluted tetramethylammonium hydroxide (TMAH, 5 wt %), doped with silicic acid (30.5 g/l) and ammonium persulphate (5.5 g/l), is suitable for CMOS silicon microprocessing. To protect Al and AlN, silicic acid has been chosen instead of pure silicon powder, because silicic acid dissolves quickly in TMAH solution. Ammonium persulphate (AP) is also added to the above-mentioned solution to reduce the surface roughness of etched silicon. The etch rate of silicon in doped TMAH is found to be 50 µm/hour. During silicon etching the Al, AlN and SiO_2 films are used as mask layers (Fig. 10). Low etch rates of Al and AlN (18-30 nm/hour) as well as SiO_2 (2.5 nm/hour) are found to be suitable for MEMS applications. Probable reason for low etch rate of Al may be the formation of a passivating layer during TMAH etching (Fujitsuka et al., 2004). Dilute TMAH is a well known etchant for AlN film (Kim et al., 2004) and Al as well. But doped TMAH shows significantly lower etch rates for AlN and Al, which are exploited for AlN based suspended microstructures. Fig. 10 (d) depicts suspended Cr/AlN/Cr/SiO_2 cantilevers fixed at one end, where in one of the microstructures is lifted up because of the stress (Kar et al., 2009).

Property	KOH	EDP	TMAH
Si etch rate (100), $\mu m/h$	150	30-35	40-60
Etch quality	high	high	medium
Selectivity (111)/(100)	1:30-100	1:20	1:10-50
Under-etch rate	0.5-1.5	1.4-1.5	0.2-1.7
CMOS compatible	no	yes	yes
Selectivity PECVD SiO_2/Si	1:100-300	1:10,000	1:100-1000
Selectivity PECVD SIN/Si	1:10,000	----------	1:150-200
Attack of aluminum	high	medium	low with Si
Etch stop	boron dope	boron dope	boron dope
Toxicity	low	high	low
Long-term stability	high	low	medium
Cost	low	high	medium

Table 2. Characteristics of important wet etchants used for silicon micromachining (French, 2001)

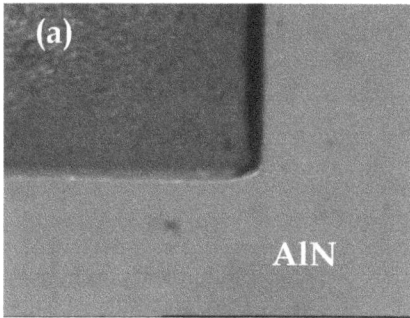

Cavity depth = 75 µm

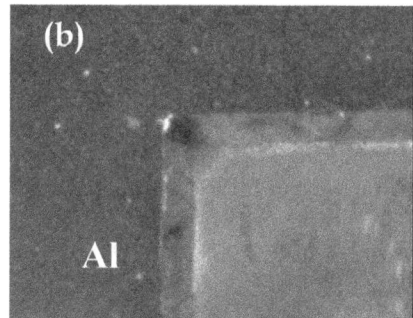

Cavity depth = 125 µm

Cavity depth = 150 µm

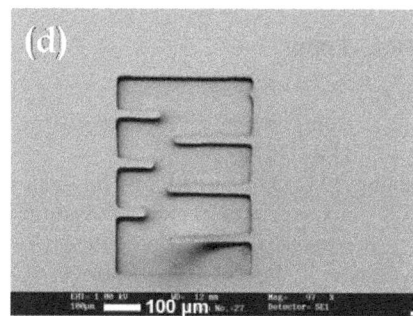

Fig. 10. Micrographs of etched silicon (a) AlN/Si, (b) Al/Si, (c) SiO_2/Si, (d) AlN based suspended microstructures

7. Conclusion

This chapter focuses on the study of RF sputtered AlN films in view of ICs, acoustic devices and MEMS applications. It is divided into two distinct parts; growth, characterization and optimization of device worthy AlN film (mostly on silicon), and demonstration of acoustic and MEMS device applications. The morphological and electrical properties of RF sputtered AlN films are studied with sputtering power, deposition temperature, sputtering pressure, gas flow ratio and target to substrate spacing (D_{ts}). Higher RF power (400 W) and nitrogen concentration (80%), moderate substrate temperature (200 °C) and sputtering pressure (6×10^{-3} mbar), lower target-substrate distance (5 cm) is suitable for the growth of smooth, highly c-axis oriented film with better electrical properties. Post-deposition (RTA and furnace) annealing has a significant impact on the morphology as well as the electrical properties. The RTA processed AlN films have relatively high c-axis (002) orientation films at 800 °C, where microcracks are appeared during RTA process at 1000 °C. Bulk charge density is increased with annealing temperature both types of annealing. A significant reduction in interface charge density is found at 600°C with RTA process, whereas it decreases with furnace annealing temperature. AlN films are deposited and characterized on different substrates by RF reactive magnetron sputtering. On SiO_2/Si substrates, (002) orientation is deteriorated and surface roughness of the films is increased with the increase in oxide thickness. c-axis (002) oriented films are observed on Si and GaAs substrates, whereas AlN (100), (002) and (102) oriented peaks are seen on InP substrates. AlN films, deposited on GaAs substrate, show bump like structures. c-axis oriented AlN films are also observed on metallic films. The AlN films, deposited on Al and Cu, are found to be rough with larger grains. Piezoelectric nature of RF deposited AlN films is ascertained from the performance of a SAW device. This device is centred around a frequency of 84.3 MHz and acoustic phase velocity is inferred to be 5058 m/ sec with K^2 of 0.34 %. The TMAH solution, doped with an appropriate ratio of silicic acid and ammonium persulphate, is developed for micromachining of AlN based structures. The etch rate of silicon is around 50 µm/hour. On the otherhand, the doped solution has negligible impact on Al, AlN and SiO_2 films. The growth of highly (002) oriented AlN films, post deposition process and micromachining method will provide an appropriate platform for the fabrication of futuristic electronic devices.

8. References

Ahmed, A.U., Rys, A., Singh, N., Edgar, J.H. & Yu, Z.J. (1992). The Electrical and Compositional Properties of AlN-Si Interfaces, *J. Electrochem Soc*, Vol. 139(4), pp. 1146-1151.

Assouar, M.B., Hakiki, M.E., Elmazria, O., Alnot, P. & Tiusan, C. (2004). Synthesis and microstructural characterisation of reactive RF magnetron sputtering AlN films for surface acoustic wave filters, *Diamond and Related Materials*, Vol. 13, pp. 1111-1115.

Belyanin, A.F., Boulov, L.L., Zhirnov, V.V., Kamenev, A.I., Kovalskij, K.A. & Spitsyn, B.V. (1999). Application of aluminum nitride films for electronic devices, *Diamond and Related Materials*, Vol. 8, pp. 369-372.

Bender, S., Dickert, F.L., Mokwa, W. & Pachatz, P. (2003). Investigations on temperature controlled monolithic integrated surface acoustic wave (SAW) gas sensors, *Sensors and Actuators B*, Vol. 93, pp. 164-168.

Biswas, K., Das, S., Maurya, D. K., Kal, S. & Lahiri S.K. (2006). Bulk micromachining of silicon in TMAH-based etchants for aluminum passivation and smooth surface, *Microelectronics Journal*, Vol. 37(4), pp. 321-327.

Cheng, H., Sun, Y. & Hing, P. (2003). The influence of deposition conditions on structure and morphology of aluminum nitride films deposited by radio frequency reactive sputtering, *Thin Solid Films*, Vol. 434, pp. 112-120.

De Los Santos, H.J. (1999). Introduction to Microelectromechanical (MEM) Microwave Systems, Artech House, pp. 83.

French, P.J. (2001). Integration of silicon MEMS devices: Materials and processing considerations, *Smart Materials Bulletin*, January, pp. 7-13.

Fujitsuka, N., Hamaguchi, K., Funabashi, H., Kawasaki, E. & Fukada, T. (2004). Silicon anisotropic etching without attacking aluminum with Si and oxidizing agent dissolved in TMAH solution, *Sensors and Actuators A*, Vol. 114, pp. 510-515.

Humad, S., Abdolvand, R., Ho, G.K. & Ayazi F. (2003). Micromechanical Piezo-on-Silicon Block Resonators, *in Proc. IEEE International Electron Devices Meeting (IEDM'03)*, Washington DC, Dec., pp. 957-960.

Kar, J.P., Mukherjee, S., Bose, G., Tuli, S. & Myoung, J.M. (2009), Impact of post-deposition annealing on the surface, bulk and interface properties of RF sputtered AlN films, *Materials Science and Technology*, Vol. 25, pp. 1023-1027.

Kar, J.P., Bose, G., Tuli, S., Myoung, J.M. & Mukherjee, S. (2009). Morphological investigation of AlN films on various substrates for MEMS applications, *Surface Engineering*, Vol. 25, pp. 526-530.

Kar, J.P., Bose, G., Tuli, S., Dangwal, A. & Mukherjee, S. (2009). Growth of AlN films and its process development for the fabrication of acoustic devices and micromachined structures, *Journal of Materials Engineering and Performance*, Vol. 18, pp. 1046-1051.

Kar, J.P., Mukherjee, S., Bose, G. & Tuli, S. (2008). Effect of inter-electrode spacing on structural and electrical properties of AlN films, *Journal of Materials Science: Materials in Electronics*, Vol. 19, pp. 261-265.

Kar, J.P. (2007). *Growth and characterization of aluminum nitride (AlN) films for electroacoustic and MEMS applications*, Ph. D. thesis, Indian Institute of Technology Delhi, India

Kar, J.P., Bose, G. & Tuli, S. (2006). A study on the interface and bulk charge density of AlN films with sputtering pressure, *Vacuum*, Vol. 81, pp. 494-498.

Kar, J.P., Bose, G. & Tuli, S. (2006). Correlation of electrical and morphological properties of sputtered aluminum nitride films with deposition temperature, *Current Applied Physics*, Vol. 6, pp. 873-876.

Kar, J.P., Bose, G. & Tuli, S. (2006). Influence of nitrogen concentration on grain growth, structural and electrical properties of sputtered aluminum nitride films, *Scripta Materialia*, Vol. 54, pp. 1755-1759.

Kar, J.P., Bose, G. & Tuli, S. (2005). Influence of rapid thermal annealing on morphological and electrical properties of RF sputtered AlN films, , *Materials Science in Semiconductor Processing*, Vol. 8, pp. 646-651.

Kim, H.H., Ju, B.K., Lee, Y.H., Lee S.H., Lee J.K. & Kim, S.W. (2004). Fabrication of suspended thin film resonator for application of RF bandpass filter, *Microelectronics Reliability*, Vol. 44, pp. 237-243.

Lakin, K. (1999). Thin film resonators and filters, *IEEE Ultrasonics Symposium*, pp. 895-906.

Lee, H.C., Park, J.Y., Lee, K.H. & Bu, J.U. (2004). Preparation of highly textured Mo and AlN films using a Ti seed layer for integrated high-Q film bulk acoustic resonators, *J. Vac. Sci. Technol. B*, Vol. 22(3), pp. 1127-1133.

Lee, S.H., Yoon, K.H., Cheong, D. S. & Lee, J.K. (2003). Relationship between residual stress and structural properties of AlN films deposited by r.f. reactive sputtering, *Thin Solid Films*, Vol. 435, pp. 193-198.

Loebl, H.P., Klee, M., Metzmacher, C., Brand, W., Milsom, R. & Lok, P. (2003). Piezoelectric thin AlN films for bulk acoustic wave (BAW) resonators, *Materials Chemistry and Physics*, Vol. 79, pp. 143-146.

Luo, J.K., Lin, M., Fu, Y.Q., Wang, L., Flewitt, A.J., Spearing, S.M., Fleck, N.A. & Milne, W.I. (2006).MEMS based digital variable capacitors with a high-k dielectric insulator, *Sensors and Actuators A*, Vol. 132 (1), pp. 139-146.

Naik, R.S., Reif, R., Lutsky J.J. & Sodini, C.G. (1999). Low-Temperature Deposition of Highly Textured Aluminum Nitride by Direct Current Magnetron Sputtering for Applications in Thin-Film Resonators, *J. the Electrochemical Society*, Vol. 146(2), pp. 691-696.

Ni, H., Lee H.J. & Ramirez, A.G. (2005). A robust two-step etching process for large-scale microfabricated SiO$_2$ and Si$_3$N$_4$ MEMS membranes, *Sensors and Actuators A*, Vol. 119, pp. 553-558.

Olivares, J., Iborra E., Clement M., Vergara, L., Sangrador, J. & Sanz-Herv´as, A. (2005). Piezoelectric actuation of microbridges using AlN, *Sensors and Actuators A*, Vol. 123–124, pp. 590-595.

Quandt, E. & Ludwig, A. (2000). Magnetostrictive actuation in microsystems, *Sensors and Actuators A*, Vol. 81, pp. 275-280.

Schreiter, M., Gabl, R., Pitzer, D., Primig, R. & Wersing, W. (2004). Electro-acoustic hysteresis behaviour of PZT thin film bulk acoustic resonators, *J. the European Ceramic Society*, Vol. 24, pp. 1589-1592.

Strite, S. & Morko, H. (1992). *GaN, AlN and InN: A review, J. Vac. Sci. Technol. B*, Vol. 10(4), pp. 1237-1266.

Wang, H.H. (2000). Properties and preparation of AlN thin films by reactive laser ablation with nitrogen discharge, *Modern Physics Letters B*, Vol. 14, 523-530.

Wang, X.D., Jiang, W., Norton, M.G. & Hipps, K.W. (1994). Morphology and orientation of nanocrystalline AlN thin films, *Thin Solid Films*, Vol. 251(2), pp. 121-126.

Xu, X.H., Wu, H.S., Zhang, C.J. & Jin Z.H. (2001). Morphological properties of AlN piezoelectric thin films deposited by DC reactive magnetron sputtering, *Thin Solid Films*, Vol. 388, pp. 62-67.

Yamada, H., Ushimi, Y., Takeuchi, M., Yoshino, Y., Makino, T. & Arai, S. (2004). Improvement of crystallinity of ZnO thin film and electrical characteristics of film bulk acoustic wave resonator by using Pt buffer layer, *Vacuum*, Vol. 74, pp. 689-692.

Yim, W.M., Stofko, E.J., Zanzucchi, P.J., Pankov, J.I., Ettenberg, M. & Gilbert, S.L. (1973). Epitaxially grown AlN and its optical band gap, *J. Appl. Phys.*, Vol. 44(1), pp. 292-296.

11

Polymer Coated Rayleigh SAW and STW Resonators for Gas Sensor Applications

Ivan D. Avramov

Georgi Nadjakov Institute of Solid State Physics, Sofia
Bulgaria

1. Introduction

Polymer coated gas-phase sensors using the classical Rayleigh-type surface acoustic wave (RSAW) mode have enjoyed considerable interest worldwide over the last two decades [1-3]. This interest is motivated by their orders of magnitude higher sensitivity and larger dynamic range compared to bulk acoustic wave (BAW) sensors, fast response times, excellent overall stability, coming close to that of quartz crystal sensors, and low phase noise of the sensor system making high-resolution measurements possible [4]. Because of these features that are difficult to achieve with other technologies, RSAW based gas sensors have found successful application in a variety of industrial implementations such as electronic noses, systems for analysis of chemical and biological gases, medical diagnostics, environmental monitoring and protection, etc. [5-11]. On the other hand, surface transverse wave (STW) based gas sensors, even though sharing the same operation principle, have not been studied so extensively yet. The purpose of this article is to present and discuss systematic experimental data with both acoustic wave modes which will prove that STW based gas-phase sensors not only successfully compete with their RSAW counterparts but also complement them in applications where RSAW gas sensors reach their limits. Successful corrosion proof RSAW sensors using gold metallization for operation in highly reactive chemical environments will also be presented.

2. Operation principle of RSAW/STW based resonant gas phase sensors

Both RSAW and STW based gas sensitive resonant sensors share the same operation principle illustrated in Fig. 1. The sensor device typically is a two-port RSAW or STW resonator on a temperature compensated rotated Y cut of quartz whose geometry has been optimized in such manner that the resonator retains a well behaved single-mode resonance and suffers minimum loss increase and Q-degradation after the gas sensitive layer, (typically a solid, semisolid or soft polymer film with good sorption properties), is deposited on its surface. On the other hand, the sensor has to have maximum active area in the centre of its geometry where the magnitude of the standing wave and deformation are maximized. Thus, strong interaction with the gas adsorbed in the polymer film occurs and maximum gas sensitivity is obtained. The sensor operation principle according to Fig. 1 is fairly simple. If a gas-phase analyte of a certain concentration is applied to its surface, gas molecules are absorbed by the sensing layer until thermodynamic equilibrium is achieved; i. e. the number of adsorbed molecules becomes equal to the number of desorbed ones. Due to adsorption, the layer becomes heavier and this increases the mass loading on the sensor surface. As a result of that, the acoustic wave propagation velocity v decreases and causes a concentration proportional frequency down shift Δf of the sensor's resonance, called sensor signal. The resonance frequency shift of RSAW gas sensors coated with a polyisobutilene (PIB) polymer film is shown in Fig. 2 a) and b) for two different concentrations of tetrachloroethilene vapors. If the vapor concentration is small (0,1% in Fig. 2 a)) then the resonance shifts down by 83 ppm without degradation in loss or Q. At large concentrations of the gas vapors (0,7% in Fig. 2 b)), the 550 ppm of observed frequency down shift is accompanied by a 2 dB loss increase due to the heavy mass loading. However, the sensor device retains a high loaded Q, (above 2000 in Fig. 2 a) versus >4000 in Fig. 2 b)) and a steep phase slope in a well behaved single-mode resonance without distortion or excitation of undesired longitudinal modes.

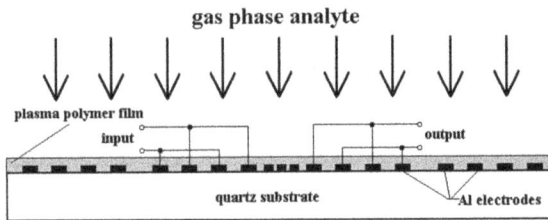

Fig. 1. Operation principle of RSAW/STW based resonant gas phase sensors

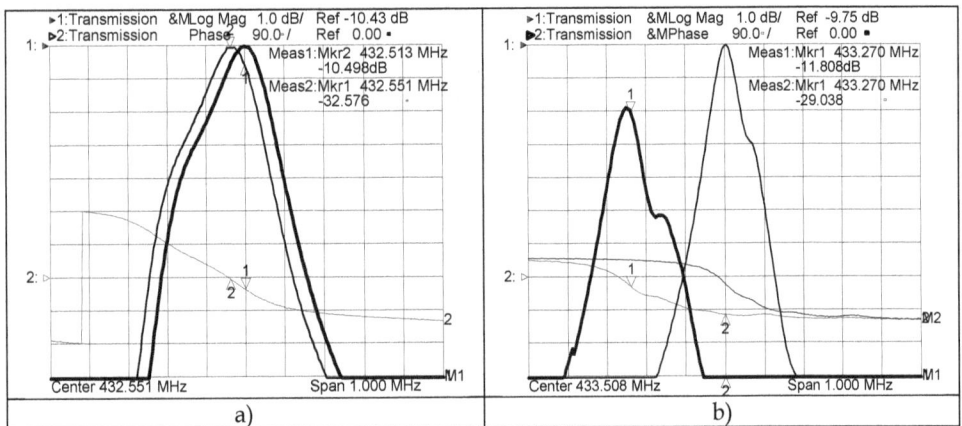

Fig. 2. Frequency (upper curves and phase (lower curves) responses of PIB coated RSAW sensors prior to (right) and after (left) tetrachloroethilene vapor probing at a) 0,1% and b) 0,7% concentration

3. Measurement resolution of RSAW/STW gas phase sensor systems

If a sensor device as the ones from Fig. 2 a) and b) is used as a frequency stabilizing element in the feedback loop of an oscillator circuit and its frequency f_0 is adjusted at the resonance frequency of the sensor (see marker positions in Fig. 2 a) and b)) then due to the high Q of the sensor device, low-noise oscillation with high short-term stability will be obtained. Any change in gas concentration will alter the resonance frequency and the output frequency f_0 of the sensor oscillator, accordingly. Thus Δf can be measured with a high precision using a high-resolution frequency counter, connected to the output of the sensor oscillator. At a given gas concentration C, measured in parts per million (ppm), the resolution R of the sensor system, also measured in ppm, will be limited only by the short-term stability of the sensor oscillator $\sigma_y(\tau)$, also called Allan's variation, for the measurement time τ. *The value of $\sigma_y(\tau)$ represents the flicker phase noise of the sensor oscillator in the time domain which is dominated by the actual flicker phase noise of the coated acoustic wave sensor.* The resolution R determines the minimum change in gas concentration that the system can detect and is, therefore, also called detection limit. It is calculated as follows:

$$R = [C\sigma_y(\tau)f_0\tau] / \Delta f \qquad (1)$$

To calculate R for a given gas concentration C, according to (1), it is sufficient to measure $\sigma_y(\tau)$ of the sensor oscillator for the time interval τ which is normally 1s for most frequency counters operating in the typical 0,3 to 1,0 GHz RSAW/STW sensor range with 1 Hz resolution. Then, according to [12], $\sigma_y(\tau)$ can be calculated from a finite number M of consecutive frequency measurements y_i of f_0 as:

$$\sigma_y(\tau) = \left[\frac{1}{2(M-1)} \sum_{i=1}^{M-1} (y_{i+1} - y_i)^2 \right]^{1/2}, \qquad (2)$$

where i is an integer. In a well stabilized against thermal transients sensor oscillator typically 20 to 50 consecutive measurements of f_0 are enough to calculate $\sigma_y(\tau)$ with sufficient accuracy for practical sensor applications.

4. Chemosensitive layers for RSAW/STW based gas sensors

The correct choice of the sensing layer suitable for the chosen acoustic mode is the key to proper sensor operation and good sensitivity and dynamic range [13, 14]. A sensing layer is considered as "good" if it has an excellent adhesion to the surface of the acoustic device for proper interaction with the acoustic wave, can easily adsorb and restlessly desorb large amounts of probing gases without chemically reacting with them, has good temperature stability and low ageing and does not change its sensitivity and sorption characteristics over thousands of measurement cycles. It is also desirable that the layer provides some selectivity to a certain chemical compound, i. e. it absorbs larger amounts of that compound than other compounds. Finally, the layer should not significantly degrade the Q, loss and the shape of the resonance after deposition onto the acoustic device.

Because of their complicated net structure, many polymers feature excellent physical sorption, as required for reproducible sensor performance and this makes them appropriate for gas sensing applications [15-19]. If some of them have also appropriate viscoelastic properties for good interaction with the RSAW or STW mode, then they will provide the required performance of the acoustic wave sensor, accordingly. Layers with appropriate viscoelastic properties are those that follow the deformation of the surface as a result of the wave propagation without causing significant propagation loss and conversion of the acoustic energy into undesired modes that decay into the bulk of the substrate and may cause degradation of sensor performance.

An important parameter of the sensing film, except for its viscoelastic properties is its solidness. On one hand, the parameter "solidness" determines the sorption properties of the film and the amount of gas that the layer can accommodate before saturation is reached. On the other hand, it determines the way in which the polymer film interacts with the acoustic wave. Therefore, the film solidness will determine the sensitivity, dynamic range and detection limit of the sensor. Based on their solidness, there are three types of polymer films that are appropriate for RSAW/STW sensors:

a. *Solid polymer films.* In fact, these films are solid as glass. That is why, they are often called "glassy polymer films" and have a stiffness value close to that of the sensor's quartz substrate that they are deposited on. If used with the STW mode, due to the lower propagation velocity, these solid films trap the wave energy to the substrate surface and the acoustic wave propagates with low loss. That is why, solid films work much better with the STW mode than with the RSAW one. When their thickness becomes too high, a second slightly faster mode, called "Love mode" gets excited and multimoding occurs. Solid polymer films feature surface sorption and become easily saturated by the adsorbed gas but on the other hand, they feature very fast response times and are very sensitive if the sensor is operated far below saturation. That is why they are appropriate for high resolution measurements at low gas concentrations, (typically below 0,1%). A typical representative of the solid polymer family is the hexamethyldissiloxane (HMDSO), obtained in a glow-discharge plasma polymerization process [19].

b. *Soft polymer films.* These films are soft and elastic just like rubber. That is why they are referred to as "rubbery" or "jelly-like" films. Typically, they are deposited using spin coating or more advanced techniques such as airbrush or electro spray methods that provide good control over film thickness and uniformity. Since these soft polymers provide profound bulk sorption, they are capable of adsorbing large amounts of gas and are appropriate for measurements at high gas concentrations, (typically above 0,1%). They are well tolerated by the RSAW mode but do not work so well with STW. The reason is that they cause energy leakage of the STW into the bulk of the soft layer which results in increased loss and Q-degradation of the sensor resonator. Polymers like polyisobutilene (PIB), poly-(2-hydroxyethylmethacrylate) (PHEMA) and poly-(n-butylmethacrylate) (PBMA) are often used in RSAW based gas sensors.

c. *Semisolid polymer films.* These light and highly elastic films are also typically obtained in a plasma polymerization process [17, 18] for good reproducibility of the film parameters and have a structure very similar to polystyrene, the material used in plastic bags. They are highly resistant to almost all aggressive chemicals such as acids, bases and organic solvents and this makes them appropriate for environmental sensing applications. They are well tolerated by both, the RSAW and STW mode and often feature sensitivities comparable to those of the soft polymer films. The two semisolid films used in this study are styrene (ST) and allylalkohol (AA) synthesized in a plasma polymerization reactor.

5. Comparative characteristics of polymer coated RSAW and STW gas sensors operating at the same acoustic wave length

To identify the advantages and disadvantages of the STW mode versus its RSAW counterpart on quartz for gas sensor applications it is necessary to compare the sensor performance of both modes under identical real-life conditions. Such a performance comparison would be correct only if it is carried out with sensor devices of both modes operating on the same acoustic wave length for the following reason: If both types of devices are fabricated on the same piezoelectric material and cut orientation (AT-cut quartz in this case), use the same device geometry, are coated with the same sensing layer of the same thickness and are probed with identical gases and concentrations, then the only factors responsible for the differences in electrical and sensor performance would be the type of motion for each mode, (elliptical for the RSAW and shear horizontal for the STW) and the way the acoustic wave interacts with the sensing layer. The results presented in the next sections were performed with RSAW and STW sensors whose electrical characteristics in the uncoated state are summarized in Table 1.

5.1 Electrical performance of STW/RSAW sensor resonators coated with solid and semisolid sensing layers

The frequency and phase responses of the STW and RSAW sensor resonators from Table 1 prior to and after coating with the solid HMDSO are compared in Fig. 3. After film deposition, the frequency of the RSAW device shifts down by about 1,5 MHz (3500 ppm), its insertion loss increases by 5,7 dB and the loaded Q decreases from 6000 to about 2000.

Acoustic wave mode	STW	RSAW
Acoustic wave length	7,22 μm	7,22 μm
Sensor resonator frequency	433 MHz	701 MHz
Device insertion loss	5-7 dB	6-7 dB
Loaded Q-factor	3000-4000	5000-6000
Side lobe suppression	> 8 dB	>12 dB
Metallisation	Al	Al

Table 1. Electrical characteristics of the uncoated STW/RSAW sensor resonators used in the comparative studies

In addition, the RSAW device retains a well behaved single-mode resonance with excellent side lobe suppression as required for stable operation of the sensor oscillator. The STW device shows a different behavior. Its frequency shifts down by 4 MHz (6100 ppm) which accounts for about 2 times higher relative mass loading sensitivity than its RSAW counterpart. The insertion loss increases by just about 3 dB versus 5,7 dB for the RSAW mode which implies that the STW mode tolerates solid films better in terms of loss increase. On the other hand, excitation of a second higher-order Love wave mode [20] about 7 MHz higher than the main STW mode at 697 MHz is observed. Since a 180 deg. phase reversal at this Love mode occurs, (see the lower data plot in Fig. 3 b)), it is not very likely to degrade the performance of the sensor oscillator. A more serious problem, however, is the distortion at the main STW mode that indeed can cause the sensor oscillator to jump onto an adjacent peak during the measurement. That is why, coating STW sensor resonators with excessively thick solid films as the 190 nm HMDSO from Fig. 3 should be stopped before distortion and multiple peak behavior on the main STW mode occurs. As far as the higher-order Love mode at 704 MHz is concerned, we have noticed that its gas sensitivity is orders of magnitude lower than the STW mode on the right side. This lack of sensitivity is explained by the fact that the Love mode scatters its energy into the bulk of the sensing layer [20].

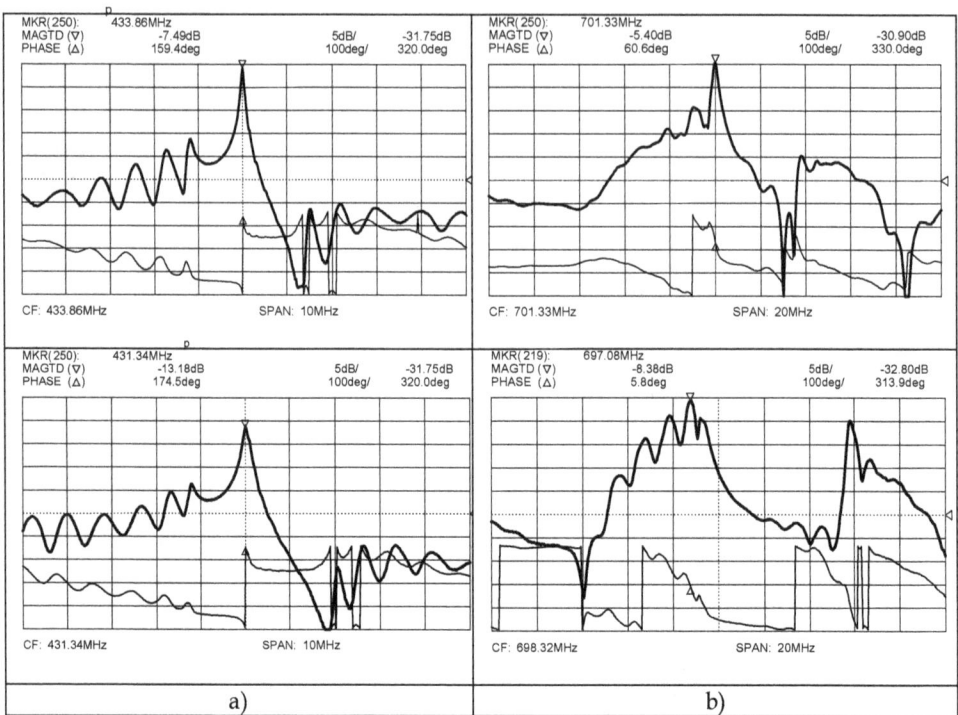

Fig. 3. Frequency (upper curves) and phase responses (lower curves) of the a) RSAW and b) STW sensor resonators from Table 1 prior to (upper plots) and after (lower plots) 190 nm HMDSO solid film deposition

5.2 Electrical performance of STW/RSAW sensor resonators coated with soft polymer films

A similar comparison between both acoustic wave modes was performed by coating the devices from Table 1 with the soft polymer film PIB using the micro drop deposition method. The data obtained shows quite the opposite tendency compared to the solid film behavior from Section 5.1. The STW devices suffered a 5 dB increase in insertion loss and rather distorted frequency responses even at fairly thin soft layers. Only a moderate frequency down shift of 1330 ppm was obtain as a result of film coating. As evident from the frequency responses in Fig. 2 the RSAW devices were found to provide a much better performance at the same film thickness. They retain a high loaded Q and low insertion loss, as well as an undistorted single-mode resonance. These data imply that RSAW sensors will work better with soft polymer films while the STW mode will provide better performance with solid films as long as they are not excessively thick to cause distortion.

6. A practical method for film thickness optimization of RSAW/STW gas sensors coated with solid and semisolid sensing layers

The most important step in designing practical RSAW/STW resonant sensors is the selection of an optimum thickness of the sensing layer. It should be selected in such manner that maximum sensor sensitivity and dynamic range are obtained at minimum degradation of the electrical resonator performance (insertion loss, loaded Q, side lobe suppression and distortion) as required for stable low-noise operation of the sensor oscillator. A very efficient method for film thickness optimization using a controlled plasma deposition of the semisolid polymer Parylene C is described in [21]. This material has viscoelastic properties very similar to practical solid and semisolid layers but has the unique feature that it can polymerize directly on the surface of the acoustic devices at room temperature, thus avoiding undesired thermal frequency drifts. Also the actual deposition is performed in a chamber separate from the plasma reactor where the devices are protected from the high electric fields of the main plasma generator. This allows direct measurement of their frequency and phase responses with a network analyzer in the process of film deposition while the film thickness is measured with a quartz crystal microbalance (QCM). The results from a Parylene C deposition on a 433 MHz RSAW resonator and real time measurements of its electrical characteristics are shown in Fig. 4 for a polymer thickness ranging from 0 to 700 nm. From this measurement it is possible to identify the thickness range in which optimum sensor performance is expected and to extract information on the behavior of important sensor parameters in the process of polymer coating as follows:

- the down shift of the resonant frequency versus film thickness;
- the loss increase with film thickness;
- the loaded Q decrease with film thickness;
- the behavior of the adjacent longitudinal modes in the process of deposition;
- the thickness range over which the sensor demonstrates maximum mass sensitivity of frequency while retaining good electrical performance. In Fig. 4 this optimum thickness range is between 100 and 300 nm with an average of 200 nm.

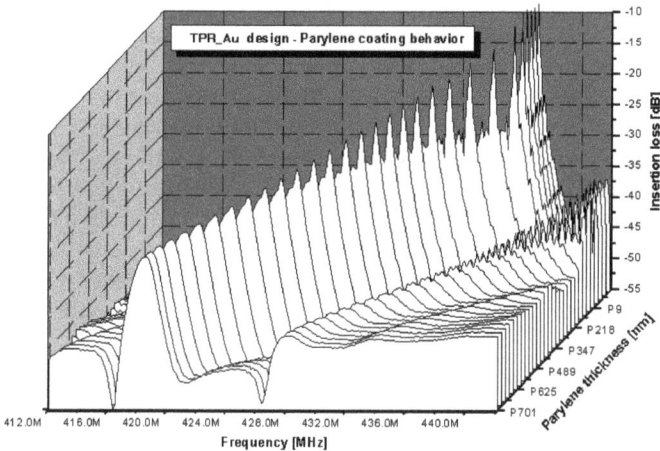

Fig. 4. Parylene C coating behavior of a 433 MHz RSAW sensor resonator in the 0 to 700 nm polymer thickness range

As evident from Fig. 4, in the optimum thickness range (100 to 300 nm) a maximum linear frequency shift, (maximum mass sensitivity), is accompanied by just about 7 dB loss increase. The Q remains high as shown by the sharp resonance while the first longitudinal mode on the left side of the resonance remains suppressed by at least 12 dB. If the sensor is intended for high-resolution measurements at low gas concentrations, then a thickness close to 100 nm should be chosen due to the highest Q and lowest loss. If measurements at higher gas concentrations are expected then a 300 nm thickness may be more appropriate since the thicker layer may adsorb larger amounts of gas without film saturation.

6.1 Critical thickness in RSAW/STW based sensor resonators coated with solid and semisolid sensing layers

As shown in the previous sections, the sensing layer does not only shift the resonant frequency down, increase the loss and decrease the loaded Q as a result of mass loading but it also influences the longitudinal modes supported by the resonator geometry that appear on the left side of the main resonance. In the uncoated resonator these modes are well enough suppressed (typically by 5 to 15 dB) and do not cause any problems when the resonator is operated in an oscillator circuit. As soon as a sensing layer is deposited on the surface, it will change the phase conditions along the device topology and this will cause the adjacent longitudinal modes to arise in magnitude at the expense of the main resonance. This situation gets worse at thick solid films for both, the STW and the RSAW mode. At a certain thickness which we call "critical thickness" the magnitude of the first adjacent low-frequency longitudinal mode on the left becomes equal to the magnitude of the higher-frequency main resonance. This creates a potential for instability in the sensor oscillator stabilized with this sensor since it can easily jump from the main resonance onto the left longitudinal mode during gas probing which will ruin the measurement. The critical thickness situation is illustrated in Fig. 5 a) and b) for a STW and a RSAW device from Table 1, accordingly, in the process of Parylene C deposition as described in Section 6. As evident from Fig. 5 a), at thickness values above 185 nm the first longitudinal mode on the left starts rapidly growing until its magnitude becomes equal to the main resonance. The critical thickness at which this happens is about 350 nm. At this thickness also a strong Love mode excitation on the right is observed. The RSAW device in Fig. 5 b) reaches its critical thickness at about 650 nm. From these data we can draw the conclusion that the devices from Table 1 can be usable as Parylene C coated sensors as long the film thickness is lower than 300 nm and 600 nm for the STW and RSAW devices, respectively. Comparing the coating behavior of both modes in Fig. 5 we see that the STW mode retains a much better behaved resonance than its RSAW counterpart until the critical thickness is reached. At that thickness the STW device has a loss of 14 dB (Fig. 5 a)), versus 35 dB for the RSAW device (Fig. 5 b)). Therefore, the STW mode tolerates solid and semisolid sensing films much better than the RSAW one and is more appropriate for operation with such films in practical gas sensors.

7. Gas sensing characteristics of RSAW/STW resonant sensors coated with solid and semisolid chemo sensitive films

In the following sections we present results from gas probing experiments on RSAW/STW sensors coated with solid HMDSO and semisolid ST and AA films. Four different chemical agents at different vapor concentrations are used for gas probing as follows:

- Dichloroethane   6500 ppm
- Ethylacetate   17600 ppm
- Tetrachloroethylene   2650 ppm
- Xylene   1400 ppm

The purpose of the gas probing tests is to identify which acoustic wave mode provides better performance in real-world gas sensing conditions.

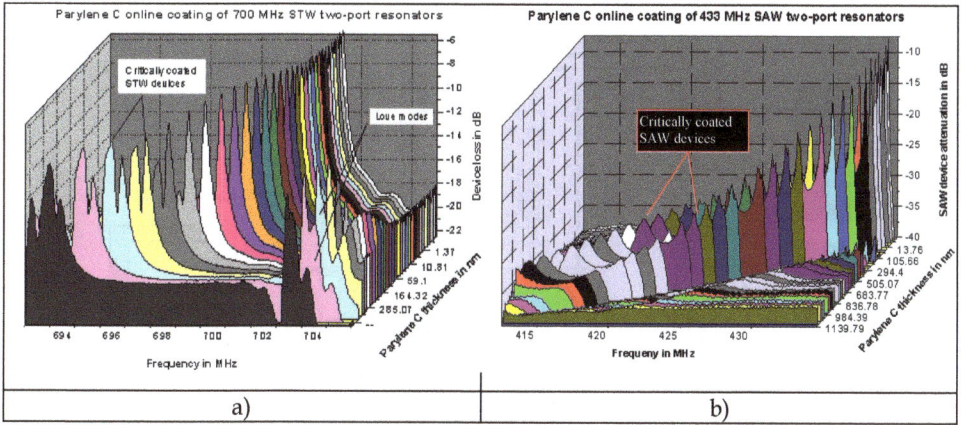

Fig. 5. Critical thickness in a) STW and b) RSAW devices in the process of Parylene C coating

7.1 Computer controlled automatic system for gas probing measurements

The block diagram of the computer controlled system for measuring the gas sensing characteristics of the RSAW/STW polymer coated sensors is shown in Fig. 6. For correct comparison of the gas probing performance of both acoustic wave modes four pairs of devices (one RSAW and one STW sensor in each pair, coated with the same polymer to the same film thickness and in the same deposition process) are mounted in open TO 92 packages and placed in the sensor head which can accommodate a total of eight sensors. Each device is connected to one of the 8 sensor oscillator circuits in the head. During gas probing each of the 8 oscillators is turned on for a short period of time to take the measurement. The oscillators are operated one at a time and multiplexed consecutively to avoid possible injection locking. Their output frequency is down converted to an intermediate frequency in the 4-9 MHz range by means of a stable heterodyne reference oscillator to allow fast high-resolution measurements with a reciprocal frequency counter. The chemical compounds 1 through 4 used for gas probing are vapors from the 4 liquid-phase analytes in the 4 containers. A permeation cell is placed on top of each container to allow a defined vapor pressure which is controlled by the rotation speed of the pump. By a switch block of valves the vapors of each analyte are then consecutively fed to the sensor head where they interact with the sensors. After the measurements at each analyte are completed the sensors are flushed with dry air passing through a silica gel integrator which provides also a homogenous air and gas flow. The entire system is controlled by a computer which performs measurements in probe-flush cycles over time and provides real-time sensor data on the computer screen. The data in Fig. 7 is the gas probing performance of a 700 MHz STW styrene coated sensor probed with dichloroethane vapors at 6500 ppm concentration in 100 s probe-flush cycles. It should be noted that prior to this measurement the sensor was probed with a different compound (xylene) for 62 hours and 40 minutes. Note the excellent reproducibility of the noise free sensor signal with a magnitude

Δf=160KHz over time indicating that prolonged xylene treatment has not had any influence on the sensor performance. This is a clear indication that styrene has very good physical sorption properties to a variety of gas phase compounds.

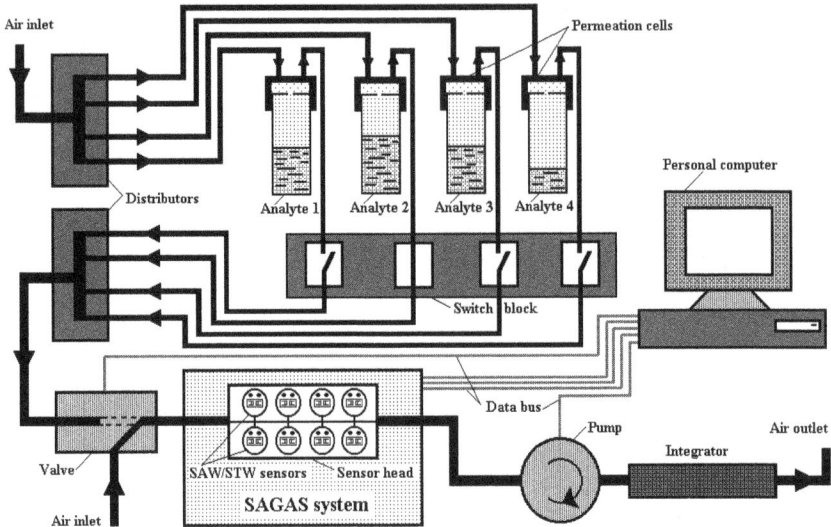

Fig. 6. Block diagram of the automated system for simultaneous gas sensitivity measurements on eight sensor devices

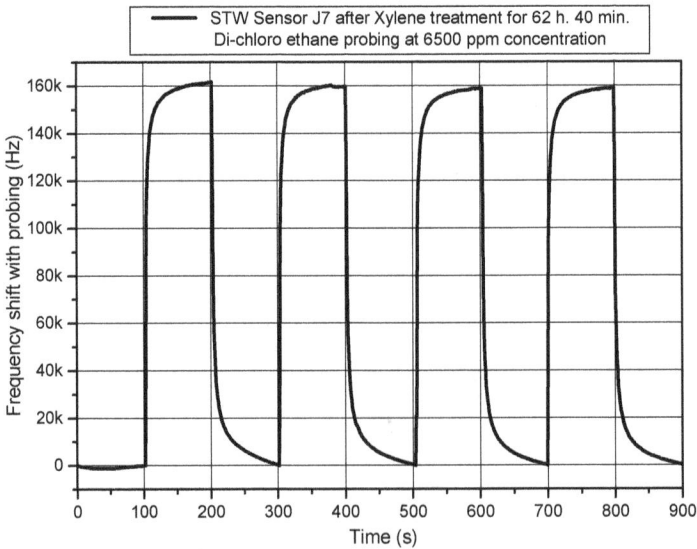

Fig. 7. Gas sensing performance of a 700 MHz styrene coated STW sensor probed with dichloroethane at 6500 ppm concentration in 100 s probe-flush cycles

7.2 Gas sensitivity comparison of RSAW/STW sensor resonators coated with solid HMDSO films

This study aims at finding out which of both acoustic wave modes provides better gas sensitivity when coated with solid HMDSO films and what is the optimum film thickness at which maximum sensitivity is achieved. For this purpose, 5 pairs of RSAW/STW devices according to Table 1 were coated at 5 different HMDSO thicknesses (50, 100, 190, 280 and 350 nm) each, in the same plasma deposition process for each pair. Figure 8 compares the gas sensing characteristics of both modes gas probed with tetrachloroethilene at 2650 ppm concentration. The results from all gas probing experiments on the 5 pairs of devices are summarized in Table 2. In this table the "sensitivity factor" is the ratio between the relative sensitivities (in ppm) for the two devices of each pair. It is given for each of the 5 film thicknesses and shows which mode is more sensitive and at which thickness. From the data in Fig. 8 and Table 2 the following important practical conclusions can be drawn:

1. *The HMDSO coated sensors have very short response times and reach adsorption-desorbtion equilibrium just a few seconds after the gas flow is applied. We attribute this behavior to the surface sorption of the HMDSO which is typical for solid sensing polymers.*

Sensor/Compound	Dichloroethane 6500 ppm	Ethylacetate 17600 ppm	Tetrachloroethylene 2650 ppm	Xylene 1400 ppm
700 MHz STW 50 nm HMDSO	2 KHz (2.9 ppm)	2.8 KHz (4 ppm)	3.5 KHz (5 ppm)	2.4 KHz (3.4 ppm)
433 MHz RSAW 50 nm HMDSO	1.5 KHz (3.5 ppm)	4 KHz (9.2 ppm)	3 KHz (6.9 ppm)	2.2 KHz (5 ppm)
Sensitivity factor (STW/ RSAW)	0.82	0.43	0.72	0.68
700 MHz STW 190 nm HMDSO	3 KHz (4.3 ppm)	4.8 KHz (6.9 ppm)	7.5 KHz (10.7 ppm)	3.7 KHz (5.3 ppm)
433 MHz SAW 190 nm HMDSO	1.8 KHz (4.2 ppm)	3.8 KHz (8.8 ppm)	3.5 KHz (8.1 ppm)	2.5 KHz (5.8 ppm)
Sensitivity factor (STW/RSAW)	1.02	0.78	1.32	0.91
700 MHz STW 280 nm HMDSO	8.3 KHz (11.9 ppm)	8.5 KHz (12.1 ppm)	7 KHz (10 ppm)	4 KHz (5.7 ppm)
433 MHz SAW 280 nm HMDSO	3.5 KHz (8 ppm)	6.5 KHz (15 ppm)	6 KHz (14 ppm)	4.3 KHz (10 ppm)
Sensitivity factor (HПAB/RSAW)	1.49	0.81	0.71	0.57
700 MHz STW 350 nm HMDSO	11 KHz (15.7 ppm)	14 KHz (20 ppm)	15 KHz (21.4 ppm)	9.5 KHz (13.6 ppm)
433 MHz SAW 350 nm HMDSO	1.8 KHz (4.2 ppm)	2.3 KHz (5.3 ppm)	5.1 KHz (11.8 ppm)	4.2 KHz (9.7 ppm)
Sensitivity factor (STW/RSAW)	3.74	3.77	1.81	1.4
700 MHz STW 100 nm HMDSO	11 KHz (16 ppm)	20 KHz (29 ppm)	37 KHz (53 ppm)	9 KHz (13 ppm)

Table 2. Gas sensitivity comparison of RSAW vs. STW devices coated with solid HMDSO film of 5 film thicknesses. Data on the 100 nm coated SAW device are not available

Compound Concentration	Dichloroethane 6500 ppm	Ethylacetate 17600 ppm	Tetrachloroethylene 2650 ppm	Xylene 1400 ppm
700 MHz STW, 100 nm HMDSO	11 KHz (16 ppm)	20 KHz (29 ppm)	37 KHz (53 ppm)	9 KHz (13 ppm)
433 MHz RSAW, 280 nm HMDSO	3.5 KHz (8 ppm)	6.5 KHz (15 ppm)	6 KHz (14 ppm)	4.3 KHz (10 ppm)
Sensitivity factor	*STW/RSAW* 2.0	*STW/RSAW* 1.93	*STW/RSAW* 3.79	*STW/RSAW* 1.3

Table 3. Gas sensitivity comparison of the RSAW and STW sensors coated at their optimum HMDSO thickness values

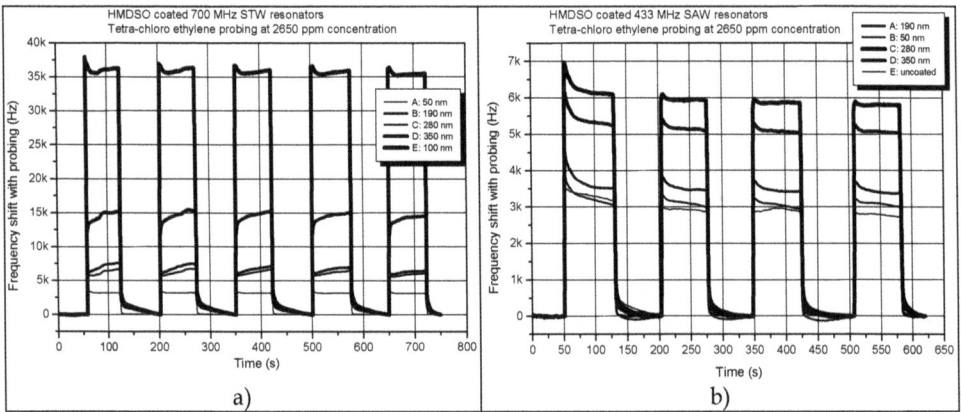

Fig. 8. Tetrachloroethilene probing data of HMDSO coated a) STW and b) RSAW sensors at 50, 100, 190, 280 and 350 nm film thicknesses

2. *Starting from very thin films (50 nm in this case) and increasing the thickness, the gas sensitivity increases in both the RSAW and STW devices, accordingly, until a thickness value is reached at which maximum sensitivity is achieved. This optimum thickness value is different for both modes (100 nm for the STW and 280 nm for the RSAW mode at the wavelength of 7,22 μm in this case). Further increase in film thickness beyond the optimum thickness value only reduces the relative frequency sensitivity and increases the loss of the gas sensor.*

3. *The optimum thickness values for both modes are far below critical thickness and are well within the thickness ranges in which the sensor devices demonstrate high mass sensitivity while retaining low insertion loss, high Q and a well behaved single-mode resonance in the Parylene C coating experiment from Fig. 5. Therefore, the practical film thickness optimization method described in Section 6 is well suited for identifying the optimum film thickness at which maximum gas sensitivity should be expected.*

4. *The relative sensitivities for both acoustic wave modes at their optimum thickness values, summarized in Table 3, demonstrate a 1,3 to 3,8 times higher sensitivity to all 4 gases of the STW mode versus its RSAW counterpart operating at the same acoustic wavelength. This suggests that the STW mode is much more appropriate for operation with solid chemo sensitive polymers.*

7.3 Gas sensitivity comparison of RSAW/STW sensor resonators coated with semisolid styrene (ST) and allylalcohol (AA) films

A similar comparative study was performed also when pairs of RSAW/STW devices from Table 1 were coated at three different thicknesses of the semisolid ST and AA polymer films in a glow discharge plasma reactor. Since no equipment was available to measure the thickness of semisolid layers directly we used the deposition time in seconds for each layer as a measure of the layer thickness. In this study the deposition times for all three thicknesses were 10, 15 and 20 s for the films with the lowest, medium and highest thicknesses, accordingly. The results from these gas probing tests are summarized in Tables 4 and 5 for the ST and AA coated sensors, accordingly. Note that in these experiments we had to reduce the concentrations of all four probing gases by a factor of 4 to avoid saturation of the sensing layers due to the much higher adsorption capacity of the ST and AA films compared to HDMSO. When comparing the data from Tables 4 and 5 an interesting behavior is observed. Styrene coated STW devices are up to 3 times more sensitive than their ST coated RSAW counterparts while with the AA coated sensors we see the opposite behavior - the AA coated RSAW devices are up to 3,6 times more sensitive than their AA coated STW counterparts. We attribute this behavior to the fact that AA is the softest of the three polymers we used in this work and this material behaves much more like a soft polymer than a semisolid one. This implies that the RSAW mode might be more suitable for operation with soft sensing films than the STW mode.

Compound/ Concentration	Tetrachloroethylene (630 ppm)		Dichloroethane (1550 ppm)		Ethylacetate (4190 ppm)		Xylene (330 ppm)	
Acoustic mode	STW	RSAW	STW	RSAW	STW	RSAW	STW	RSAW
Deposition time 10 s/styrene	119 ppm	74 ppm	100 ppm	92 ppm	94 ppm	76 ppm	71 ppm	44 ppm
Deposition time 15 s/styrene	254 ppm	85 ppm	206 ppm	171 ppm	190 ppm	95 ppm	140 ppm	65 ppm
Deposition time 20 s/styrene	239 ppm	108 ppm	219 ppm	201 ppm	206 ppm	122 ppm	111 ppm	76 ppm
Sensitivity factor 10s / 15s / 20s	STW/RSAW 1.61 / 3.0 / 2.21		STW/RSAW 1.09 / 1.2 / 1.09		STW/RSAW 1.24 / 2.0 / 1.69		STW/RSAW 1.61 / 2.15 / 1.46	

Table 4. Gas sensitivity comparison of semisolid ST coated RSAW and STW devices

Compound/ Concentration	Tetrachloroethylene (630 ppm)		Dichloroethane (1550 ppm)		Ethylacetate (4190 ppm)		Xylene (330 ppm)	
Acoustic mode	STW	RSAW	STW	RSAW	STW	RSAW	STW	RSAW
Deposition time 20 s/AA	12.3 ppm	20.8 ppm	22.8 ppm	66.4 ppm	32.6 ppm	69.9 ppm	7.1 ppm	25.2 ppm
Deposition time 25 s/AA	14.3 ppm	25.9 ppm	23.1 ppm	83.1 ppm	31.5 ppm	90.5 ppm	8.3 ppm	28.4 ppm
Sensitivity factor 20s / 25s	RSAW/STW 1.69 / 1.81		RSAW/STW 2.91 / 3.18		RSAW/STW 2.14 / 2.87		RSAW/STW 3.55 / 3.42	

Table 5. Gas sensitivity comparison of semisolid AA coated RSAW and STW devices

For the semisolid film coated RSAW/STW sensors we can make the following conclusions:

1. *Semisolid sensing films improve gas sensitivity of the RSAW and STW modes dramatically compared to the solid films. ST coated devices demonstrate one to two orders of magnitude higher relative gas sensitivities compared to HMDSO coated ones.*

2. *As observed with the solid film, also semisolid layers seem to have an optimum film thickness at which maximum gas sensitivity is achieved. Here these optimum thicknesses are achieved at 15 s for ST and 20 s for AA with the STW mode, while the RSAW mode needs somewhat higher optimum thicknesses – 20 s for ST and 25 s for the AA film.*

3. *There is a significant film type dependent difference in gas sensitivities for both modes. ST provides much better gas sensitivity compared to AA and this is attributed to both – the different sorption properties and different viscoelastic properties of the films which determine how the wave interacts with the film and the gas sorbed in it.*

4. *The RSAW mode operates better with the relatively soft AA than with the semisolid ST.*

8. Noise and measurement resolution (detection limit) of RSAW/STW resonant sensors operating with gas sensing polymer layers

As shown in Section 3, if the sensor resonator is connected in the feedback loop of a sensor oscillator whose short-term stability over the time τ has been measured as $\sigma_y(\tau)$, then measuring the sensor signal Δf which is the response of the sensor oscillator to the gas with the concentration C, the measurement resolution R, also called detection limit of the sensor system, can readily be calculated with Equation (1). As an example, let us determine the measurement resolution of the sensor system with which the dichloroethane measurement at C=6500 ppm ($6,5\times10^{-3}$) from Fig. 7 was performed. With $\sigma_y(\tau)$ measured as $1,17\times10^{-9}$/s at the oscillator frequency f_0=700 MHz (7×10^8 Hz), for the sensor resolution over the measurement time τ=1 s we obtain R=33,3 ppb (parts per billion). This means that this sensor system can detect changes in the dichloroethane concentration as small as 33 ppb. Such high sensor resolutions are extremely difficult to achieve with other sensor technologies. They are attributed to the fact that RSAW/STW resonant sensors retain excellent resonance characteristics, low loss, high Q and low flicker phase noise when coated with solid and semisolid chemo sensitive polymer films at optimum thickness.

The data in Tables 6 and 7 represent the detection limits of the styrene coated RSAW and STW sensors, respectively, during the measurements on the 4 gas-phase analytes used in this work. The $\sigma_y(\tau)$ values for the sensor oscillators were measured as $5,6\times10^{-9}$/s and $3,04\times10^{-9}$/s, respectively. The "Resolution factor" in Table 7 indicates the resolution improvement of the STW mode versus its RSAW counterpart in these measurements. In

Compound/ Concentration	Tetrachloroethylene (630 ppm)	Dichloroethane (1550 ppm)	Ethylacetate (4190 ppm)	Xylene (330 ppm)
Acoustic mode/f_0	RSAW/433 MHz			
Dep. time/polymer	20 s/styrene			
Sensor signal Δf	47 KHz (108 ppm)	87 KHz (201 ppm)	53 KHz (122 ppm)	33 KHz (76 ppm)
$\sigma_y(1s)$	$5,6\times10^{-9}$/s			
Resolution R	32,5 ppb	43,2 ppb	192 ppb	24 ppb

Table 6. Detection limits of the styrene coated RSAW sensors at the 4 gas-phase analytes

Compound/ Concentration	Tetra-chloro ethylene (630 ppm)	Di-chloro ethane (1550 ppm)	Ethyl acetate (4190 ppm)	Xylene (330 ppm)
Acoustic mode/f_o	STW/700 MHz			
Dep. time/polymer	15 s/styrene			
Sensor signal Δf	178 KHz (254 ppm)	144 KHz (206 ppm)	133 KHz (190 ppm)	98 KHz (140 ppm)
$\sigma_y(1s)$	$3{,}04 \times 10^{-9}$/s			
Resolution R	7,5 ppb	22,9 ppb	67 ppb	7,2 ppb
Resolution factor (STW/RSAW)$^{-1}$	4,3	1,9	2,9	3,3

Table 7. Detection limits of the styrene coated STW sensors at the 4 gas-phase analytes

Table 7 this resolution improvement is by a factor of 1,9 to 4,3. We attribute this advantage of the STW mode to the following factors:

- *The styrene coated STW sensors feature lower flicker noise values than their RSAW counterparts. This is evident from the $\sigma_y(\tau)$ measurement ($5{,}6 \times 10^{-9}$/s vs. $3{,}04 \times 10^{-9}$/s for the RSAW and STW sensors, respectively). This suggests that the STW devices tolerate the styrene film better than the RSAW ones;*
- *The STW mode features better relative gas sensitivity than its RSAW counterpart at the same acoustic wave length and type of semisolid polymer film (styrene) as evident from Table 4.*

8.1 External factors that may degrade measurement resolution in practical RSAW/STW sensor systems

The data in Tables 6 and 7 represent the physical detection limits that can be achieved with practical RSAW/STW based sensor systems. If all other factors that may have a negative effect on the measurements are excluded then the only limiting quantity to the measurement resolution remains the electrical flicker phase noise of the sensor oscillator represented by its Allan's variation $\sigma_y(\tau)$. Unfortunately, in practical sensor systems there are several external factors that may seriously degrade sensor resolution and therefore care should be taken to eliminate them or to reduce their influence to acceptable values. The closer the system is brought to its $\sigma_y(\tau)$ limit, the better it has been designed.

The major factors that may degrade sensor resolution in practical sensor systems will be briefly discussed next.

a. *Gas flow homogeneity.*
    This is one of the key disturbances that may seriously degrade sensor noise and should be eliminated first. If the gas flow applied to the sensor head from Fig. 6 is not homogeneous then the sensor devices will sense a variable gas concentration during the probe-flush cycles. Since the measurement cycle is much longer than the time over which inhomogeneities occur, amplitude fluctuation of the sensor signal at or close to equilibrium will occur. This situation is illustrated in Fig. 9 showing results from a tetrachloroethilene measurement with the setup from Fig. 6 when the integrators are removed. This causes a serious turbulence of the gas flow in the system which results in strong noise levels on top of the sensor signals. Even a periodicity on the noise signal is observed which is attributed to the pump rotation. Fortunately, gas flow

inhomogeneities are easily eliminated. After the integrators are placed back at the air inlet and outlet a noise free measurement similar to the one from Fig. 7 is obtained.

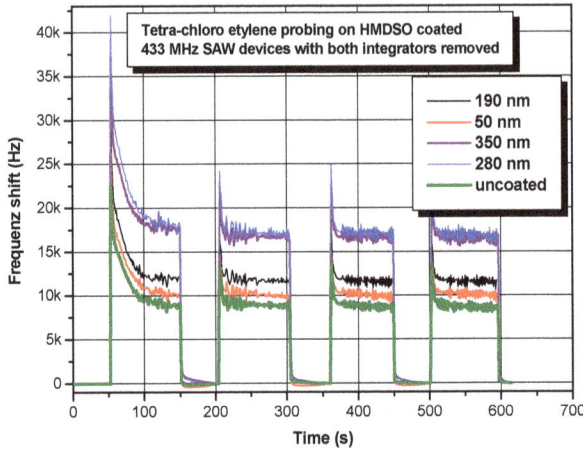

Fig. 9. Tetrachloroethilene probing with 5 HMDSO coated RSAW sensors with the integrators from the setup in Fig. 6 removed

b. *Gas saturation of the sensing films.*

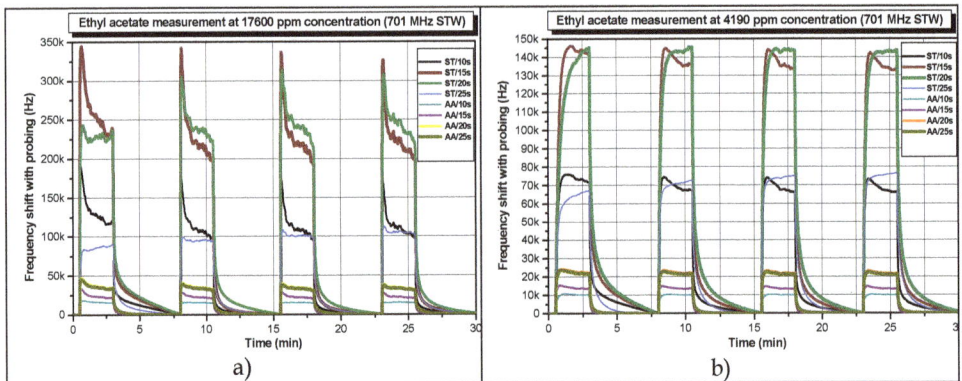

Fig. 10. Ethylacetate probing on ST and AA coated STW sensors at a) 17600 and b) 4190 ppm vapour concentration

Saturation of the sensing films occurs when gas concentrations become so high that sorption limit of the layer is reached. A situation like this during an ethylacetate measurement at 17600 ppm concentration is illustrated in Fig. 10 a). In this case strong peaks of overpressure in the analyte container as well as noise and distortion on the sensor signals are observed as a result of film saturation. When the gas concentration is reduced by a factor of 4 to 4190 ppm, the sensor signals become much more uniform and noise fluctuations are greatly reduced. The behaviour in Fig. 10 a) has the following explanation. When the films become saturated dynamic equilibrium is disturbed and a very turbulent adsorption-desorbtion process takes place, the films get lighter and heavier in a stochastic way and this generates noise on top of the sensor signals. Once the gas concentration is reduced, equilibrium occurs and the adsorption-desorbtion process returns back to normal (see Fig. 10 b)).

c. *Adsorption-desorbtion noise (ADN) in soft film coated RSAW sensors operated far below gas saturation.*

When RSAW sensors are coated with soft polymer films featuring profound bulk sorption these films can accommodate large amounts of gas without being driven into saturation. The larger the amount of adsorbed gas, the more turbulent the adsorption-desorbtion process at equilibrium becomes even though the film is operated far below its saturation limit. This results in ADN evident in Fig. 11 which presents results from tetrachloroethilene probing on RSAW/STW sensors coated with the soft PIB film. ADN is visible on top of all sensor signals regardless of how strong they are. It should be noted that ADN levels depend entirely on the sorption characteristics of the soft polymer films and this makes elimination of this type of noise very difficult. In practical sensor systems one should either cope with ADN or, in critical situations, a different type of polymer film with lower ADN should be used. For example, the magnitude of the sensor signals in Fig. 11 is comparable with those in Fig. 7 where ST was used and no measurable ADN levels were observed. Therefore, results as the ones from the PIB film coated sensors from Fig. 11 but free of ADN could readily be obtained with ST coated sensors.

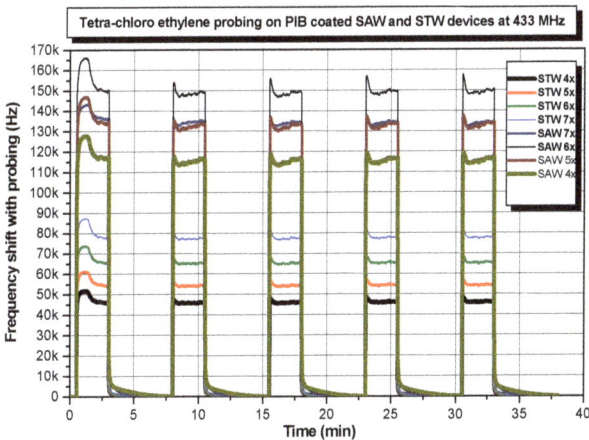

Fig. 11. Tetrachloroethilene probing on RSAW/STW sensors coated with the soft PIB polymer film and operated far below gas saturation

9. Corrosion proof RSAW resonant sensors using gold electrode structure

Typically, RSAW/STW sensors are fabricated with Al electrode structure using a well established photolithographic process. Al metallization is cheap and provides excellent electrical performance in almost all SAW devices fabricated to date. However, a major problem occurs if SAW devices with Al metallization are coated with sensing layers and used as gas sensors. Very often, the chemical gas-phase compounds to be detected form corrosive bases and alkalis with the ambient humidity and attack the Al electrode structure by entering into a chemical reaction with the Al film. The problem is further aggravated by the presence of the sensing layer which, by absorbing large amounts of gas, greatly increases the concentration of the aggressive analyte that comes in contact with the Al electrodes. As a result of corrosion, the sensors suffer irreversible performance degradation, provide inconsistent data and even dye within a limited number of measurement cycles. The solution to that problem is the implementation of SAW devices with corrosion proof gold (Au) metallization that can successfully stand severe corrosion attacks by chemically reactive substances. The design of RSAW sensor resonators with Au electrode structure is not so straight forward as with Al metallization. Due to the fact that Au has a 7 times higher density than Al and is much softer several side effects, such as excitation of strong SSBW modes and transverse waveguide modes occur that can cause serious loss and Q-degradation as well as distorted characteristics at the main resonance. However, by careful selection of the Au thickness and choosing proper device geometry, these side effects can be kept under control and very good resonance characteristics appropriate for gas sensor applications can be achieved [22], [23]. In the next sections we will discuss the performance of RSAW sensors with Au electrode structure intended for operation as gas sensors in highly reactive chemical environments. These sensors were designed to replace their predecessors using the problematic Al electrode structure in a practical sensor system operating at 433 MHz.

9.1 Electrical performance comparison of Au vs. Al RSAW sensor resonators

Generally two types of resonator devices are used in a practical resonator system – two-port resonators (TPR) featuring a single-mode resonance and coupled resonator filters (CRF) that have a two-pole resonance achieved with a coupling grating in the centre of the resonant cavity. As evident from Fig. 12 a) and b), the CRF devices have twice the phase slope of a TPR in their filter pass bands and generally provide better stabilization of the sensor oscillator than the TPR, especially in measurements at high gas concentrations. In a real-world sensor system, the sensor oscillator is designed to operate on the right CRF resonant mode since the left one vanishes when the device is coated with a polymer film [6]. Typically, the sensor oscillator provides stable oscillation on the right mode and never jumps onto the left one since a 180 deg. phase reversal makes oscillation impossible at that mode (see Fig. 12 b)). The frequency and group delay responses of the Al RSAW devices previously used in the sensor system are shown in Fig. 12 c) and d) while a) and b) represent the electrical performance of their Au substitutes designed in [23]. The insertion loss, Q and group delay data from these devices at resonance are compared in Table 8. Es evident from that data and also from Fig. 12 the Al and Au devices feature very similar frequency responses, insertion loss and loaded Q values and the replacement of the Al devices with their Au successors was made without any changes or adjustments of the sensor circuitry. The Au devices are slightly inferior to the Al ones in terms of loss and loaded Q. This is attributed to the loss of energy as a result of the heavy Au loading on the quartz surface.

Parameter / Device (433 MHz)	Al-CRF	Au-CRF	Al-TPR	Au-TPR
Unmatched insertion loss [dB]	6.5	10.5	6.5	7.5
Group delay (50Ω load) [μs]	4.01	3.44	3.94	2.83
Loaded Q	5450	4430	5350	3890
Unloaded Q	10400	6190	10160	6730

Table 8. Comparison of the insertion loss, Q and group delay data at resonance of the uncoated RSAW sensors with Al and Au metallization characterized in Fig. 12

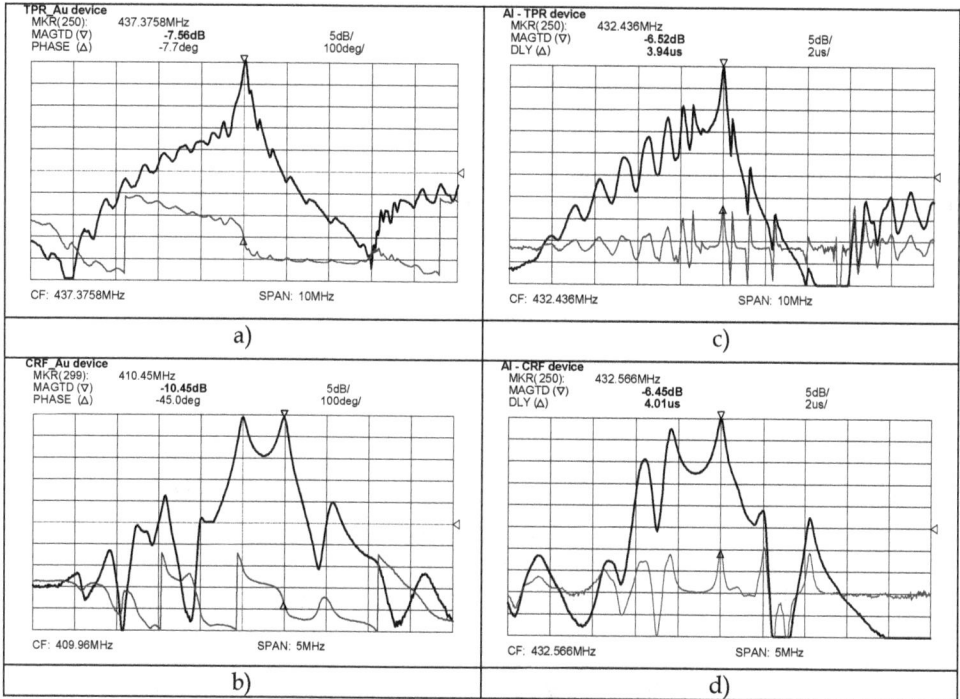

Fig. 12. Frequency (upper curves) and phase/group delay (lower curves) responses of RSAW single mode two-port resonators (upper row) and coupled resonator filter (lower row) using (a) and (b) Au and (c) and (d) Al electrode structure

9.2 Chemo sensitive polymer films and deposition methods used

This section compares the new Au vs. the old Al sensor devices in their sensor characteristics to find out if the replacement causes any performance degradation of the sensor system the Au devices were expected to operate in. To check the sensor performance we again used two types of polymer films: (A) solid Parylene C to simulate coating behaviour with solid and semisolid films as described in Section 6 and (B) a soft polymer called poly[chlorotrifluoroethylene-co-vinylidene fluoride] (PCFV) to test sensor performance at high gas concentrations. Since Parylene C coating was discussed in Section 6 already, here we will briefly discuss a relatively novel soft polymer deposition method, which is called electro spray method and is described in [24] in detail. We applied it successfully to all 4 devices from Fig. 12 to obtain very uniform high-quality soft PCFV films for reproducible sensor performance. According to this method, the holder with the SAW devices mounted on it, spins in a cloud of very small liquid-phase polymer droplets coming out from a narrow capillary tube and directed by an electrostatic field towards the sensor surface. The droplets settle onto the device surface, stick together and form a uniform film. Its thickness depends on the deposition time. Since the SAW device loss increases with film thickness, a certain insertion loss value, as necessary for optimum sensor performance, can be obtained simply by adjusting the deposition time. Except for excellent control over film thickness and uniformity [24], major advantage of this polymer coating method is that, even at 433 MHz, the droplets are much smaller than the acoustic wavelength of about 7 μm at this frequency. Because of the small droplet size, the electro spray films cause much less propagation loss for the SAW, compared to films of the same type and thickness, deposited in an older airbrush coating technique. In contrast to electro spray films, airbrush coatings have a rough textured structure and the droplet size typically exceeds the acoustic wavelength. The two optical microscope pictures in Fig. 13 a) and b) compare the film structures of PCFV deposited with the two methods on identical 433 MHz RSAW devices. Fig. 13 a) shows part of the electrode structure and bus bars at the centre of the electro spray coated SAW device. For better visibility of the textured film structure obtained in an airbrush technique, the picture in Fig. 13 b) has been taken with a slightly higher magnification and shows part of the reflector with some free substrate area on which the large drops are clearly visible.

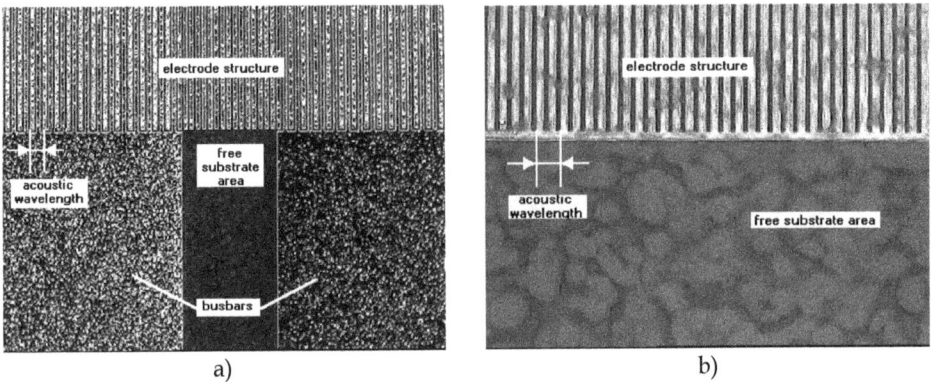

a)   b)

Fig. 13. Comparison of two identical RSAW devices PCFV coated with (a) the electrospray and (b) the airbrush method

9.3 Polymer coating behaviour of Au vs. Al sensor resonators

The three dimensional data plots in Fig. 14 compare the Parylene C coating behaviour of the Au and Al CRF devices from Fig. 12 b) and d) accordingly with a film thickness varying from 0 to 700 nm. In both cases the magnitude of the left longitudinal mode decreases with thickness until it disappears completely at about 300 nm for the Au and 450 nm for the Al device. Above these thickness ranges both devices demonstrate a smooth and well behaved single-mode resonance. In the 300 to 500 nm range the Au CRF reaches its highest mass sensitivity, accompanied with a gradual increase in insertion loss while the loss of the Al device decreases more rapidly. The critical for the system operation loss value of 20 dB is reached at 370 nm vs. 450 nm for the Au and Al devices, respectively.

An identical comparative Parylene C coating test, (not shown here), was performed also with the Au vs. Al TPR devices from Fig. 12 a) and c). In these tests again the Au devices were found to tolerate the solid Parylene C better than their Al counterparts. The increase in device insertion loss with Parylene C thickness for all four tested devices from Fig. 12 is shown in Fig. 15. At film thicknesses up to about 180 nm, all devices yield the same loss increase. Above 200 nm the loss behaviour starts to diverge. The two Al devices keep the same insertion loss up to about 550 nm thickness. Above 200 nm thickness, the loss of the Au devices increases at a much lower rate indicating that these devices can tolerate up to

40% thicker solid films than their Al counterparts for the same amount of loss increase. Finally, Fig. 16 compares the frequency sensitivities of the four tested devices with Parylene C film thickness which is also an indication of their gas probing sensitivity with solid films. Up to about 300 nm thickness, the sensitivity slope of the devices is nearly identical with a small advantage of the Al TPR device, followed by the Al CRF. Above 300 nm, the sensitivity slope of the Al TPR device increases but in view of its strong loss degradation, its sensitivity advantage gets lost. The other three devices keep a nearly constant sensitivity slope up to 500 nm thickness. Below the practical 370 nm thickness at which the critical for this particular system 20 dB of loss is reached for the Al devices, the sensitivity of all four devices differs by less than 20%. This difference is insignificant for practical sensor systems.

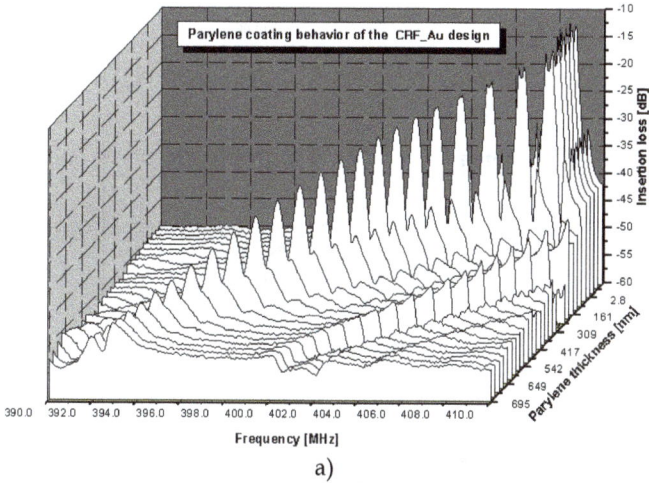

a)

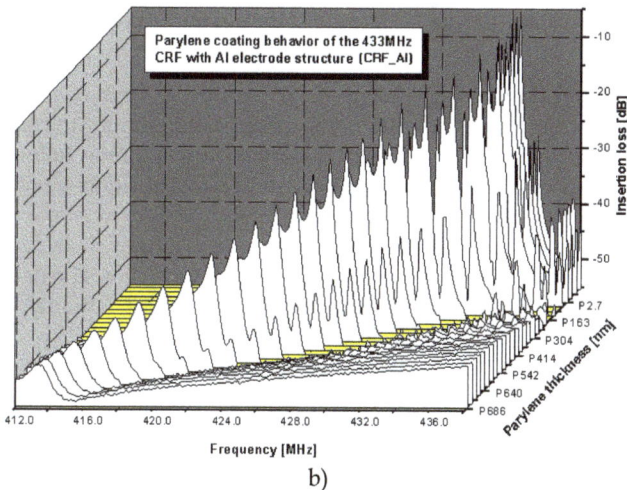

b)

Fig. 14. Parylene C coating behaviour of the RSAW CRF devices from Fig. 12 using (a) Au and (b) Al metallization

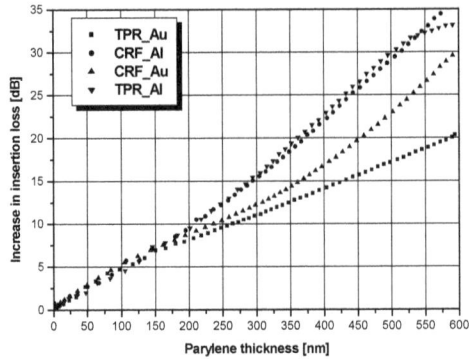

Fig. 15. Insertion loss behaviour of the Au and Al devices from Fig. 12 vs. Parylene C thickness

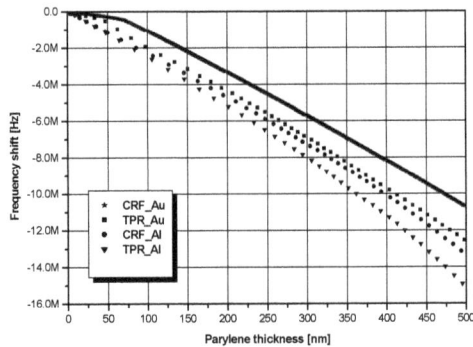

Fig. 16. Frequency (mass) sensitivity behaviour of the Au and Al devices from Fig. 12 vs. Parylene C thickness

The soft PCFV polymer coating experiments were performed on the two TPR devices from Fig. 12 a) and c) since they have about the same amount of loss prior to coating, as required by the sensor system. Since the devices are mounted on a spinning holder, monitoring of their electrical performance in the process of electro spray deposition is not possible. That is why we recorded the frequency and phase responses of each device prior to and after the deposition to obtain the frequency shift and loss increase as a function of the deposition time which we used as a measure for the thickness of the soft PCFV film. The data plots in Fig. 17 illustrate the PCFV coating behaviour of an Au TPR device in a 7,5 minutes deposition time. As a result of film loading the device loss increases by about 8 dB to 17,9 dB while its frequency shifts down by 3 MHz. The coated device on the left retains a well behaved single-mode resonance with a smooth phase response in the resonance region. The loss increase vs. thickness proportional deposition time for the Al and Au devices is shown in Fig. 18. For the Au device this dependence is linear while the Al device shows a small loss increase up to about 10 min. of deposition time and after that its loss degrades very rapidly. We attribute this behaviour again to the difference in Au vs. Al densities. Once the Au device has been optimized for operation under the heavy Au film load, it tolerates much better the polymer film which is much lighter than Au. Finally, Fig. 19 compares the frequency sensitivity of both devices to increased PCFV thickness. The curve for the Au device is steeper meaning that its mass sensitivity is higher than its Al counterpart.

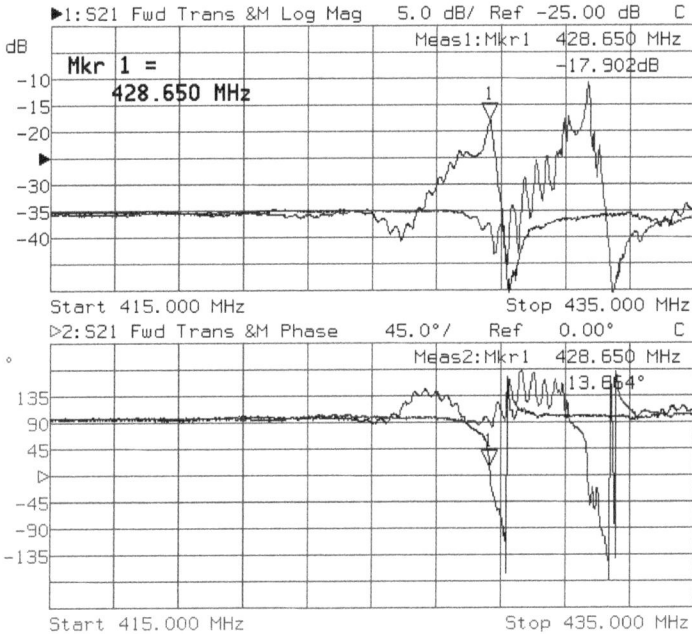

Fig. 17. Frequency responses (upper plots) and phase responses (lower plots) of an Au TPR device before (data on the right) and after (data on the left) 7.5 minutes of PCFV deposition using the electro spray method

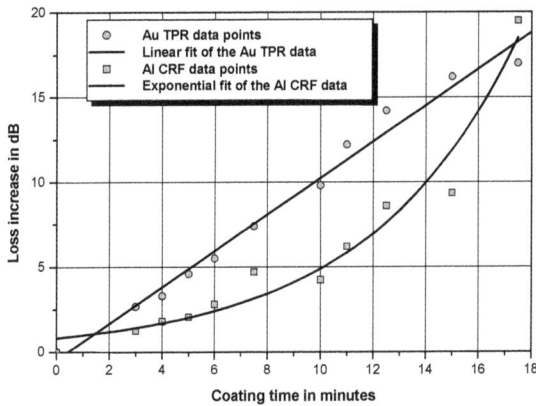

Fig. 18. Loss increase vs. deposition time for Au and Al devices electro spray coated with soft PCFV film

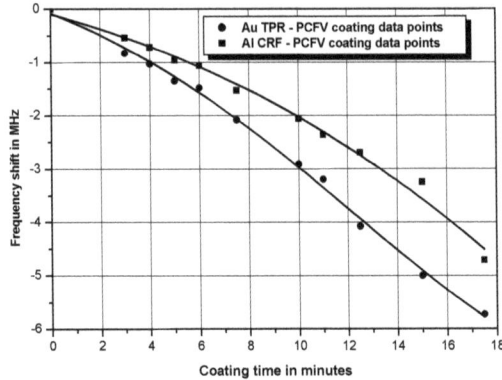

Fig. 19. Frequency downshift vs. deposition time for Au and Al devices electro spray coated with soft PCFV film

9.4 Gas probing behaviour of PCFV coated Au and Al TPR and CRF RSAW sensors

To compare the gas sensitivities of Au vs. Al sensors, pairs of devices of the same type according to Fig. 12 were PCFV coated in the same electro spray deposition method and probed with cooling agent, octane and tetrachloroethylene at different concentrations. Cooling agent and octane were among those gases that the sensors were intended to operate with in a specific application. After coating, the Au/Al pairs were selected to have nearly the same loss increase as a result of PCFV deposition to simulate identical mass loading. The gas probing results for 4 different loss increase values (thicknesses) are summarized in Table 9. As expected from the mass sensitivity data in Fig. 19, the Au devices demonstrate higher gas sensitivity than their Al counterparts. Another important conclusion evident from Table 9 is that soft polymer coating also requires an optimum film thickness for maximum gas sensitivity. In this case the optimum PCFV thickness is achieved when both the Au and Al devices are coated to about 6 dB loss increase, (numbers in bold in Table 9).

Probing gas → Loss increase values for each device pair ↓	Cooling agent 40000 ppm concentration		Octane 1100 ppm concentration		Tetrachloroethylene 1000 ppm concentration	
PCFV coated device→	Al CRF	Au TPR	Al CRF	Au TPR	Al CRF	Au TPR
Al_8/Au_8,7 dB	12,5 kHz	15 kHz	13 kHz	15 kHz	23 kHz	25 kHz
Al_6,2/Au_5,5 dB	**11 kHz**	**21 kHz**	**14 kHz**	**19 kHz**	**26 kHz**	**28 kHz**
Al_4,6/Au_4,6 dB	8 kHz	11 kHz	9 kHz	18 kHz	17 kHz	27 kHz
Al_3,5/Au_3,3 dB	6 kHz	8 kHz	11 kHz	13 kHz	17 kHz	17 kHz

Table 9. Summary of the gas probing performance of Au vs. Al device pairs PCFV coated to nearly identical loss increase values in the same electro spray deposition process

10. Summary and conclusions

This chapter has highlighted important practical aspects for the design and operation of chemical gas detection systems using STW and RSAW resonant devices. It has been shown that both acoustic wave modes provide excellent gas sensitivity and low detection limits, down to a few ppb when coated with solid, semisolid and soft polymer sensing layers. Furthermore, the RSAW and STW modes do not only compete but rather complement each other in different measurement tasks. The STW mode operates better with solid and semisolid sensing films featuring surface sorption and is better suited for high-resolution measurements at low gas concentrations (<1%) while the RSAW mode tolerates much better thick soft sensing layers with profound bulk sorption that operate better at high gas concentrations (>1%). Carefully designed RSAW sensors with Au metallization provide an excellent corrosion proof substitute of their Al counterparts when operated in highly reactive gas-phase environments, thereby greatly increasing system reliability and measurement reproducibility over time and a large number of measurement cycles. All gas sensors, regardless of acoustic wave mode, design, metallization and type of sensing polymer requires a careful thickness optimization to provide highest gas sensitivity, maximum dynamic range and lowest detection limit.

11. Acknowledgments

The author wishes to gratefully acknowledge Dr. E. Radeva from the Georgi Nadjakov Institute of Solid State Physics, Bulgarian Academy of Sciences in Sofia, Bulgaria for expert preparation of the HMDSO films used in this study as well Professor Shigeru Kurosawa and his research associates from the National Institute of Materials Chemistry in Tsukuba, Japan for the deposition of the semisolid ST and AA films. Special thanks are directed to Dr. Michael Rapp and his research team at the Research Centre Karlsruhe in Germany for the opportunity to perform a substantial part of this work at those laboratories.

12. References

[1] R. M. White, Acoustic sensors for physical, chemical and biochemical applications, *Proc. IEEE 1998 International Symposium on Frequency Control*, pp. 587-594.

[2] R. M. White, Surface acoustic wave sensors, *Proc. IEEE 1985 Ultrasonics Symposium*, pp. 490-494.

[3] H. Wohltjen, Mechanism of operation and design considerations for surface acoustic wave device vapour sensors, *Sens. Act.*, vol. 5, p. 307, 1984.

[4] R. Chung, R. A. McGill and P. Matthews, "Phase noise characterization of polymer coated SAW-gas sensors: Implications for the performance of an oscillator circuit", *Proc. 1997 IEEE Int. Freq. Control Symp.*, pp. 169-174.

[5] S. J. Martin, G. C. Frye, J. J. Spates, and M. A. Butler, Gas sensing with acoustic devices, *Proc. IEEE 1996 Ultrasonics Symposium*, pp. 423-434.

[6] M. Rapp, J. Reibel, S. Stier, A. Voigt, and J. Bahlo, SAGAS: Gas analyzing sensor systems based on surface acoustic wave devices—An issue of commercialization of SAW sensor technology, in *Proc. IEEE Int. Freq. Contr. Symp.*, 1997, pp. 129–132.

[7] E. J. Staples, Dioxin/furan detection and analysis using a SAW based electronic nose, *Proc. IEEE 1998 Ultrasonics Symposium*, pp. 521-524.

[8] E. J. Staples, T. Matsuda, and S Viswanathan, Real Time Environmental Screening of Air, Water and Soil Matrices Using a Novel Field Portable GC/SAW System, Environmental Strategies for for the 21st Century, *Asia Pacific Conference*, pp. 8-10 April 1998.

[9] United States Patent No. 5,289,715, Vapour Detection Apparatus and Method Using an Acoustic Interferometer.

[9] G. C. Frye, S. J. Martin, R. W. Cerenosek, K. B. Pfeifer, and J. S. Anderson, Portable acoustic wave sensor systems, in *Proc. IEEE Ultrason. Symp.*, 1991, pp. 311–316.

[10] G. C. Frye and S. J. Martin, Dual output acoustic wave sensor for molecular identification, in *Proc. IEEE Transducers*, 1991, pp. 566-569.

[11] T.Wessa, S. Kueppers, G. Mann, M. Rapp, and J. Reibel, "On-line monitoring of process HPLC by sensors," *Organ. Process Res. Develop.*, vol. 4, no. 2, pp. 102–106, 2000.

[12] D. A. Howe, D.W. Allan, and J. A. Barnes, Properties of Signal Sources and Measurement Methods, *U.S. Dept. Commerce, NIST Tech. Note 1337*, 1990.

[13] D. S. Ballantine, Jr. and H.Wohltjen, Elastic properties of thin polymer films investigated with surface acoustic wave devices, in *Chemical Sensors and Microinstrumentation*, R.W. Murry, R. E. Dessy,W. R. Heinemann, J. Janata, and W. R. Seitz, Eds. Washington, DC: Amer. Chem. Soc., 1989, pp. 222–236.

[14] J. W. Grate and E. T. Zellers, The fractional free volume of the sorbed vapor in modeling the viscoelastic contribution to polymer-coated surface acoustic wave vapor sensor responses, *Anal. Chem.*, vol. 72, no. 12, pp. 2861–2868, July 1, 2000.

[15] F. L. Dickert and A. Haunschild, Sensor materials for solvent vapor detection-donor-acceptor and host-guest interactions, *Adv. Mater.*, vol. 5, no. 12, pp. 887–895, Dec. 1993.

[16] N. Barie, M. Rapp, and H. J. Ache, UV crosslinked polysiloxanes as new coating materials for SAW devices with high long-term stability, *Sens. Actuators B, Chem.*, vol. B46, pp. 97–103, 1998.

[17] H. Yasuda, *Plasma Polymerization*. New York: Academic, 1985, p. 11 and 294.

[18] C. Hamann and G. Kampfrath, Glow discharge polymeric film: Preparation, structure, properties and applications, *Vacuum*, vol. 34, pp. 1053–1059, 1984.

[19] E. I. Radeva, Thin Plasma-Polymerized Layers of Hexamethyldisiloxane for Acoustoelectronic Humidity Sensors, *Sensors and Actuators, B*, vo1.44, pp.275-278, 1997.

[20] G. Kovach, G. W. Lubking, M. J. Vellekoop, and A. Venema, Love waves for (bio) chemical sensing in liquids, in *Proc. IEEE Ultrason. Symp.*, 1992, pp. 281–285.

[21] I. D. Avramov, M. Rapp, A. Voigt, U. Stahl and M. Dirschka, Comparative studies on polymer coated SAW and STW resonators for chemical gas sensing applications, *Proc. 2000 IEEE International Frequency Control Symposium*, pp. 58-65.

[22] I. D. Avramov, A. Voigt and M. Rapp, "Rayleigh SAW Resonators Using Gold Electrode Structure for Gas Sensor Applications in Chemically Reactive Environments", *Electronics Letters*, 31-st March 2005, Vol. 41, No. 7, pp. 450-452.

[23] I. D. Avramov, Design of Rayleigh SAW resonators for applications as gas sensors in highly reactive chemical environments, *Proc. 2006 IEEE International Frequency Control Symposium*, 5-7 June 2006, Miami, Florida, USA, pp. 381-388.

[24] F. Bender, L. Waechter, A. Voigt and M. Rapp, Deposition of High-Quality Coatings on SAW Sensors using Electrospray, *IEEE Sensors Conference*, 2003, Paper ID 1022 (on CD).

12

Surface Acoustic Wave Devices for Harsh Environment

Cinzia Caliendo

Istituto dei Sistemi Complessi, ISC-CNR, Area della Ricerca Roma 2, Rome
Italy

1. Introduction

There is an increasing demand of electronic components for aerospace, aircraft industries, sensors, automotive, chemical and material processing applications, to name just a few, able to operate reliably and for long time at high-temperature. Measurements reliability requires the electronic components to be placed directly inside the extreme environment, and to withstand temperatures of several centigrade degrees with lifetimes of several hours. The device mounting and packaging, but first of all the device materials must be stable with the working temperature, otherwise temperature-induced stress may result in device's failures. Electroacoustic devices based on surface and bulk acoustic wave (SAW and BAW) technology must satisfy the requirements of low cost, high frequency, high-Q, low loss, large piezoelectric coupling and zero temperature coefficient of delay (TCD) to be key devices in the communication and sensor fields. The temperature stability of the piezoelectric crystal is an essential characteristic because of its direct link with the temperature sensitivity of the electroacoustic device operation frequency. The high operation frequency is an essential characteristic for SAW and BAW devices to be used in mobile phones, cordless headphones, alarm and security systems, military equipment, sensors, etc. The temperature stability and the high operation frequency demands can be met through a proper choice of the piezoelectric substrate crystal cut, new piezoelectric materials and/or multilayer configurations. The use of temperature stable cuts of single crystal bulk piezoelectric materials or temperature compensated multilayers represents two possible solutions to the temperature stability requirement. The use of high-resolution lithography techniques and/or of high SAW velocity materials is required in order to extend the upper limit of the electroacoustic device frequency range. Submicron feature sized interdigital transducers (IDTs) are required to implement GHz range SAW devices on *slow* piezoelectric materials, while micron feature sized IDTs can still be used on *fast* materials, since the SAW device centre frequency, $f = v/\lambda$, depends on both the phase velocity of the propagating medium, v, and on the acoustic wavelength λ, being the IDT's period $p = \lambda/2$. Conventional piezoelectric substrates, such as quartz, lithium niobate (LiNbO$_3$), and lithium tantalate (LiTaO$_3$) crystals, cannot be used above 500°C. Quartz ST cut is a temperature stable material but it shows an alpha-beta transition at 573°C, which causes the loss of piezoelectricity, and results in a non-operable device. SAW devices implemented on LiNbO$_3$ have been studied for a temporary usage at 400°C [1]; however the LiNbO$_3$ acoustic wave properties are highly dependent on temperature since it is a pyroelectric material and has a TCD as high as ~75 ppm/°C. $LiTaO_3$ shows properties, such as a low Curie temperature (607°C), sensitivity to temperature variations (TCD = 22 ppm/°C for the X-112°Y cut) and a strong pyroelectricity, which limit its operation at elevated temperatures. Piezoelectric bulk single crystals such as $GaPO_4$, LGS ($La_5Ga_5SiO_{14}$) and its isomorphs (called LGX family group) substrates are widely investigated for the realization of SAW-based devices able to work at high temperature. LGS belongs to the trigonal class 32 group as quartz but it has no α-β transitions and can operate up to its melting temperature of 1470°. It shows zero or very low TCD cuts with zero power flow angle and higher electromechanical coupling coefficient than that of quartz [2]. Langasite based SAW devices are not suitable for operation in the GHz range as a consequence of their low phase velocity and high acoustic losses (from 1 to 0.01 dB/wavelength [3]). $GaPO_4$ has twice the sensitivity of quartz and many its physical constants are stable up to about 900°C, but the accessible frequencies are limited to values of 1 GHz as a consequence of quite high acoustic losses.

The technology of thin piezoelectric films (such as AlN) offers the opportunity of combining the properties of the substrate with those of the film: thus a composite arrangement of fast materials with opposite sign TCDs and a proper design of the electroacoustic configuration enable achieving a thermally stable SAW device operating in the GHz range.

Aluminium nitride (AlN) is a piezoelectric material that shows interesting properties, such as excellent thermal conductivity (180W/mK), low coefficient of thermal expansion (CTE, 4.1×10^{-6} °C^{-1}), and good resistance to thermal shock and caustic chemicals [4], that make it useful as protective coating and guarantee the stability of the AlN-based devices when they are in contact with extreme environments. It is currently being investigated due to its promising potentialities for high-temperature, high-power, and high-frequency electronics. It has demonstrated to be an ideal candidate for packaging SiC-devices for high-temperature applications [5] thanks to its CTE that closely matches those of Si (3.5×10^{-6} °C^{-1}) and SiC (3.7×10^{-6} °C^{-1}), high electrical resistivity, high mechanical strength, and chemical inertness. Reactively sputtered AlN films have been used as an effective encapsulant for GaN [6] at an annealing temperature of 1100°C substituting the standard dielectric encapsulants, such as SiO_2 and Si_3N_4, that are not viable at so high temperatures. AlN maintains its piezoelectricity up to 1200°C in vacuum and shows very high BAW and SAW velocities (~6000 m/s and 11300 m/s for transversal and longitudinal BAWs propagating along the z direction, 5607 m/s for SAWs propagating in the z plane) that make it the ideal candidate for microwave electroacoustic devices implementation. Furthermore, AlN can be grown in thin film form onto non piezoelectric substrates by techniques as simple as the rf reactive sputtering. Both the structural properties of the substrate and the experimental sputtering parameters (such as the reactive gas flow rate, the partial pressure of reactive and inert gasses, the substrate temperature, the rf power, and the substrate-target distance) affect the morphological and structural properties of the sputtered thin films. The requirements for a suitable substrate include also a thermal coefficient-of-expansion compatible with that of the film, high-temperature stability, machinability, good adherence of the AlN film: among the available substrates, silicon, platinum and sapphire satisfy these requirements. Al_2O_3, substrates have a wide range of industrial applications as structural ceramic and optical materials. Al_2O_3 is extensively used as a high temperature, corrosion resistant refractory material due to its hardness, chemical durability, abrasive resistance, mechanical strength, and good electrical insulation. Al_2O_3 shows good thermal conductivity (24 W/mK), high SAW velocity (in the range 5555 to 5706 m/s in the c-plane), positive TCD (~70 ppm/°C) and a CTE that closely matches that of AlN. The AlN/Al_2O_3 –based multilayers can withstand temperatures up to

900°C, thus allowing the realization of high frequency, temperature compensated dispersive electroacoustic devices for high temperature applications. Electroacoustic devices implemented on Si substrates offer the opportunity to integrate the device with the surrounding electronic circuitry on the same chip. Moreover, the opposite TCD of Si (~30 ppm/°C) and AlN (~ -30 ppm/°C) allows the realisation of zero-temperature-coefficient acoustic devices at the proper film thickness to be used as sensors and actuators where low loss, low thermal drift, high sensitivity and high signal-to-noise ratio are demanded [7, 8]. Pt is the material of choice for metallic components that have to withstand oxidation, thank to its high temperature coefficient of resistance: Pt can be grown in thin film form and both the IDTs and ground electrodes can be easily defined by lift off technique.

In the present chapter the sustainability of Pt and AlN films on sapphire and Si substrates for high temperature applications is assessed.

2. AlN-based SAW devices

Bulk piezoelectric single crystals, such as LGS and GaPO$_4$, can be used for the implementation of non dispersive SAW devices, such as delay lines, filters and resonators, and the SAW propagation characteristics, such as phase velocity, electroacoustic coupling efficiency K^2 and TCD, depend on the crystal cut and SAW propagation direction, as well as on the geometry of the IDTs . The SAW propagation is excited by IDTs located at the free surface of the piezoelectric substrate and directly exposed to the surrounding environment, as shown in figure 1.

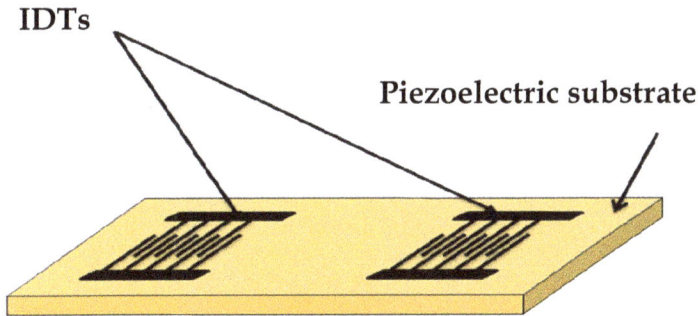

Fig. 1. SAW delay line on a piezoelectric substrate

AlN can be grown in thin film form onto non piezoelectric substrates, such as silicon or sapphire, thus allowing the realization of dispersive electroacoustic devices. Moreover, if the AlN film is sandwiched between the IDTs and the ground electrode, four piezoelectric coupling configurations can be obtained by placing the IDTs at the substrate/film interface or at the film surface, with and without the floating electrode opposite the IDTs. These four structures will be mentioned hereafter as substrate/film/IDT (SFT), substrate/IDT/film/metal (STFM), substrate/IDT/film (STF) and substrate/metal/film/IDT (SMFT), respectively. Figure 2a shows the top view of a dispersive SAW delay line, and figure 2b shows the cross sections of the four coupling configurations.

When the IDTs are located at the substrate/film interface, the piezoelectric film plays the role of both the acoustic wave transductor and protective layer of the underlying IDTs.

These four configurations show frequency dispersive SAW propagation characteristics, that are no longer solely determined by the geometry of the IDTs, the crystals cut, and the SAW propagation direction, but also by the film thickness and the electrical boundary conditions. For SAW propagating along layered structures, the achievable K^2 value is sometimes larger than that of the individual piezoelectric materials; it is frequency dispersive and depends on the type and orientation of the piezoelectric material, and it is drastically affected by the location of the IDTs and counter electrode with respect to the piezolectric layer. As an example, figures 3a and 3b show the K^2 vs film thickness to wavelength ratio, h/λ, for SAW propagation along zx- and zy-Al_2O_3/AlN for the four coupling configurations: the highest K^2 values obtainable are about 0.50 and 0.67% for STF x and y propagation (at $h/\lambda \sim 0.65$ and 0.60), 0.49 and 0.64 % for STFM x and y propagation (at $h/\lambda \sim 0.67$ and 0.62), being 0.3% the AlN K^2 value.

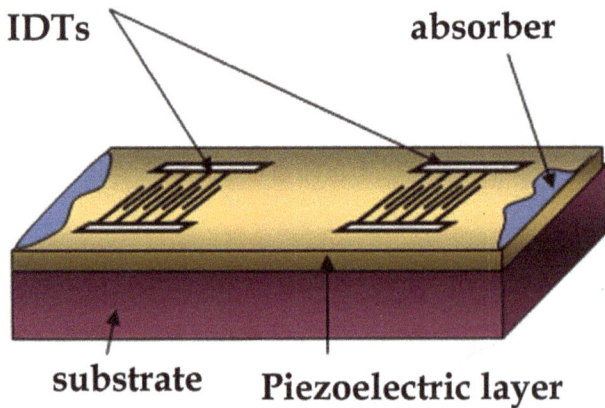

Fig. 2a. Dispersive SAW delay line

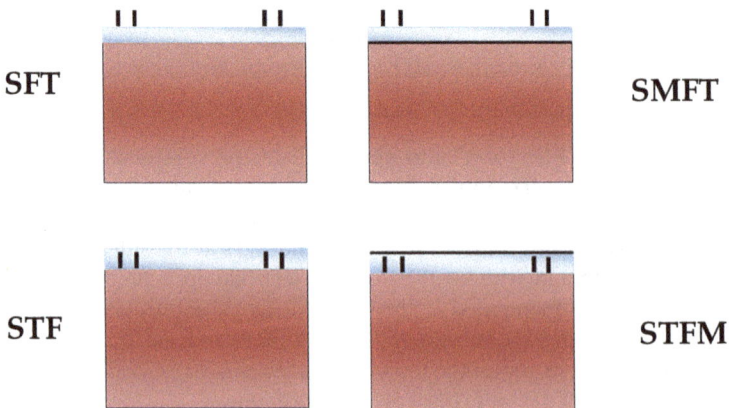

Fig. 2b. The four electroacoustic coupling configurations

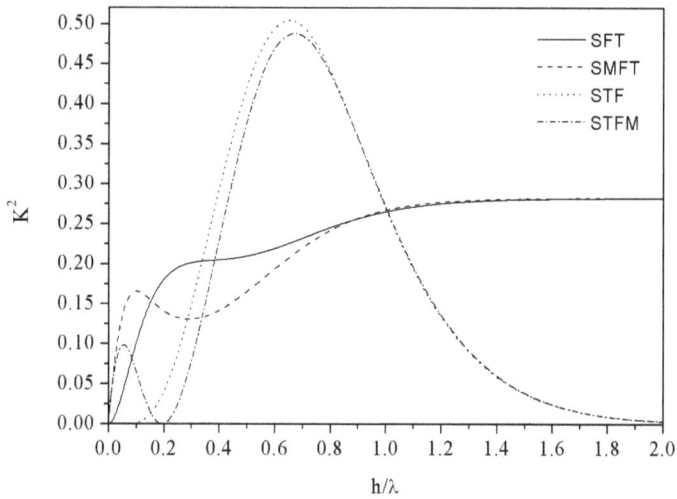

Fig. 3a. The K² vs h/λ for SAW propagating along zx-Al₂O₃/AlN for the four coupling configurations

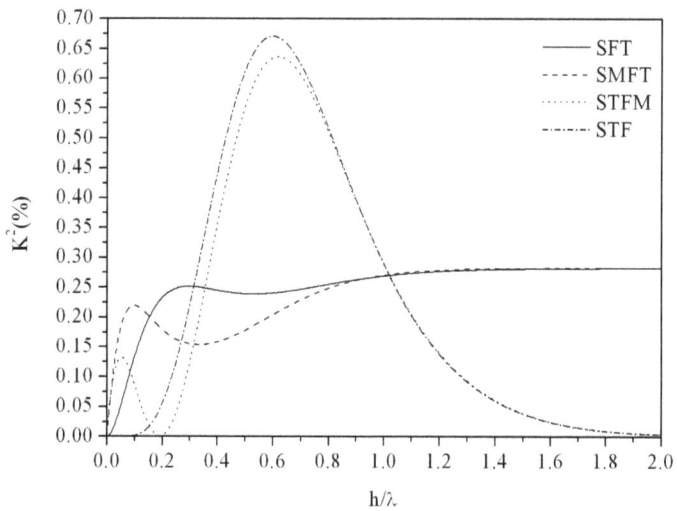

Fig. 3b. K² vs h/λ for SAW propagating along zy-Al₂O₃/AlN for the four coupling structures

Figures 4a and 4b show the K² vs h/λ for SAW propagation along Si/Pt/AlN and Si/AlN/Pt, being the Pt thickness the running parameter.

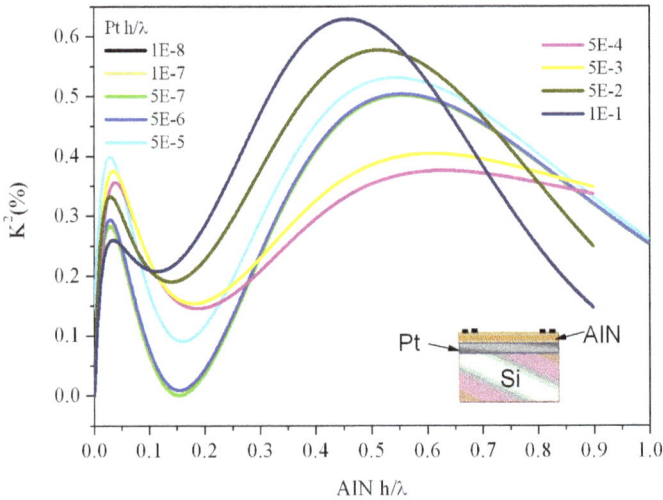

Fig. 4a. The K² vs AlN h/λ for SAW propagation along Si/Pt/AlN, for different Pt thickness values normalized to the acoustic wavelength

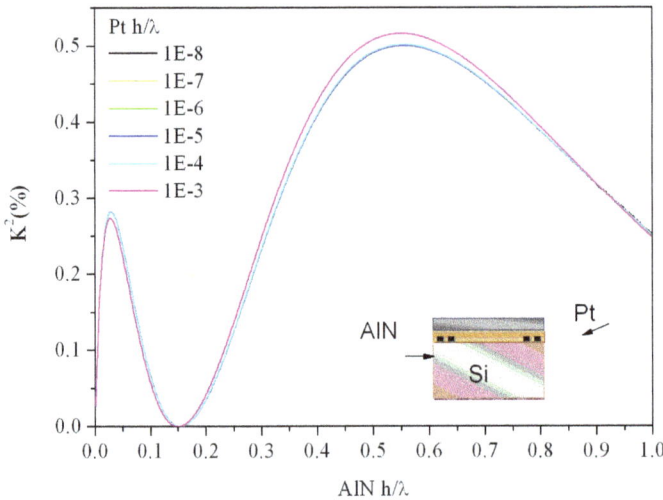

Fig. 4b. The K² vs AlN h/λ for SAW propagation along Si/AlN/Pt, for different Pt thickness values normalized to the acoustic wavelength

In figure 4a the highest K² values obtainable are in the range 0.5 to 0.63% at ~h/λ = 0.55 for Pt h/λ = 10^{-8} to 10^{-1}. In figure 4b the highest K² values obtainable are in the range 0.5 to

0.52% at ~h/λ = 0.555 for Pt h/λ = 10^{-8} to 10^{-3}. The theoretical data shown in figures 3 and 4 have been evaluated using the PC SAW software developed by Mc Gill University [9]. The theoretical K^2 has been approximated as $2 \cdot \dfrac{v_{ph}^{f} - v_{ph}^{m}}{v_{ph}^{f}}$ where v_{ph}^{f} and v_{ph}^{m} are the SAW phase velocities along the free and electrically short-circuited surfaces of the AlN film. The phase velocity v_{ph}^{m} is obtained by the insertion of a perfectly conductive and infinitesimally thin film at the interfaces where the IDTs and the ground plane are located in each of the four coupling structure. The physical data relative to the elastic, piezoelectric and dielectric constants of AlN film are extracted from [10] and [11] and refer to single crystal AlN thin films grown on the basal plane of Al$_2$O$_3$(0001) by metalorganic vapor deposition. The *TCD* of the bulk piezoelectric crystal depends only on the crystal cut and the SAW propagation direction, while that of a layered structures is frequency dispersive. Because the SAW penetration depth inside the propagating medium is about one wavelength, in a layered medium, for h/λ << 1, the most of the SAW energy is confined to the substrate, while, as h increases with respect to λ, more and more of the SAW energy is confined to the film. For small AlN film thickness (h/λ < 1) the *TCD* value of the multilayer corresponds approximately to that of the substrate. With increasing the AlN film thickness respect to the acoustic wavelength (h/λ ≥ 1) the *TCD* reaches the AlN *TCD* value. If the film and the substrate show opposite sign TCD values, there will be a h/λ value at which *TCD* = 0 ppm/°C (the temperature compensated point, TCP). This h/λ value represents the film thickness for which the two opposite sign *TCDs* of the film and of the substrate equilibrate to form a thermally compensated structure. Thus high-frequency, enhanced coupling, and thermally compensated elctroacoustic devices can be designed at the proper AlN films thickness values [12].

3. Materials and methods

Highly c-axis oriented AlN films were grown at 180°C and at 200°C by rf reactive sputtering technique on the polished surface of (0001) oriented single crystal Al$_2$O$_3$ substrate, on SiO$_2$/Si(100) and Pt/SiO$_2$/Si(100) substrates. The AlN deposition process parameters were the following: gas atmosphere of 100% of N$_2$, high purity (99.999%) 4″ diameter Al target disc, RF power 200 watt, background vacuum $5 \cdot 10^{-8}$ Torr and pressure during the deposition process 3×10^{-3} Torr. Before starting the sputtering process, a 30 minute pre-sputtering was performed. The substrate temperature was held at 180 °C during the deposition process. The optimized sputtering parameters ensure AlN films showing a high adhesion to the substrate, c-axis orientation, a columnar growth, smooth surface, and high piezoelectricity; the films are also uniform, stress-free and extremely adhesive to the substrates. The Pt sputtering process parameters were the following: substrate temperature 200°C, gas atmosphere of 100% of Ar, high purity (99.99%) 4″ diameter Pt target disc, RF power 150 watt, background vacuum 10^{-7} Torr and pressure during the deposition process 5×10^{-3} Torr. The deposition process of both Pt and AlN films is performed subsequently without breaking the vacuum in order to avoid any oxidation effects of the layers. Then the obtained samples were heated at 900°C in air at ambient pressure by a quartz tube furnace, for different lengths of time. The *cold* (20°C) sample was abruptly put inside the furnace pre-heated at 900°C and the annealing time was measured from the set temperature was reached; then the sample was removed from the furnace and brought abruptly to room temperature. A temperature ramp of 1°C/s was measured by a thermocouple after the insertion of the *cold* sample inside the furnace. The furnace tube was not hermetically sealed, so ambient air was present during the loading and unloading of samples [14]. The structural properties of annealed Pt and AlN films were investigated by by X-ray diffraction measurements   (XRD) a Seifert XRD 3003P performed on a a Seifert XRD 3003P diffractometer operating in the Bragg-Brentano geometry using Cu-Kα radiation (λ = 1.5418 Å) and the diffracted intensities were collected in θ-2θ scan mode in the range 20° <2θ < 80° with step size 0.04°. The reflection peaks of the diffractograms were compared with the standards of the JPCDS database. The crystallite size D was calculated from the Scherrer formula $D = \dfrac{0.9 \cdot \lambda}{B \cdot \cos \vartheta}$ where λ is the wavelength, B the θ-2θ full width at half maximum (FWHM) of the (0002) peak in rad and θ the Bragg angle. The c and a lattice parameters of the AlN and Pt films were calculated from the angular position of the AlN (002) and Pt (111) diffraction peaks of the θ-2θ scan and compared to the value from ref. 13 for Pt and powder AlN (a = 3.9231Å, c = 4.979Å). Since the electrical resistivity of the thin conducting films influences the device characteristics (such as insertion loss and Joule heating), the electrical resistance and the surface morphology of the outer Pt electrode were investigated at room temperature after each thermal annealing. The annealing effects on the piezoelectric constant d_{33} of the AlN films were also estimated [14].

3.1 Pt/AlN/Pt/SiO₂/Si

AlN films, 3.15 µm thick, were sputtered on bare and Pt (2200 Å thick) -covered SiO_2/Si(100) substrates, being ~2 µm the silicon oxide thickness; a Pt film (2200 Å thick) was sputtered on the AlN free surface and then the Pt/ AlN/Pt/SiO₂/Si multilayers were heated at 900°C in air for lengths of time ranging from 1 to 32 hours. Figure 5 shows the XRD patterns of the as grown and annealed multilayers [14].

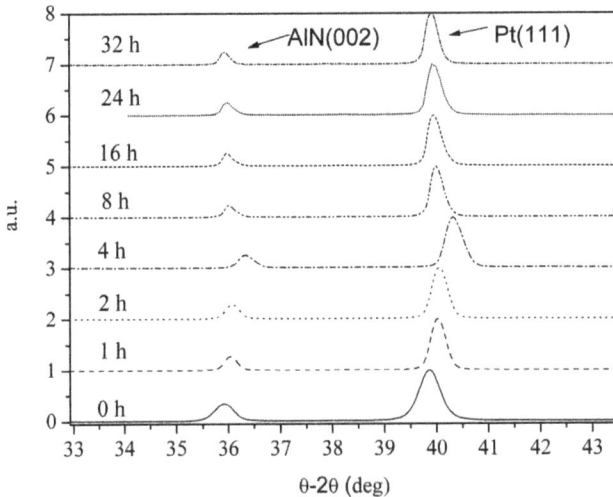

Fig. 5. XRD patterns of Pt/ AlN/Pt/SiO₂/Si structures: the running parameter represents the annealing time

The piezoelectric AlN film is c axis oriented perpendicularly to the growth plane: in all the samples the AlN (002) and (004) peaks, at ~ 36° and at ~ 76°, are visible even after 32 hours annealing. The peak at 2ϑ ~40° corresponds to the Pt film strongly oriented along the (111) plane; with increasing the annealing time, a very small stress-induced shift in the Pt (111) peak position can be observed, while the same peak becomes narrower indicating a growing (111) fiber texture. As a consequence of the temperature-induced improvement of the Pt local epitaxy, the alignment precision of the AlN film crystalline planes also improves. The XRD data of the outer Pt film showed two peaks at 2θ~ 40° and 46.3° corresponding to the (111) and (200) orientations, and also a small platinum oxide (220) peak, at ~66°, clearly visible after the 1st annealing, that does not increase in percentage after the successive thermal cycles [14]. The AlN and Pt FWHM of the $\vartheta-2\vartheta$ scan before and after the annealing [14] resulted improved, starting from 0.391 and 0.415° of the as grown AlN and Pt films, up to 0.24 and 0.28° after 32 hours annealing. The c and a lattice parameter of AlN and Pt films of the as-grown samples, calculated from the angular position of the (002) and (111) diffraction peaks, are respectively larger (4.997 Å) and smaller (3.914Å) than the values reported for bulk single crystal AlN and Pt [13], as a consequence of the lattice mismatch between the two materials. Figures 6 and 7 show the FWHM of the AlN(002) and Pt(111) peaks, and the c and a lattice parameters of AlN and Pt films vs the annealing time.

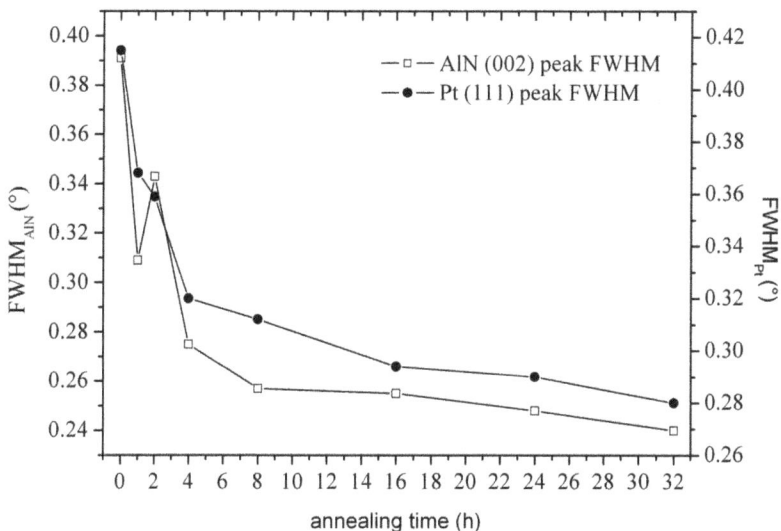

Fig. 6. FWHM of the AlN(002) and Pt(111) peaks vs annealing time

After the first annealings the AlN c parameter relaxes to the bulk while the Pt a parameter slightly decreases. Further annealings result in a permanent in-plane compressive stress for both the AlN and Pt films. The AlN(002) and Pt(111) peaks' FWHM decreases with increasing the annealing time indicating a decrease of inhomogeneous strain distribution.

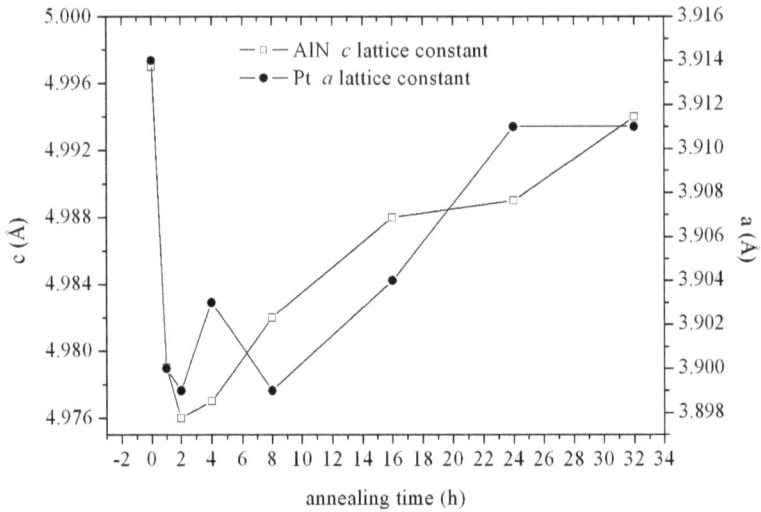

Fig. 7. c and a lattice parameters of AlN and Pt films vs the annealing time

The measurement of the longitudinal piezoelectric coefficient d_{33f} of the AlN film was done at ambient temperature on the same type multilayer without the outer Pt films, before and after the thermal annealing, with a method described in ref. 15 and based on the direct piezoelectric effect: a longitudinal acoustic wave perturbs the sample via a special probe and the electrical voltage induced in the piezoelectric film is measured. The probe consisted of a metal rod in contact with a Pb(Zr,Ti)O$_3$ (PZT)-based low frequency transducer that was connected to a pulse generator (pulse width 0.1-1.0 ns) to produce longitudinal bursts propagating along the metal rod. The contact between the rod and the piezoelectric film surface resulted in the application of a stress on the surface of AlN films. Stress-induced electrical charges were collected at the piezoelectric film surfaces by electrodes (the metal rod and the conducting substrate) and observed on a scope. The piezoelectric strain constant, d_{33f}, of the tested films was evaluated comparing the film response with the response of a thin single crystal reference sample, whose d_{33} was known. All the tested films showed to be piezoelectric with a difference in the d_{33f} obtained values not appreciable with this measurement technique because of an error of about 15-20 %. The estimated mean value is in the range from 6.2 and 7.4 pC/N, for both the as grown and all the annealed samples: these values well agree with the corresponding value reported in the available literature [16] that is about 6 pC/N.

The sheet resistivity of the Pt layer deposited on the AlN/Pt/SiO$_2$/Si multilayer free surface was measured at ambient temperature by the four point method, before and after the thermal annealing. It was observed that the sheet resistance values decreases with increasing the annealing time, starting from ~0.6 Ω/sq, refereed to the unannealed samples, to ~0.5 Ω/sq, referred to the 32 hours annealed samples [14]. Scanning electron microscopy (SEM) investigations revealed an average grain size increased with increasing the annealing lasting. The as deposited Pt films have small grain size (about 400 nm), as shown in figure 8, and this high density of grain-boundary affects the high resistivity of the films.

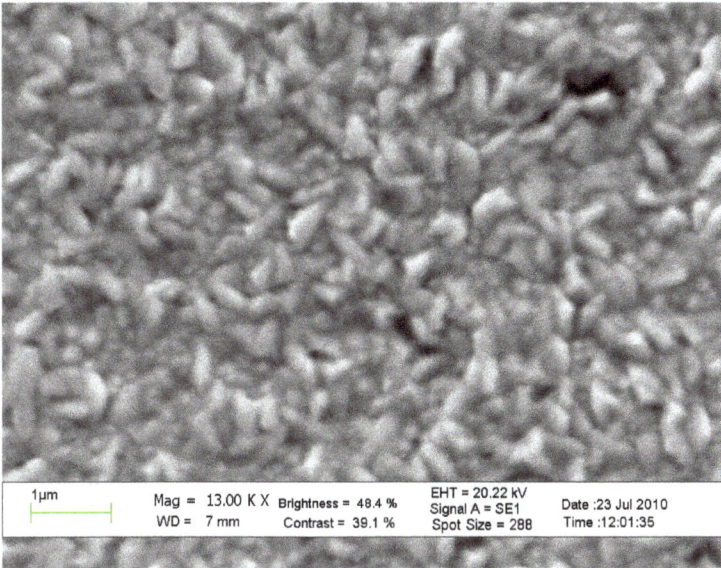

Fig. 8. SEM photo of the as deposited Pt film

Fig. 9. SEM photo of the 32 hours annealed Pt film

Moreover, the as-grown films contain a number of structural defects that anneal out when heat treatment is carried out and, as a result, the film's resistivity decreases. The Pt annealing results in the relaxation of intrinsic stresses, as well as in the redistribution of structural imperfections: grains coalescence take place and the film sheet resistivity decreases. SEM photo of the 32 hours annealed sample, shown in figure 9, demonstrates that recrystallization of Pt surface occurs [14].

The unannealed outer Pt film was silvery while the annealed films were matt: this color change can be explained with the increase of the Pt grain size. The adhesion strength of the Pt films was good enough to pass a rudimentary "tape test" with a transparent tape after each anneal as well.

3.2 c-AlN/(0001)-Al$_2$O$_3$

Highly c-axis oriented AlN films were grown by rf reactive sputtering technique on the polished surface of (0001) oriented single crystal Al$_2$O$_3$ substrate. Then the as grown films, 0.26 to 4.7 μm thick, were thermally annealed at 900°C in air for 1 to 18 hours and the AlN structural characteristics were evaluated after each thermal cycle. References are available in the literature concerning the high thermal annealing (HTA) of AlN on Si performed at 700 – 1200 °C for 2 to 12 hours in controlled atmosphere (in oxygen or nitrogen flux) or in high vacuum; AlN films on Al$_2$O$_3$ were heated at 950 °C in air for 30 minutes [20] to 1 hour [21], at 900 to 1200 °C for 10 s in flowing N$_2$[22], and at 800 °C for 90 minutes in air [23]; bulk AlN was annealed in oxygen at 900 – 1150 °C for 6 hours [24]. In the present work, the HTA of the AlN films on Si and on Al$_2$O$_3$ substrates have been performed up to 32 and 18 hours, respectively. No damage was observed on the surface of the AlN/Al$_2$O$_3$ film even after 18 hours annealing: the AlN films were still clear, uniform, and extremely adhesive to the substrate. The impact of the annealing on the films structural properties was investigated by XRD before and after undergoing the thermal annealing. The D and c parameter of the AlN films thermally annealed for different time periods, are listed in tab. 1.

AlN thickness (μm)	Time (hours)	FWHM$_{\theta-2\theta}$ (deg)	D (Å)	c (Å)
4.7	0	0.384	218	4.995
	4	0.281	297	4.977
	8	0.289	288	4.978
	12	0.290	290	4.978
4	0	0.270	540	4.976
	1	0.269	548	5.024
	3	0.264	555	5.026
	5	0.262	593	5.027
	9	0.261	611	5.037
	18	-	-	-
1.5	0	0.289	252	4.978
	4	0.332	288	4.984
	8	0.350	281	4.989
	10	0.410	303	5.001
0.26	0	0.500	220	4.986
	3	0.422	198	5.048
	4	-	-	-

Table 1. The AlN (002) FWHM, the D and c parameter of the AlN films thermally annealed for different time periods

Figures 10, 11 and 12 show the diffraction patterns of the AlN films, 4.0, 1.5 and 4.7 μm thick, annealed for different time.

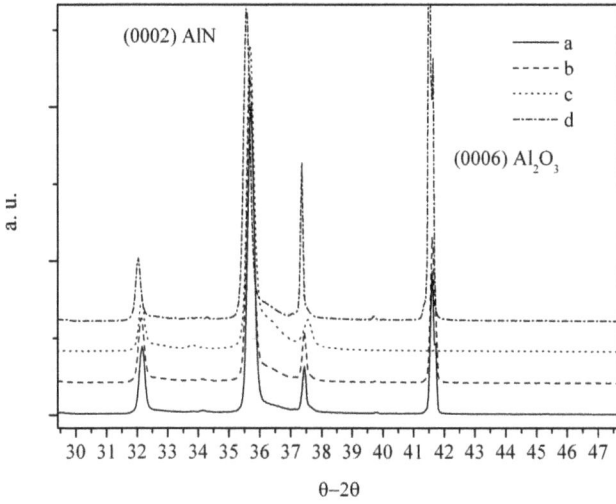

Fig. 10. XRD pattern of the AlN, 4 μm thick, annealed for 1 (a), 3 (b), 5 (c), and 9 (d) hours

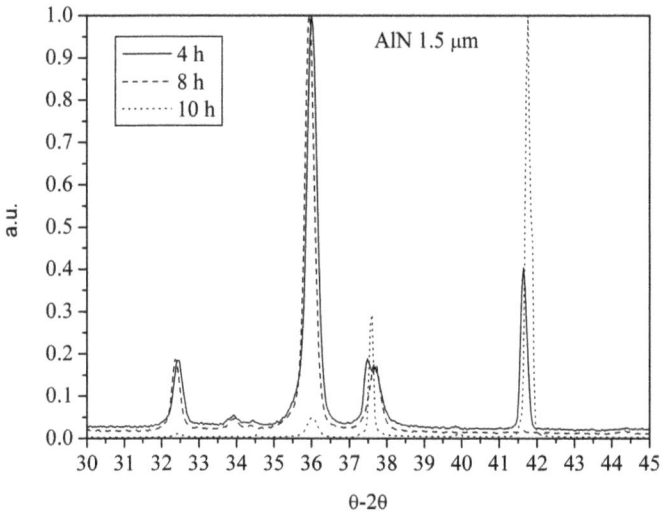

Fig. 11. XRD pattern of the AlN, 1.5 μm thick, annealed for 4, 8 and 10 hours

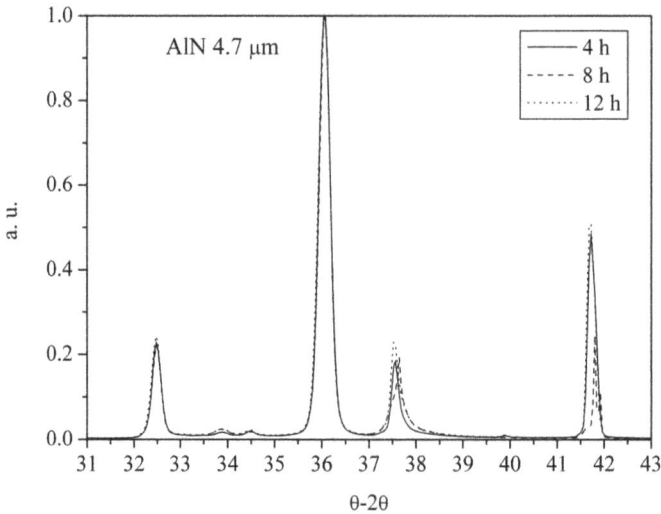

Fig. 12. XRD pattern of the AlN, 4.7 μm thick, annealed for 4, 8 and 12 hours

The AlN (0002) peak at approximately $2\theta = 36°$ of the as deposited films, 0.26 to 4.7 μm thick, showed a FWHM (from θ-2θ scan) in the range 0.5° to 0.27°. Both the as grown and the annealed samples exhibit a strong peak at $2\theta \sim 42°$ due to the (0006) reflection of the Al_2O_3 substrate, and one small peak at 76°, corresponding to the (0004) reflections of the wurtzite AlN structure, not shown in figures 10 to 12. The 2nd order peaks of AlN(0002) and Al_2O_3 (00001) are clearly visible at ~32.4° and ~37.5°. After 4 hours at 900°C in air, the (0002) peak of the thin AlN film (0.26 μm thick) broadens completely as well as that of the 4 μm thick AlN film after 18 hours annealing. No other AlN phases are present, nor AlN-oxide traces are evident in the spectra after the annealing; only a decrease in the 2θ value of the (0002) peak is observed, resulting in an increase in the c lattice parameter whose values are larger than the bulk lattice parameter [13], indicating the presence of compressive stress in the surface plane. This lattice elongation, perpendicular to the growth plane, is usually associated with the intrinsic compressive stress caused by the lattice mismatch between the AlN and Al_2O_3, $(a_{sapphire}-a_{AlN})/a_{sapphire}$, that is approximately equal to 30%. Our previous results [25] have shown that the c-axis of the as grown AlN films on sapphire relaxes to the bulk value with increasing the film thickness: an interface layer, strained because of the large lattice mismatch, is formed on the sapphire surface and is followed by a columnar AlN layer which runs through the entire thickness of the film. The thinner the AlN film and more is strained and unable to survive to the HTA.

4. Conclusions

Highly c-axis oriented AlN films and thin film stacks of Pt/AlN/Pt were sputtered at 180°C on (0001)Al_2O_3 substrates and at 200°C on oxidized Si substrates. The multilayers were heated at 900 °C in air up to 32 hours to test their resistance to high temperature. The structural investigation of the annealed films showed that the thermal annealing improved the crystal quality of the AlN films sandwiched between the Pt films, as confirmed by the decreased FWHM of the rocking curves, and the film piezoelectric d_{33f} coefficient resulted unaffected by the temperature. The study of the electrical, morphological and structural characteristics of the Pt electrode revealed a dense surface with a hillock-free morphology, confirming that Pt is the material of choice when a high oxidation resistance is required for metallic components within devices operating at elevated temperatures. The AlN films on sapphire show a lattice elongation, perpendicular to the growth plane, that is usually associated with the compressive stress caused in the growth plane by the lattice mismatch between the film and the substrate. After the first few hours annealing, the FWHM of the (002) AlN peak decreases showing an improvement in the film texture; further annealing results in the FWHM broadening, whose magnitude depends on the film thickness. The structural investigations of thin (0.26 μm) and thick (from 2 to 4.7 μm) annealed AlN films revealed that the film behaviour in harsh environment is strongly affected by the film properties. Thick films, whose structure is more relaxed than the thin one, is able to survive to high temperature without suffering significant deterioration for longer annealing times than the thin one. The obtained results confirm that, during the HTA, the Pt film on the substrate surface is protected by the AlN film, while the Pt film directly exposed to the ambient conditions acts as a protective layer with respect to the AlN film; thus the AlN-based STFM coupling configuration to be implemented on sapphire or silicon substrates is an attractive alternative to langasite and $GaPO_4$ for the development of microwave electroacoustic devices able to work at high temperatures.

5. Acknowledgements

The author wishes to thank Mr. P.M. Latino for his technical support in the development of the technological processes and the HTA.

6. References

[1] Hauser, R. Reindl, L. Biniasch, J., ú Ultrasonics Symposium, Vol. 1, 192 (2003).

[2] M. Pereira da Cunha, M.P. Saulo de A. Fagundes, ú Ultrasonics Symposium, Vol. 1, 283 (1998).

[3] Maurício Pereira da Cunha, Eric L. Adler and Donald C. Malocha, ú Ultrasonics Symposium, Vol.1 169 (1999).

[4] Kar, J.P., Mukherjee, S., Bose, G., Tuli, S., Myoung, J.M., Materials Science and Technology, 25 1023 (2009).

[5] Zhigang Lin and Yoon, R.J., International Symposium on Advanced Packaging Materials: Processes, Properties and Interfaces, 156 (2005).

[6] C. Zolper, D. J. Rieger, and A. G. Baca S. J. Pearton and J. W. Lee R. A. Stall, Appl. Phys. Lett. 69, 538 (1996).

[7] S.M.Middelhoek, S.A.Audet, Silicon sensors, Academic Press, London 1989.

[8] M.J.Vellekoop, E.Nieuwkoop, J.C.Hsaartsen, A.Venema, ú Ultrasonics Symposium, Vol. 1, 375 (1981).

[9] E.L. Adler, G.W. Farnell, J. Slaboszewicz, C.K. Jen, ú Ultrasonics Symposium, Vol. 1, 103 (1982).

[10] K. Tsubouchi, K. Sugai, N. Mikoshiba, ú Ultrasonics Symposium, Vol. 1, 375 (1981).

[11] J.G. Gualtieri, J.A. Kosinski, A.Ballato, ú *Ulltrasonics Symposium*, Vol. 1 403 (1992).

[12] C. Caliendo, *Appl. Phys. Letters* 92 103501 (2008).

[13] JPCDS cards No. 25-1133 Powder Diffraction File, Joint Committee on Powder Diffraction Standards, ASTM, Philadelphia, PA, 1967, Card 25-1133

[14] C. Caliendo, P. M. Latino, *Thin Solid Films* 519 (2011), pp. 6326-6329

[15] C.K.Xu, V.N.Umashev, I.B.Yakovkin, *Sov. Phys.* PTE6, 192 (1986).

[16] Landolt-Bornstein, *Numerical Data and Functional Relationships in Science and Technology, Group III: Crystal and Solid State Physics*, (Springer-Verlag Berlin, Heidelberg-New York, 1979), Vol. 11 (1979).

[17] C.-Y. Lin, F.-H. Lu, *J. Eur. Ceram. Soc.* 28 691 (2008).

[18] E. A. Chowdhury, J. Kolodzey, J. O. Olowolafe, G. Qiu, G. Katulka, D. Hits, M. Dashiell, D. van der Weide, C. P. Swann, K. M. Unruhm, *Appl. Phys. Lett.* 70 2732 (1997).

[19] F. Jose, R Ramaseshan, S. Dash, S. Bera, A. K. Tyagi, B. Raj, *J. Phys. D: Appl. Phys.* 43 075304 (2010).

[20] T. Aubert, O. Elmazria, B. Assouar, L. Bouvot, M. Oudich, *Appl. Phys. Lett.* 96 2035031 (2010).

[21] S.-K. Tien, C.-H. Lin, Y.-Z. Tsai and J.-G. Duh, *J. Alloys Compd.* 489 237 (2010).

[22] Z. Gu, J.H. Edgar, S.A. Spekman, D. Blom, J. Perrin, J. Chaudhuri, *J. Electron. Mater.* 34 1271 (2005).

[23] T. Aubert, M. B. Assouar, O. Legrani, O. Elmazria, C. Tiusan, S. Robert, *J. Vac. Sci. Technol. A* 29 021010 (2011).

[24] B. Liu, J. Gao, K.M. Wu, C. Liu, *Solid State Commun.* 149 715 (2009).

[25] C. Caliendo, P. Imperatori, *J. Appl. Phys.* 96, 2610 (2004).

13

Applications of In–Fiber Acousto–Optic Devices

C. Cuadrado-Laborde[1,2], A. Díez[1], M. V. Andrés[1],
J. L. Cruz[1], M. Bello-Jimenez[1], I. L. Villegas[1,3],
A. Martínez-Gámez[3] and Y. O. Barmenkov[1,3]

1. Introduction

Nowadays, in-fiber acousto-optic devices are increasingly used as frequency shifters, multiplexers, modulators, and tunable filters. They can be easily spliced into optical fiber systems, and the consequent low insertion loss, make them an attractive alternative to bulk optics devices. Our group, established at the Institute of Materials Science, Department of Applied Physics, of the Valencia University (ICMUV, Valencia, Spain), has been involved in this field for the last ten years, and our fabrication facilities allow us the development of new in-fiber acousto-optic devices for novel applications in different fields such as sensors, microwave photonics, lasers, and optical communications. Our aim here is to present the great potential shown by in-fiber acousto-optic devices for different photonic applications. Although this chapter is focused in our latest developments, the reader will also find discussed the work done by other research groups in the field. This chapter is divided in two main sections: Section 2 is focused on novel applications of acousto-optic fiber devices based on flexural acoustic waves, and Section 3 is focused on applications of the acousto-optic devices based on the interaction of longitudinal acoustic waves with fiber Bragg gratings (FBG). Finally, our conclusions are shown in Section 4.

2. Applications of acousto-optic devices based on flexural acoustic waves

In a standard single mode fiber, when a flexural acoustic wave propagates along an optical fiber a periodic perturbation is introduced in its refractive index, and it can induce coupling between the fundamental mode guided by the core and the modes supported by the cladding (Kim et al., 1997). This acousto-optic interaction can be seen as the dynamic counterpart of a long period grating (LPG). LPGs are usually fabricated by creating a periodic perturbation of the refractive index by UV radiation; this of course fixes its spectral characteristics. On the other hand, when the perturbation is introduced by an acoustic wave, its period and strength can be controlled through the frequency and amplitude of the acoustic wave, respectively. Thus, the spectral properties of the optical device can be controlled dynamically through the characteristics of the acoustic perturbation. The optical

[1]*Departamento de Física Aplicada y Electromagnetismo, ICMUV,
Universidad de Valencia, Burjassot, Valencia, Spain*
[2]*CONICET La Plata, Buenos Aires, Argentina,*
[3]*Centro de Investigaciones en Óptica, León, Guanajuato, Mexico*

coupling is resonant in wavelength; it takes place at the optical wavelength that verifies the phase-matching condition between the beat length of the two optical modes and the acoustic wavelength. At the output of the acousto-optic device, only the light that remains guided by the core mode is transmitted. Thus, the coupling of power from the fundamental mode to a cladding mode results in the appearance of an attenuation notch in the spectrum, whose amplitude is fixed when travelling acoustic waves are used. The characteristics of the fundamental flexural acoustic mode of an optical fiber fulfill the above requirements for the development of efficient in-fiber acousto-optic devices. In subsection 2.1 we will describe a Q-switched fiber laser where cavity loss modulation is reached by using this operation principle. As an interesting alternative, if standing –instead of travelling– acoustic waves are used, light transmitted at the resonance wavelength will experience an amplitude modulation at twice the frequency of the acoustic wave. In subsection 2.2 we describe a mode-locked fiber laser based on the use of this operation principle.

2.1 *Q*-switching by intermodal acousto-optic modulation in an optical fiber

Q-switched fiber lasers have attractive applications in different fields, such as in remote sensing and material processing. The mechanism of short pulse emission is based on the modulation of the Q factor of the cavity, which can be done either passively or actively (Siegman, 1986). In the former the setups are simpler, but the repetition rate only varies with the pump power of the medium gain. Further, they usually show long-term instability, and frequently the amplitude of the pulses is randomly modulated in time (Vicente et al., 2004). On the other hand, active Q-switching is independently and accurately controlled by an electrical signal, which triggers the modulator. The bulk approach, such as electro-optic (Kee et al, 1998) or acousto-optic modulators (Álvarez-Chávez et al., 2000), is not adapted to compact fiber laser systems required nowadays; further they have large optical coupling losses and stringent alignment requirements. For these reasons, the all-fiber approach is of permanent interest, being advantageous in terms of cost, loss, packaging, robustness, and simplicity. One solution widely investigated has been enabled by the use of FBGs as cavity mirrors. In this way, the tuning of the wavelength of one of the FBGs has been used to achieve active Q-switching (Andrés et al., 2008). Among the typical tuning methods, we could mention the stretching of fiber Bragg gratings by magnetostrictive materials (Pérez-Millán et al., 2005a; Andersen et al., 2006), by piezoelectric actuators (Imai et al., 1997; Russo et al., 2002), or by the interaction of longitudinal acoustic waves with the FBG (Delgado-Pinar et al., 2006; Cuadrado-Laborde et al., 2007). Other technique used for active Q-switching relied in the cavity loss modulation by the core-to-cladding mode-coupling in optical fibers induced by travelling flexural acoustic waves (Huang et al., 2000; Zalvidea et al., 2005). As opposed to the last two referred works –which were focused at the erbium lasing wavelength–, here we experimentally demonstrate the possibility of using this modulating technique at the technologically relevant ytterbium lasing wavelength (Tuchin, 1993). Very few actively Q-switched ytterbium-doped lasers with an all-fiber configuration have been reported (Andersen et al., 2006). The main difficulties arise from a relatively long time response and a limited modulation depth of all-fiber amplitude modulators. Here, we present the experimental results that we have obtained in a research work focused on the exploitation of in-fiber acousto-optics. In our proposed Q-switched fiber laser, when the acoustical signal is switched-off, the optical power losses within the cavity are reduced, and then a laser pulse is emitted (Villegas et al., 2011b).

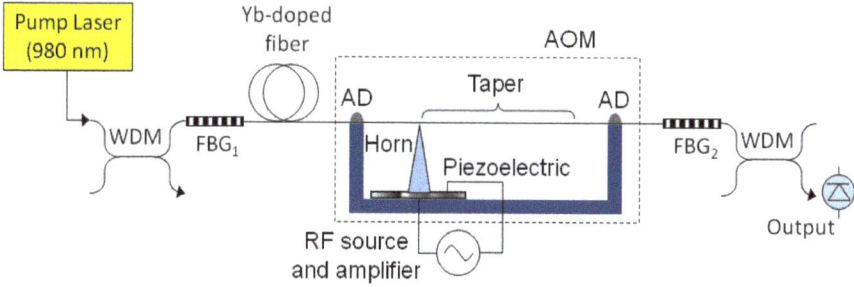

Fig. 1. Q-switched fiber-laser setup; the acousto-optic modulator is defined by the elements inside the dashed line; AD stands for acoustic dumper

The setup used for our Q-switched fiber laser is schematically illustrated in Fig. 1. The gain was provided by 0.65 m of a heavily-doped ytterbium-doped single-mode fiber –Nufern SM-YSF-HI, cut-off wavelength of 860 ± 70 nm, numerical aperture of 0.11, and fiber absorption of 250 dB/m at 975 nm–. The active fiber was pumped through a WDM coupler by a pigtailed laser diode emitting at 980 nm, providing a maximum pump power of 110 mW. The in-fiber acousto-optic modulator (AOM) was spliced between FBG$_1$ –Bragg wavelength at 1064 nm, FWHM of 0.23 nm, and 99.6 % of maximum reflectivity– and FBG$_2$ –Bragg wavelength at 1064 nm, FWHM of 70 pm, and 44 % of maximum reflectivity–, defining in this way a Fabry-Perot cavity. The AOM in turn is composed of an RF source, a transversal-mode piezoelectric disk, an aluminum horn, and a tapered single-mode optical fiber –Fibercore SM980 of low numerical aperture (0.13-0.15) –. The optical fiber was tapered down by the fusion and pulling technique using a travelling flame, which produces a taper waist with a uniform diameter of 76 µm and 0.1 m length, for this specific case. The tip of the aluminum cone –with the piezoelectric disc fixed to its base–, was glued to an uncoated section of fiber near the taper, see Fig. 1. Finally, the optical fiber in the AOM was acoustically dumped in both extremes; see Fig. 1, in order to prevent unwanted acoustical reflections.

When an RF signal is applied to the piezoelectric disc, a travelling flexural acoustic wave is launched through the taper. If the acoustic wavelength matches the beat-length between the fundamental mode guided by the core and one of the optical modes supported by the cladding, then light coupled to the later remains in the cladding downstream of the taper, being finally absorbed by the fiber coating (Birks et al., 1994). Thus, the coupling of power from the fundamental mode to a cladding mode results in the appearance of an attenuation notch in the spectrum. When the acoustic frequency is varied, the periodicity of the perturbation also does, and hence the phase-matching condition is shifted to a different optical wavelength. Figure 2(a) shows the tunability of the notches caused by the coupling between the fundamental core mode and the first three cladding modes LP$_{1m}$. The selected operating point was at an optical wavelength of 1064.1 nm for an RF signal applied to the piezoelectric of 825 kHz. These measurements were made by illuminating the taper with a broadband light source and detecting the light transmitted through the taper with an optical spectrum analyzer. As an example, Fig. 2(b) shows one of the measured transmission spectra for an applied voltage to the piezoelectric of 26 V and a frequency of 885 kHz. The transfer of optical power from the fundamental core mode to one of the cladding modes behaves periodically as a function of the acoustic power, which in turn is a function of the applied voltage to the piezoelectric. Figure 2(c) shows this effect for the selected operation point marked in Fig. 2(a). As it can be observed, the transmittance decays by a maximum of 12 dB for an applied voltage to the piezoelectric of 50 V (peak-to-peak measurement). Beyond this point, further increment in the applied voltage raises the transmittance again. Therefore, and bearing in mind its use as a Q-switching device, cavity loss modulation between 0 dB and 12 dB can be achieved by applying to the piezoelectric a sinusoidal signal at the frequency of 825 kHz fully-modulated by a rectangular signal, see the inset of Fig. 2(c). This modulation produces on–off periods of the acoustic wave travelling down the fiber, which results in a modulation of the cavity losses at the resonance wavelength. In passing, we should mention that several tapers were fabricated in order to better optimize this laser system, by using Fibercore SM980 of high numerical aperture (0.17-0.19), and Nufern SM-YSF-LO (a moderately ytterbium-doped fiber), and in turn with different taper waists. Despite this, the minimal decay in transmittance was always around the reported values, i.e. between 10 and 16 dB.

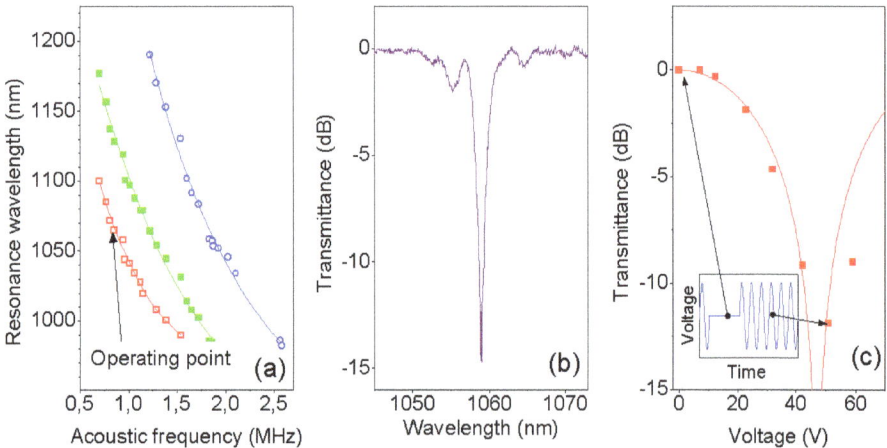

Fig. 2. (a) Resonant optical wavelengths as a function of the acoustic frequency for the first three mode-couplings LP_{01} -LP_{1m}. (b) Typical transmission notch caused by the first mode-coupling by applying to the piezoelectric a sinusoidal signal at 885 kHz and 26 V. (c) Transmittance at the selected operation point marked in (a) –i.e. 1064.1 nm and 825 kHz– as a function of the applied voltage to the piezoelectric (solid scatter points), the curve represents a theoretical fitting according to a $\sin^2$ function

The switching time is one of the key parameters for any modulator intended to be used as a Q-switching device in a fiber laser. Preferably, it should be as short as possible. For this reason we measured the temporal response of this device, by detecting the transmitted light, while simultaneously registering the modulating signal in an oscilloscope. It takes 25 μs to increase/decrease the transmitted optical power through the taper when the voltage applied to the piezoelectric is switched-off/on, respectively. This time corresponds reasonably well with the time it takes a flexural acoustical wave to travel down the taper, i.e. 0.1 m / 3764 m/s ≈ 26 μs. This value is short enough to allow the use of this modulator as a Q-switching device in a fiber laser, as it will be demonstrated in the following section. In principle, and for a given taper waist, shorter switching times could be achieved by decreasing the interaction length. Unfortunately, this simultaneously increases the acoustical power needed to reach the same optical coupling power, therefore there is a trade-off between both parameters.

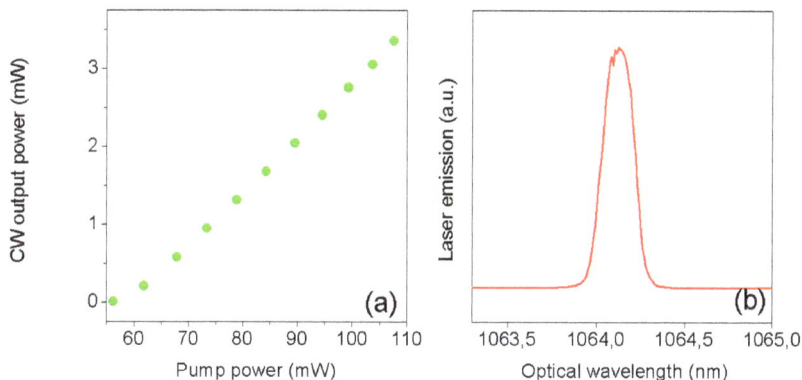

Fig. 3. (a) Output optical power in CW emission as a function of the pump power. (b) Emission spectrum

In this setup, both gratings are permanently tuned to the same wavelength by using two translational stages, one for each FBG; see the setup in Fig. 1. In this way, for zero voltage applied to the piezoelectric, this laser emits in CW, since then the cavity losses are minimal, see the transmittance in Fig. 2(c) for 0 V. Figure 3(a) shows the optical power in CW emission as a function of the pump power. There is a pump power threshold of 56 mW, about which the laser starts to emit, reaching a maximum output power of 3.4 mW, which in turn is determined by the maximum pump power available in our setup (110 mW). Figure 3(b) shows the spectrum of CW emission; its linewidth is of 0.21 nm at a center optical wavelength of 1064.1 nm.

Now, we discuss the Q-switched operation of this laser. To this end, we modulated the cavity losses by applying to the piezoelectric a fully-modulated sinusoidal signal at the frequency of 825 kHz and 50 V. A rectangular wave was used to modulate the RF voltage that generates the acoustic wave. At a given frequency of the modulating signal, we found always a maximal duty cycle –i.e. the fraction of time that the signal is in its high level– able to perform Q-switch correctly. If we decrease this duty cycle, then the cavity would stay in its high Q state longer, and more than one Q-switched pulse would be emitted in each time slot. The Q-switch repetition rate becomes determined by the frequency of the modulating signal; with this configuration we reached continuous tuning of the Q-switch repetition rate in the range 1-10 kHz. Figure 4(a) shows, as an example, a Q-switched optical pulses train at 1 kHz, together with the corresponding modulating signal, for a pump power of 59 mW. Figure 4(b) shows a detail of a single Q-switched optical pulse of the train shown in Fig. 4(a) with a time width (FWHM) of 3.72 μs. The pulses have a quasi-Gaussian profile; the fitting by this function is also shown in Fig. 4(b). The effect of pump power on the Q-switched pulses, for different repetition rates in the range 1-10 kHz, is shown in Fig. 5. For each Q-switching frequency there is a pump power threshold. Above threshold, the peak power increases with pump power, and there is a corresponding reduction of pulse width. For a given frequency there is also a pump power level beyond which extra pulses appear, the curves are truncated at that point. In order to overtake this limitation, a modulator with improved modulation depth is required. In addition, our laser produces easily multiple pulse emission, which is a well-known problem in actively Q-switched lasers with relatively large time responses. A large time response makes critical the adjustment of the duty cycle of the modulation voltage when the pump power is increased, particularly at low repetition rates. Consequently, a faster switching response of the acousto-optic modulator is required in order to improve the operation of the laser.

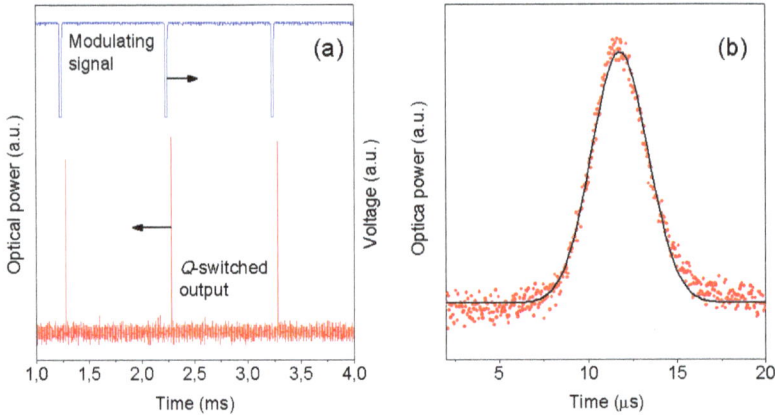

Fig. 4. (a) Modulating signal (above) at 1 kHz repetition rate together with the generated Q-switched laser output (below). (b) Detail of a single Q-switched pulse of (a) (scatter points) together with its corresponding fitting by a Gaussian function (solid curve)

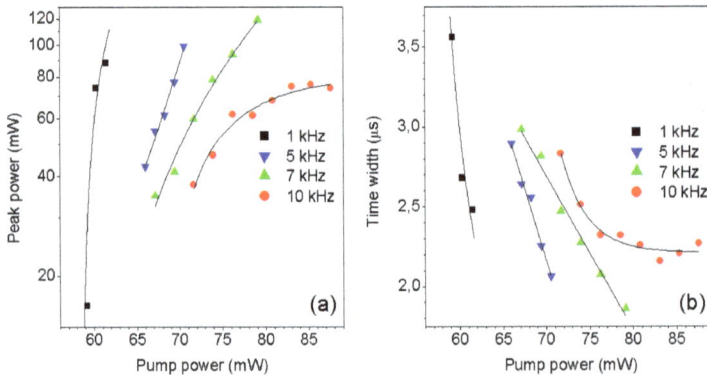

Fig. 5. Peak power and time width in Q-switched operation as a function of the pump power, (a) and (b), respectively, for several repetition rates

2.2 Mode-locking by intermodal acousto-optic modulation in an optical fiber

Mode-locking lasers have a vast number of applications, ranging from the telecommunications industry to medical surgery (Bonadeo et al., 2000; Haus, 2000; Yu et al., 2000; Schaffer et al., 2003). The shortest pulses are obtained by passive schemes. However, they also have some drawbacks, among them, they are generally unstable, they have higher timing-jitter between pulses and usually self-starting cannot be assured. An alternative to using passive schemes that reduces jitter and allows for synchronization is to use an active scheme. To this end, a modulator is driven at the cavity's fundamental repetition rate; therefore, timing-jitter is reduced since each pulse is triggered by the modulator, which in turn is accurately driven by an electrical signal. In this way, in active mode-locking one has the possibility to synchronize a laser to an external clock or electronic signal (French, 1995). Among all these lasers, the all-fiber solution has been the subject of intensive research due to their compact structure, free alignment problems, and low cost, between other merits. Earlier and subsequent active modelocked fiber lasers were not strictly all-fiber lasers, since they currently included a bulk modulator (Geister & Ulrich, 1988; Phillips et al., 1989a; Hudson et al., 2005). This not only induces extra losses into the laser cavity but of course also destroys the benefits of an all-fiber configuration. On the contrary a strictly all-fiber modulator can overcome these limitations and it can be used for actively mode-lock rare-earth-doped silica fiber lasers (Phillips et al., 1989b; Culverhouse et al., 1995; Jeon et al., 1998; Costantini et al., 2000; Myrén & Margulis, 2005; Cuadrado-Laborde et al., 2009a, 2010a, 2010b; Bello-Jiménez et al., 2010, 2011). Table 1 shows the major achievements towards the obtaining of all-fiber actively mode-locked lasers. As it can be observed, acousto-optic devices have been by far the preferred technique, with exception of the Myrén and Marguli`s work relying in a electro-optic effect. Our group was involved in works described in the last two rows of this table, which will be discussed in detail in this chapter.

In Jeon`s et al. work (1998), mode-locking is achieved by frequency shifting based on travelling flexural acoustic waves in a two-mode optical fiber; such a frequency shifter could be considered an improved version of the one reported in (Kim et al., 1986). Unlike Jeon (1998) and Kim (1986), here we propose using a standard single-mode optical fiber in which core-to-cladding mode-coupling is induced by a standing flexural acoustic wave. In this way, the device does not work as a frequency shifter, but as an amplitude modulator, which is used to actively mode-lock a fiber laser. In the following, we demonstrate that this simple acousto-optic modulator proves to be particularly well-suited for active mode-locking purposes (Bello-Jiménez et al., 2010, 2011).

The setup of the actively modelocked fiber ring laser is schematically illustrated in Fig. 6. The medium gain was provided by 4.36 m of erbium-doped fiber (EDF) containing 300 ppm Er^{3+}, with a cut-off wavelength of 939 nm, and a numerical aperture of 0.24. The active fiber was pumped through a WDM by a 980 nm pigtailed laser diode, providing a maximum pump power of 600 mW. Next –and following a clockwise direction– it was inserted the acousto-optic modulator between two polarization controllers (PCs), followed by a delay line fusion-spliced to port one of a polarization-independent three-port optical circulator (OC). The OC not only forces the unidirectional operation within the ring but simultaneously incorporates the FBG filtering in the cavity by the second port. Finally, the ring cavity is closed by connecting port three of the OC to the WDM. The output of this laser is obtained by transmission through the FBG –Bragg wavelength at 1549.5 nm, FWHM of 0.45 nm, and 50 % of maximum reflectivity–. In this way, the entire laser is spliced. The AOM in turn is composed of an RF source, a piezoelectric disk, an aluminum conical horn, and a 0.2 m long standard optical fiber stripped of its polymer coating; in order to prevent the attenuation of the acoustic wave. The horn is attached to the piezoelectric, and it focuses the vibrations into the fiber through its tip, which is glued to the uncoated fiber. In order to allow the generations of a standing flexural acoustic wave, the uncoated optical fiber of the AOM was firmly clamped at one extreme, whereas in the other it was damped, see Fig. 6.

Year	Group	Mode-locking mechanism	Major achievements
1989	Phillips et al.	Acousto-optically induced intra-fiber phase modulation	First all-fiber actively mode-locked laser, 200 ps time width pulses
1995	Culverhouse et al.	Acousto-optically induced frequency shifting in an all-fiber four-port null coupler	18 ps time width pulses (chirped)
1998	Jeon et al.	Acousto-optically induced mode-coupling in a two-mode optical fiber	Harmonic mode-locking, 3.8 ps time width pulses
2000	Costantini et al.	Acousto-optically induced intra-fiber phase modulation	Visible up-conversion laser, 550 ps time width pulses
2005	Myrén and Margulis	Electro-optic effect in an all-fiber Mach–Zehnder interferometer	1 ns time width pulses
2009	Cuadrado-Laborde et al.	Acousto-optic super-lattice effect in a fiber Bragg grating	630 ps un-chirped time width pulses. First to demonstrate double-active all-fiber Q-switching and mode-locking
2010	Bello-Jiménez et al.	Acousto-optically induced mode-coupling in a single-mode optical fiber	34 ps time width pulses

Table 1. Major achievements in the development of all-fiber actively mode-locked lasers

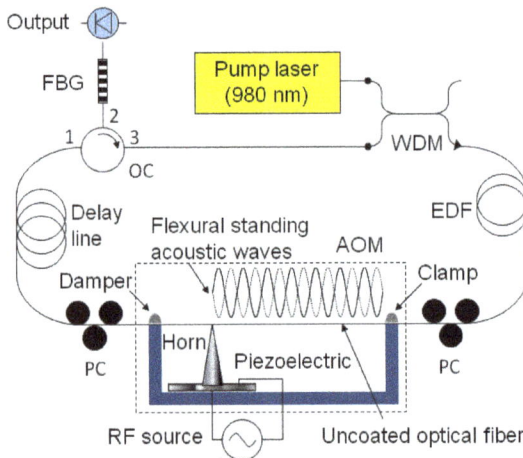

Fig. 6. Modelocked fiber-laser setup; where OC stands for optical circulator, EDF for erbium doped fiber, PC for polarization controller, FBG for fiber Bragg grating, and WDM for wavelength division multiplexer. The acousto-optic modulator is defined by the elements inside the dashed line

Figure 7(a) shows the modulation depth as a function of light wavelength for an acoustic frequency of 2.37315 MHz. The voltage applied to the piezoelectric was 18 V (whenever we refer to voltages, it is a peak-to-peak measurement). The measurement was performed by illuminating the AOM with a tunable laser and detecting the transmitted light at each wavelength. The modulation is maximal at the phase-matching wavelength of 1551 nm, and symmetrically decreases for longer and shorter wavelengths. It should be emphasized the high modulation depth achieved in this technique, reaching a maximum of 72 %, together with the relatively broad operative bandwidth of 1.5 nm (FWHM) –i.e. 187 GHz at the operation wavelength–. Moreover, the insertion loss is as low as 0.7 dB, corresponding to the maximum transmittance value attained, which is 86 %. When the acoustic frequency is varied, the periodicity of the perturbation also does, and hence the phase-matching condition is shifted to a different wavelength. Figure 7(b) illustrates this effect; the rate of change for the wavelength shift is of –0.169 nm/kHz; the maximum modulation reached at each specific operation point is also shown to the right. The differences between the modulation depths at different acoustic frequencies are mainly originated from the combination of the non-flat frequency response of the piezoelectric and the acoustic fiber resonator.

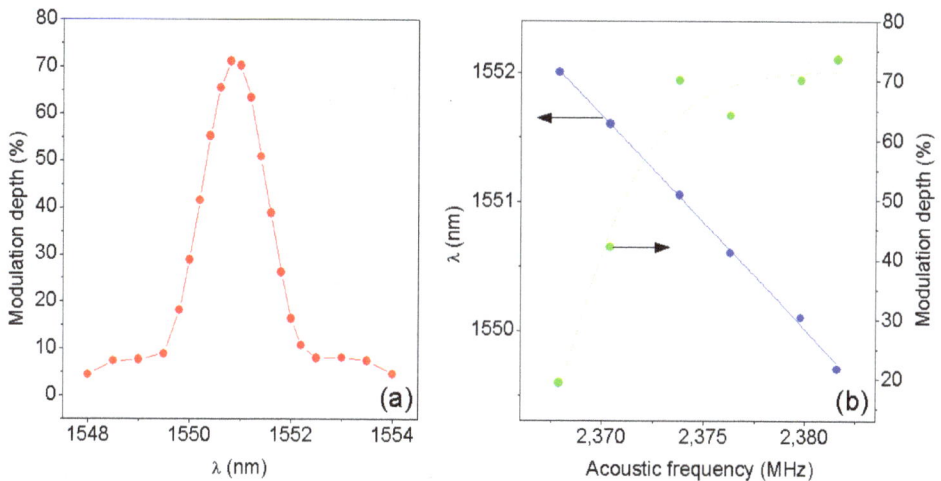

Fig. 7. (a) Modulation depth as a function of the optical wavelength for an RF signal applied to the piezoelectric of 2.37315 MHz and 18 V. (b) Resonant optical wavelengths as a function of the frequency of the RF signal, its corresponding modulation depth is also shown to the right ordinate

Now we discuss the operation of the modelocked laser shown in Fig. 6. The delay line length must be selected to match the round-trip time with the reciprocal of the optical modulation frequency, which in turn is twice the electrical frequency applied to the piezoelectric. The selected operation point for the piezoelectric is the same as in Fig. 7(a), i.e. 2.37315 MHz, and then the cavity length results in 43.7 m. However, fine tuning is always necessary to match the frequency of the modulation signal to the inverse of the round-trip time of the cavity. A translation stage was also used for matching the reflection band of the FBG to the operative wavelength (1551 nm). Regarding the dispersion of this laser; we measured the dispersion of each fiber and component within the cavity by the frequency-domain modulated-carrier method. The laser cavity is a mixed-dispersion fiber ring with an average cavity dispersion of –1 ps/nm/km (Bello-Jímenez et al., 2010). We measured also both the dispersion of the OC and WDM, both resulted normal also with a group delay per wavelength unit of −0.3 ps/nm and −0.04 ps/nm, respectively. In this way, we can assure that this laser works under the normal dispersion regime. Figure 8(a) shows the RF signal used to drive the piezoelectric together with the modelocked train of pulses generated (50 GHz bandwidth oscilloscope). As it can be observed, the frequency of the optical train (4.756 MHz) is twice the frequency of the signal used to drive the piezoelectric (2.37315 MHz). The optical pulses were best fitted by a sech² function; neither Gaussian nor Lorentzian functions improved the fitting. This is an expected profile for the optical pulses in this type of dispersion-mixed cavities, in which both kind of dispersion coexist, i.e. normal and anomalous (Bélanger, 2005). On the other hand, according to active mode-locking theory, the time width τ of the optical pulse can be expressed as (Kuizenga & Siegman, 1970):

$$\tau = \frac{\sqrt{\sqrt{2}\ln 2}}{\pi} \sqrt[4]{\frac{g_0}{\delta_m}} \frac{1}{\sqrt{f_m \Delta f_m}}, \tag{1}$$

where f_m is the modulation frequency, g_0 is the gain of the active medium, Δf_m is the modulator bandwidth, and δ_m is the modulation depth. As a rough estimation, let us compare these results with the results reported in Cuadrado-Laborde et al. (2009a), where optical pulses of 780 ps were obtained by using also AM mode-locking, but with a different modulator and setup. Thus, by comparing them, here there is a higher modulation depth (×7), modulation bandwidth (×240), and gain (×3), but a lower modulation frequency (×0.5) −where the additional assumption that gain scales with the EDF length was made−. Therefore, there is a narrowing factor for the time width of $(3/7)^{1/4} \times (0.5 \times 240)^{-1/2} = 7.4 \times 10^{-2}$. In this way, comparatively, it should be possible to reach optical pulses around $7.4 \times 10^{-2} \times 780$ ps ≈ 58 ps in this laser, which could be taken roughly as a lower limit for this configuration. One key element, out of the preceding discussion, is the FBG bandwidth, which acts as a filter −see the setup in Fig. 6 −whose bandwidth must be selected narrower than the AOM bandwidth; otherwise modes not amplitude modulated would interfere in the mode-locking process. This evidently limits the effective bandwidth and raises the minimum pulse width able to be reached with this configuration, since modulator bandwidth and time width are inversely related. The optical pulse's peak power and temporal width as a function of the pump power are shown in Fig. 8(b). The time width as well as the peak power monotonically increases with the pump power; at the lowest pump powers, pulses as narrow as 95 ps were obtained with a pump power of 110 mW; see the inset of Fig. 8(b). Pump powers below 110 mW precludes mode-locking; whereas the upper limit is reached by the maximum pump power available in our setup.

Next, we analyzed the dependence of the pulse's parameters with the applied voltage to the piezoelectric, see Fig. 9(a). Both the transmittance and modulation depth are functions of the applied voltage to the piezoelectric through a periodic relationship. However, we can consider that the relationship is linear within the range of voltages of Fig. 9(a). When the voltage increases, the modulation also does, and pulses become narrower, as expected from AM mode-locking theory, see Eq. (1). Finally, it is of practical interest to quantify the maximum allowable detuning (Δv_{max}), i.e. the maximum difference between the reciprocal of the cavity round trip's time and modulation frequency that sustain mode-locking. Figure 9(b) shows the variation of the pulse's parameters under frequency detuning, resulting in $\Delta v_{max} = \pm 300$ Hz at half the maximum peak power. According to theory of detuning in AM mode-locking (Li et al., 2001), the maximum allowable detuning can be expressed as:

$$\Delta \nu_{max} = \frac{c\sqrt{m}}{4n\Delta f_m L},$$ (2)

with c the speed of light in vacuum, L is the cavity length, and n the modal effective index. Thus, the minimum locking range is attained when the whole modulation bandwidth is available, by replacing we obtain $\Delta \nu_{max}$ = 42 Hz. This value represents a lower limit in this setup, since we are not using the whole modulator bandwidth –due to the reflection bandwidth of the FBG–. Thus, a higher measured locking range is an expected result.

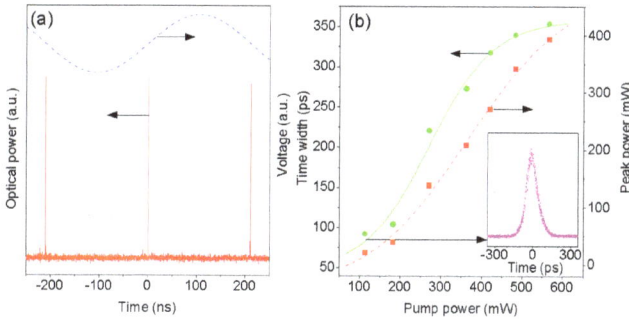

Fig. 8. (a) Voltage signal used to drive the piezoelectric and mode-locked train of pulses generated at 4.75 MHz repetition rate with 271 mW of pump power (dashed and solid curves, respectively). (b) Time width (FWHM) and peak power of the optical pulses as a function of the pump power (solid and open scatter points, respectively). The inset shows a single pulse of 95 ps time width

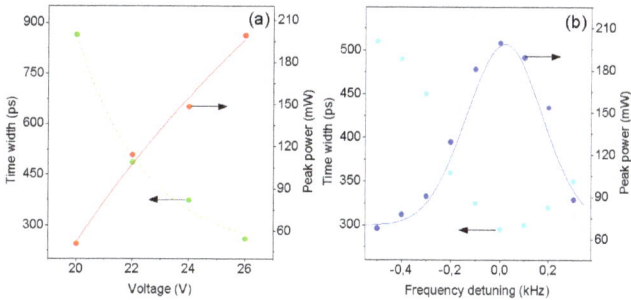

Fig. 9. (a) Time width (FWHM) and peak power as a function of the applied voltage; for a fixed frequency of 2.3731 MHz (open and solid scatter points, respectively). (b) Same as before, but as a function of the frequency detuning, for a fixed voltage of 26 V. In both cases, a pump power of 350 mW was used

Next, we carried out an experimental study of this actively mode-locked all-fiber ring laser, looking towards an improvement in its performance. To this end, we slightly modified the setup shown in Fig. 6, by replacing the FBG used before with a near 100 % reflective FBG, with a FWHM bandwidth of 0.3 nm, which should be compared with the preceding FBG with FWHM of 0.45 nm, and 50 % of maximum reflectivity. Together with this, the EDF length was shortened up to 2.85 m. As a result of these changes, output light pulses were obtained by one of the ports of a 3 dB coupler incorporated within the ring (Bello-Jímenez et al., 2011). With this new configuration, output light pulses were obtained with a maximum peak power of 380 mW and pulse width of 90 ps. By comparing the narrowest pulses obtained with this configuration with the narrowest pulses reported before (95 ps) where the same type of AOM was used, we find nearly no difference. However, from the laser operation point of view, there is an important improvement, since the present arrangement is rather more stable, and the laser can be adjusted more easily. We attribute this improvement to the FBG used, which has a flat spectral reflectivity, making less critical the spectral matching between the FBG reflectivity with the resonant dip of the AOM. On the contrary, the FBG used before, was of nearly the same FWHM but with a lower reflectivity (50 %). As a consequence, any spectral detuning resulted in a larger variation in reflectivity, which in turn induced a higher difficulty to stabilize the laser operation. Next, we analyzed the change in the pulse's parameters as the EDF length was varied, see Fig. 10 (a). Changes in the EDF length induced a slight shortening in the output light pulses, which is in accordance with Eq. 1, if we assume that gain scales with the EDF length. By this trend, the shortest light pulses (65 ps) were obtained at 2.2 m of EDF length. Further narrowing by this trend was not possible, since no mode-locked output pulses were observed for EDF lengths shorter than 2.15 m.

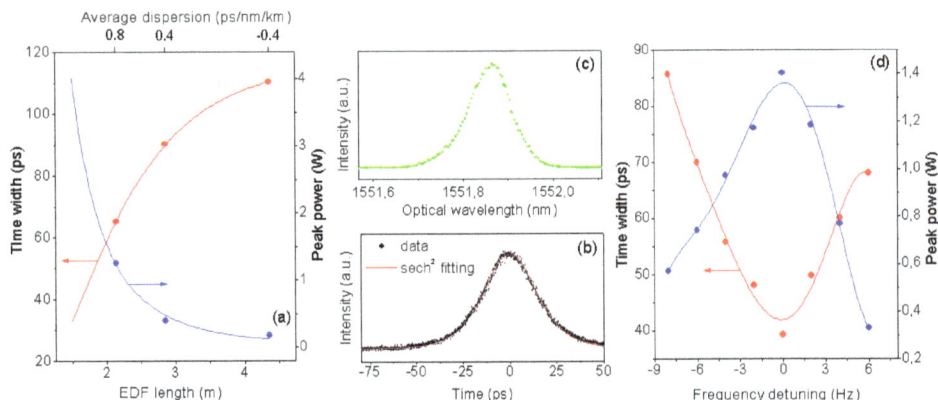

Fig. 10. (a) Time width (FWHM) and peak power as a function of the EDF length, using a FBG with 0.3 nm bandwidth. Using a 0.7 nm bandwidth: (b) single mode-locked pulse at 4.73 MHz repetition frequency, together with its corresponding fitting by a sech² function; (c) optical spectrum of the laser; (d) time width and peak power versus the frequency detuning

An important parameter of our laser arrangement is the FBG bandwidth, which acts as a spectral filter. Its bandwidth has to be selected narrower than the AOM bandwidth; in order to insure that the modes are properly amplitude modulated. For this reason we used first a 0.3 nm bandwidth FBG, which fulfills this requirement, since the AOM bandwidth is of 1.5 nm. However, it can be easily understood that a larger number of modes locked –i.e. amplitude modulated– produce the narrowest mode-locked train of pulses, which is in accordance with Eq. 1. In this respect, there is margin for improvement, since the bandwidth of the FBG used is far to be close to the AOM bandwidth. For this reason we replace the flat-top FBG with another FBG, which also has a flat-unit reflectivity, but a broader optical bandwidth of 0.7 nm. Fig. 10(b) shows the shortest pulse obtained with this configuration, by using an EDF length of 2.45 m at a pump power of 265 mW. The measured temporal width was 34 ps (FWHM), with a maximum peak power of 1.4 W. For this pulse the spectral linewidth was measured to be 110 pm (i.e. 13.7 Ghz at 1551.9 nm), using a 50 pm resolution optical spectrum analyzer, see Fig. 10(c). On the other hand, a Fourier-transform limited sech^2 pulse should have a spectral linewidth of 9.3 GHz. From the comparison between this last value and the linewidth measurement, we conclude that the optical pulses of our mode-locked laser could have some moderate degree of chirp. Finally, we present the variation of the pulse parameters as a function of the frequency detuning for this configuration, see Fig. 10(d); the measured allowable frequency detuning was of 8 Hz. According to theory of detuning in AM mode locking, the maximum allowable detuning is inversely proportional to the modulator optical bandwidth, in this case replaced by the FBG bandwidth, see Eq. 2. Since the later has been increased noticeably, i.e. from 0.3 nm to 0.7 nm, a reduction in the allowable detuning is an expected result.

3. Applications of acousto-optic devices based on the interaction of longitudinal acoustic waves with fiber Bragg gratings

When a travelling axially-propagating longitudinal acoustic wave is launched through a FBG, the periodic strain field of the acoustic wave perturbs the grating in two different ways. First, the average index changes in response to the stress-optical effect and second, the otherwise uniform Bragg grating pitch changes being modulated by the acoustical signal (Russell & Liu, 2000). As a consequence of both effects the reflectivity changes, the main features of these changes depend on the ratio between the acoustical wavelength and the grating length (Andrés et al., 2008). Thus, we can distinguish two well-different situations: the long-wavelength and the short-wavelength regimes. In the first, the Bragg grating pitch is homogeneously perturbed along its length. The successive cycles of compression and expansion generated by the longitudinal wave will shift periodically in time the spectral response of the grating as a whole to longer and shorter wavelengths (Cuadrado-Laborde et al., 2007; Andrés et al., 2008). On the contrary, in the short-wavelength regime the acoustic wave generates many compressed and expanded sections, which gives rise to a superstructure within the grating. In this case, the spectral response of the original FBG shows new and narrow reflection bands symmetrically at both sides of the original Bragg wavelength (Liu et al., 1997, 1998). The position and strength of these sidebands can be controlled by varying the frequency and voltage applied to the piezoelectric, respectively. Subsections 3.1 and 3.2 show the use of these effects to modulate the Q-factor of a fiber laser and as an active mode-locker; i.e. long and short-wavelength regimes, respectively. Subsection 3.3 shows the use of the acousto-optic interaction in a FBG in the short-wavelength regime as a tunable photonic true-time-delay line based on the group delay change of the light reflected from the grating sidebands. Finally, when the acoustic perturbation is not a harmonic wave but a single pulse, its passage through the grating creates a defect in the FBG, which can be used to actively Q-switch a distributed feedback fiber laser, which is discussed in subsection 3.4.

3.1 *Q-switching of an all-fiber laser by acousto-optic interaction in a fiber Bragg grating in the long-wavelength regime*

As an example of the use of the long-wavelength regime in FBGs, we present an all-fiber acousto-optic modulator suitable for Q-switching applications (Cuadrado-Laborde et al.,

2007). It consists of a short-length fiber Bragg grating modulated by a standing longitudinal acoustic wave, whose wavelength is much longer than the grating length. Periodic stretching and compression of the grating due to the elastic wave, causes that the Bragg wavelength is harmonically, continuously and repeatedly tuned in time over a given wavelength range. Figure 11 illustrates the operation principle. In the long-wavelength regime, the acoustic wavelength is much larger than the grating length.

However, the amplitude of the acoustic waves that can be achieved in a realistic arrangement is rather small to produce a significant perturbation of the FBG. We found that this limitation could be overcome by using an acoustic cavity. Thereby, if the grating is placed in the proper position within the acoustic cavity, a low-frequency standing longitudinal elastic wave will produce periodic stretching and compression of the grating. As a consequence, the reflection band will shift periodically in wavelength around the original Bragg wavelength, sweeping a given wavelength range in a harmonic way. The elastic wave may also introduce some chirp in the grating due to non-uniformity of the strain induced along the grating. To achieve efficient modulations this effect must be minimized. Thus, the grating length must be short compared with the acoustic wavelength, and the positioning of the grating within the cavity must match an anti-node of the standing acoustic wave.

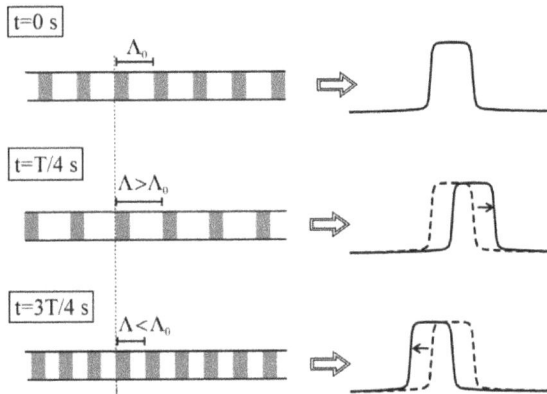

Fig. 11. Interaction of acoustic waves in the long-wavelength regime and FBGs; T is the period of the acoustic wave, Λ is the instantaneous pitch of the Bragg grating, and Λ_0 is the pitch of the Bragg grating in absence of acoustic wave. The effect on the reflection band is shown qualitatively. The standing acoustic wave stretches and compresses the grating periodically. During the time that the grating is stretched, the reflection band of the grating shifts towards longer wavelengths, whereas it shifts towards shorter wavelengths when it is compressed

Figure 12(a) shows a diagram of the laser setup, together with a detail of the acousto-optic modulator. The Fabry-Perot cavity becomes defined by FBG_1 (which also acts as modulator) and FBG_2. A piezoelectric disk (15 mm in diameter and 2 mm thickness) was used to generate the longitudinal acoustic waves, which were launched to FBG_1 through a silica horn. The base of the horn was glued to the piezoelectric and the tip reduced to ~125 μm and then fusion-spliced to FBG_1. At a distance of 210 mm from the piezoelectric disk, the fiber was clamped allowing standing acoustic waves to be created in the fiber section between the horn and the clamp. FBG_1 was written in the core of a photosensitive fiber by

UV irradiation using a phase-mask technique. The photosensitive fiber was previously tapered to enhance the acousto-optic interaction; the taper waist was 50 mm long, 90 μm in diameter, and the transition lengths were 13 mm. The grating was recorded in the taper waist (uniform diameter section); the FBG$_1$ length was 22 mm. The highest frequency used in our experiments was 61 kHz, and according to the velocity of sound in silica fibers for longitudinal elastic waves (5760 m/s), the wavelength for this frequency is 94 mm. Thus, we can verify that FBG$_1$ length is short enough to guaranty a standard long-wavelength regime. An apodization profile was introduced also during the grating inscription to reduce side-lobes in the spectrum. Figure 12(b) shows the transmission spectrum for both FBGs, the bandwidth of FBG$_1$ was 0.15 nm and the reflectivity was higher than 99.9%. On the short-wavelength side, the transmission drops sharply (~ 0.72 dB/pm). The side-lobes have been reduced entirely on this side, while a significant side-lobe still remains on the long-wavelength side of the transmission band. On the other hand, FBG$_2$ acted as the output coupler, its Bragg wavelength was 1543.27 nm, the reflectivity 50%, and the bandwidth was 40 pm.

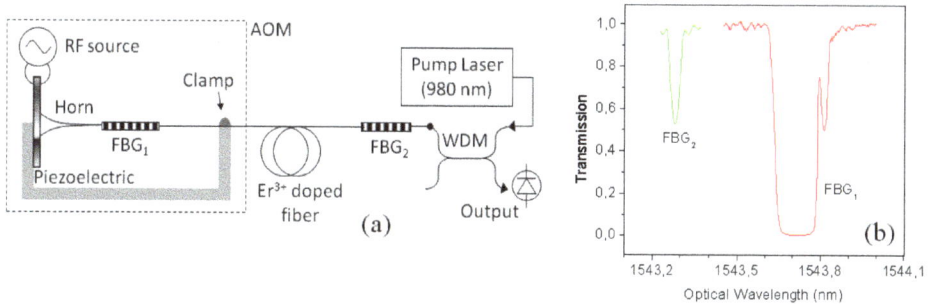

Fig. 12. (a) Laser setup, the dashed line defines the acousto-optic modulator. (b) Transmission spectra of the two fiber Bragg gratings

Next, we discuss the modulator´s performance. We detect the intensity of the light reflected by the mid-reflection point of the short-wavelength band-edge, when an AC electric signal of a given frequency was applied to the piezoelectric. To this end a tunable laser diode was used. The modulation of the grating caused by the elastic wave generates periodic modulation of the light intensity reflected by the grating. An example is shown in Fig. 13(a). At low voltage, a sinusoidal modulation was observed since the grating shifts little and the laser reflects at the band-edge at any time during the period of the electric signal. As the electric signal amplitude increases, the grating shifts further and then, the laser reflects also at the top and bottom of the band-edge, generating a square signal. We observed strong acoustic resonances at frequencies of 18 kHz, 37 kHz and 61 kHz. The peak-to-peak amplitude was measured for these frequencies as a function of the voltage applied to the piezoelectric, see Fig. 13(b). The most efficient frequency was 61 kHz, and the responses at 18 kHz and 37 kHz were quite similar. Taking into account the results shown in Fig. 13(b) and the width of the band-edge, 42 pm, the modulation efficiency (i.e. wavelength shift per unit volt) was estimated, giving 7 pm/V at 18 kHz and 37 kHz, and 16 pm/V at 61 kHz.

The switching speed is one of the key parameters of any modulator intended to be utilized as Q-switching element in a fiber laser, being desirable to be as high as possible. In this modulator, the switching time depends on the AC voltage amplitude, and it becomes shorter as the voltage is increased, as shown in Fig. 13(a). Switching times of the order of a few microseconds were achievable. Finally, although the acoustic generator was designed for the excitation of longitudinal waves, we also observed flexural waves in the resonator when large voltages were applied to the piezoelectric. A large number of flexural standing waves, spaced fractions of kHz, were observed in the range from 0 to 100 kHz. These resonances gave rise to light modulated at double frequency of the electrical excitation and exhibited relatively low amplitude. The aforementioned modulator uses an acoustic resonator, which enhances the interaction between acoustic waves and the FBG. The fundamental mechanical resonance at 18 kHz had a Q factor of 1300. With respect to an un-clamped configuration based on traveling acoustic waves, the use of an acoustic cavity increases the modulation efficiency, but it restricts the operation of the modulator at the resonant frequencies of the cavity. A modulator based on traveling acoustic waves would allow, in principle, tuning continuously the modulation frequency, but a large voltage would be required and important thermal effects would damage the acoustic transducer.

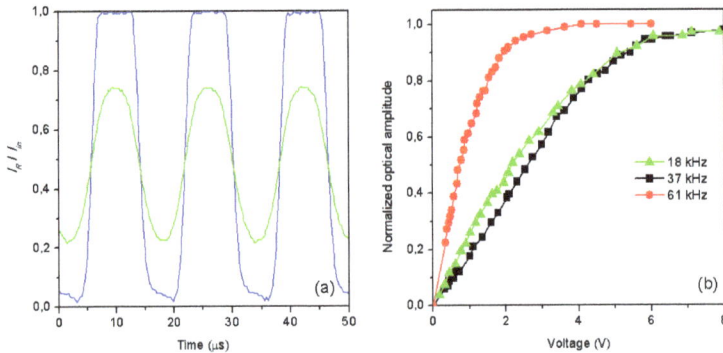

Fig. 13. (a) Light reflected from the band-edge of the grating when an AC electric signal of 61 kHz and peak-to peak amplitude of 4.6 V (green) and 14 V (blue) were applied to the piezoelectric. (b) Light modulation amplitude as a function of voltage for three different resonant frequencies

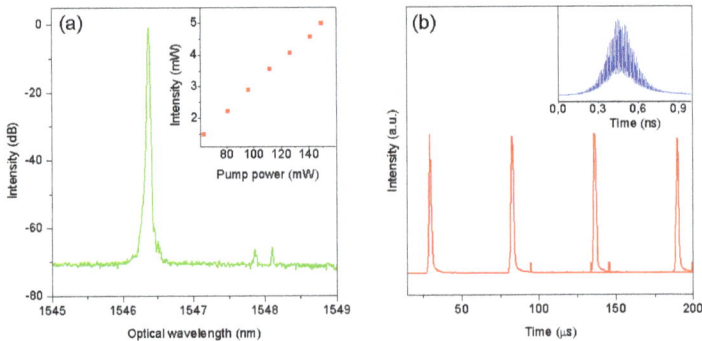

Fig. 14. (a) CW emission spectrum at 135 mW pump power; the inset shows the output CW power versus pump power. (b) Pulse train at 18 kHz repetition rate; the inset shows a single light pulse measured with a 1 GHz bandwidth photodetector, pump power of 80 mW

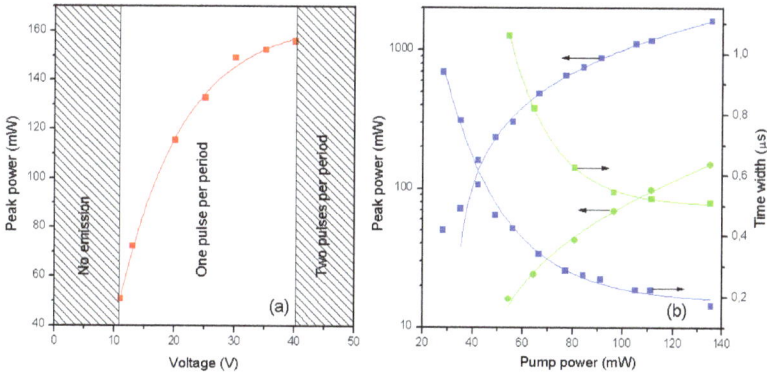

Fig. 15. (a) Pulse peak power as a function of the applied voltage to the piezoelectric (75 mW pump power, 18 kHz repetition rate). (b) Pulse peak power and pulse width as a function of pump power, for repetition rates of 18 kHz (blue) and 37 kHz (green). Voltage applied to the piezoelectric 24 V

The gain of the Q-switched all-fiber laser was provided by 0.3 m of an erbium-doped fiber containing 1000 p.p.m. Er^{3+}, with a cut-off wavelength of 965 nm, and a numerical aperture of 0.23. The active fiber was pumped through a WDM coupler by a pigtailed laser diode emitting at 979 nm, providing a maximum pump power of 140 mW. The Fabry–Perot cavity length was of 1.2 m. A translation stage was used for tuning the reflection band of FBG$_2$ closer to the short-wavelength band-edge of FBG$_1$, allowing full overlapping between both bands with moderate voltages being applied to the piezoelectric. In this arrangement, the longitudinal standing elastic wave causes periodic optimization of the Q factor of the laser cavity and, as a result, strong pulses are emitted at the repetition frequency of the acousto-optic modulation, when the fiber is pumped. The CW features of the fiber laser are shown in Fig. 14(a). The laser line was narrower than 20 pm (resolution limit of our measurement system), a signal-to-noise level as high as 70 dB was obtained. The efficiency and the threshold were 3 % and 25 mW, respectively. Figure 14(b) shows a pulse train emitted by the laser at a repetition rate of 18 kHz. At high pump powers each Q-switched pulse breaks into a train of narrower, 8 ns wide, sharp pulses (see inset). The frequency of this modulation, 83 MHz, coincides with the frequency mode-spacing of the laser cavity, which indicates beating between longitudinal cavity modes. We observed that the modulation amplitude of this high frequency component was affected by variations in pump level and repetition rate. This apparent passive mode-locking response has been previously reported repeatedly in both, passively and actively Q-switched fiber lasers (Philippov et al., 2004; Andersen et al., 2006). Figure 15(a) shows the peak power of the Q-switched pulse as a function of the applied voltage to the piezoelectric. Three regions can be clearly distinguished. For low voltages applied to the piezoelectric there is not emission, since the wavelength shift of FBG$_1$ is not enough to overlap the reflection band to FBG$_2$. As the applied voltage increases, there is a Q-switched pulse per period. For voltages close to the lower limit, the overlapping between both gratings is only partial and the laser is only a small amount above threshold, so weak pulses are emitted. As the voltage is increased, the gratings overlap further and the peak power increases. A steady state is reached when the voltage amplitude is high enough to obtain complete overlapping between the two reflection bands. If the applied voltage is increased further, the cavity stays in the high Q state longer and more than one Q-switched pulse is emitted per period. When changing the pump power and/or repetition rate, the upper and lower AC voltage limits vary, but the trend is preserved. Finally, the effect of pump power on the Q-switched pulses, for repetition rates of 18 and 37 kHz, is shown in Fig. 15(b). The pump power threshold at each frequency was 27 mW and 53 mW, respectively. Above threshold, the peak power increases with pump power, and there is a corresponding reduction of the pulse width. No evidences of peak power saturation were observed within the pump power range of our experiments. Pulses of 1.6 W peak power and 172 ns width were obtained at 18 kHz repetition rate and 135 mW pump power.

3.2 Mode-locking and Q-switching mode-locking by acousto-optic interaction in a fiber Bragg grating in the short-wavelength regime

In this subsection we discuss another mechanism developed by our group to mode-lock an all-fiber laser (Cuadrado-Laborde et al., 2009a, 2010a). This is based on the acousto-optic super-lattice modulation, and it is an example of the interaction of longitudinal acoustic waves and Bragg gratings in the short-wavelength regime (Liu et al., 1997, 1998). In the following, we start by reviewing the behavior of the acousto-optic super-lattice modulator, subsection 3.2.1. Then, we show the use of this device as mode-locker, by showing two different mode-locked lasers, either by driving the acousto-optic modulator by standing or travelling acoustic waves, subsections 3.2.2 and 3.2.3, respectively. The possibility to reach simultaneous Q-switching and mode-locking is also discussed in these subsections.

3.2.1 The fiber Bragg grating based acousto-optic modulator

The spectral response of the original FBG in this acoustical regime shows new –and narrow– reflection bands symmetrically at both sides of the original Bragg wavelength (Liu et al., 1997, 1998). The position of these sidebands can be controlled by varying the frequency at a slope of 0.15 nm /MHz and 0.30 nm /MHz for the first and second order sideband, respectively. The strength of the reflection bands, on the other hand, can be controlled independently by varying the voltage applied to the piezoelectric. Since these sidebands can be regarded as weak ghosts of the strong permanent Bragg grating, its FWHM bandwidth is that of a weak Bragg grating of the same length (Liu et al., 1997). Figure 16 shows the setup for a typical reflectivity measurement on an AOSLM. The AOSLM in turn is composed of an RF source, an electrical RF amplifier, a piezoelectric disk, a silica horn, and a FBG. The tip of the silica horn was reduced by chemical etching to the same diameter of FBG –125 μm– and subsequently fusion-spliced to the fiber. The uniform and non-apodized grating was written in photosensitive fiber using a doubled argon laser and a uniform period mask; the FBG was 120 mm long. The reflection properties of the AOSLM were investigated by illuminating the FBG through an optical circulator with a broadband light source, and detecting the reflected light with an optical spectrum analyzer (OSA). Figure 17(a) shows the reflectivity of the unperturbed FBG –i.e. without electrical signal applied to the piezoelectric– and with an electrical signal applied to the piezoelectric of 4.55 MHz and 16 V (whenever we refer to voltages throughout this chapter, it is a peak-to-peak measurement). The presence of the sidebands symmetrically positioned around the Bragg wavelength is clearly discernible. Further, these sidebands are produced either by standing or travelling acoustic waves. However, the light reflected by the sidebands in each case behaves differently. When travelling acoustic waves are used –by dumping the end of the FBG opposite to the silica horn for example with a drop of oil–, the light reflected on these sidebands is completely downshifted or upshifted by the frequency of the acoustical signal, depending if the reflection was in the long or short-wavelength sideband, respectively (Liu et al., 1997, 1998). Together with this and as a function of the instantaneous phase of the acoustic signal, there is also present an amplitude modulation at the frequency of the acoustical signal. Figure 17(b) shows the measurement performed by tuning a laser diode to the center of the short-wavelength sideband and measuring the reflected light. As expected, the amplitude modulation is at the same frequency of the electrical signal used to drive the piezoelectric. On the other hand, when standing acoustic waves are used –by clamping the end of the FBG opposite to the silica horn– the sidebands raise and fall at twice the frequency of the electrical signal (Cuadrado-Laborde et al., 2009a). Further, in principle, for a perfect acoustical reflection, the light reflected by the sidebands does not experiment any Doppler shift, as opposed to the previous case when travelling acoustic waves are used. Figure 17(c) shows the measurement performed by tuning a laser diode to the center of the short-wavelength sideband and measuring the reflected light. Now the optical modulation frequency is two times the acoustical signal frequency. To summarize, an AOSLM can be driven in two different regimes, either by using travelling or standing acoustic waves. In both cases we can use it as an amplitude modulator. However in the first case –travelling acoustic waves– the light reflected on the sidebands is modulated at the same frequency of the acoustical signal, whereas in the second case –standing acoustic waves– it is modulated at two times this frequency, respectively.

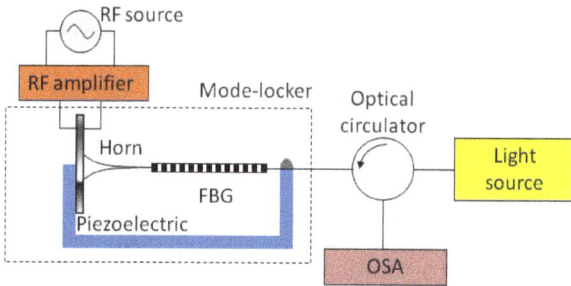

Fig. 16. Setup for the characterization of the acousto-optic super-lattice modulator; OSA stands for optical spectrum analyzer, FBG for fiber Bragg grating and RF for radio-frequency

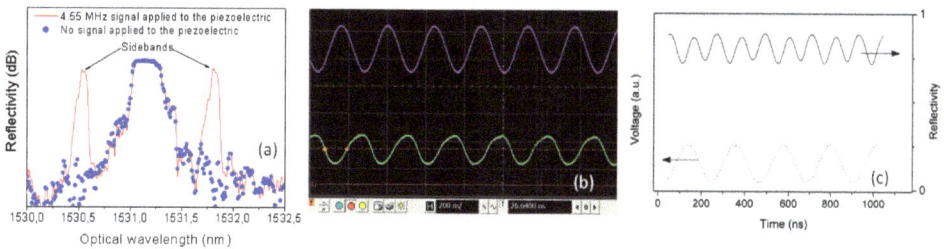

Fig. 17. (a) Reflectivity of the fiber Bragg grating with and without electrical signal applied to the piezoelectric (4.55 MHz and 16 V peak-to-peak). (b) Optical signal reflected by the short-wavelength sideband when travelling acoustic waves are used (lower trace) and RF voltage applied to the piezoelectric (4.11 MHz and 16 V, upper trace). (c) Same as in (b), but for standing acoustic waves (4.55 MHz and 16 V, dotted curve)

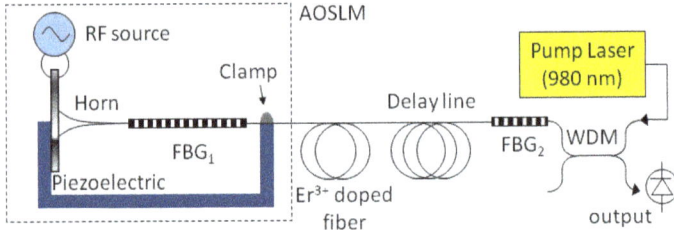

Fig. 18. Setup of the mode-locked all-fiber laser by using an acousto-optic supper-lattice modulator (AOSLM) as mode-locker

3.2.2 All-fiber actively mode-locked laser with a fiber Bragg grating based acousto-optic modulator driven by standing acoustic waves

Now, we show the use of the amplitude modulation induced by standing acoustic waves in the AOSLM to mode-lock an all-fiber laser (Cuadrado-Laborde et al., 2009a, 2010a). The setup proposed for the standard mode-locking laser is schematically illustrated in Fig. 18. The gain was provided by an erbium-doped fiber (EDF) containing 300 parts per million (ppm) of Er^{3+}, with a cut-off wavelength of 939 nm, and a numerical aperture of 0.24. The active fiber was pumped through a WDM coupler by a pigtailed laser diode emitting at 976 nm, providing a maximum pump power of 160 mW. The acousto-optic super-lattice modulator and a short delay line followed by a second fiber Bragg grating FBG_2 were fusion-spliced at each end of the active fiber. FBG_2 (10 mm long and with a Bragg wavelength of 1530.2 nm) was written with a uniform period in a photosensitive fiber using a doubled argon laser and a uniform period mask. The Fabry-Perot cavity was established once the reflection band of FBG_2 is made to match the short wavelength sideband of the FBG_1 of the AOSLM, i.e. 1530.5 nm, by straining it with a translational stage (see Fig. 18). The delay line length must be selected to match the round-trip time with the reciprocal of the optical modulation frequency, which in turn is twice the electrical frequency applied to the piezoelectric. Since the selected piezoelectric operation point for the piezoelectric was 4.55 MHz, this results in a cavity length of 11.4 m. However, fine tuning of the electrical frequency was required to achieve mode-locking.

Figure 19(a) exemplifies the laser behavior showing the train of optical pulses generated, at a frequency rate of 9 MHz (50 GHz bandwidth oscilloscope). Figure 19(b) shows a single optical pulse; the timing jitter was measured to be 40 ps (RMS). The pulses of this laser are best fitted by a $sech^2$ function, which is in accordance with earlier amplitude modulated mode-locking theory. Further, the dispersion measurements of the different fibers we used to form the laser cavity show that both kind of dispersion coexist, from the anomalous dispersion of the delay line to the normal dispersion of the EDF. For this type of dispersion-mixed cavities, the analytical steady-state solution is expected to be also of the $sech^2$ type (Belanger, 2005), which adds further evidence in the same direction. The emission linewidth was measured using a high resolution optical spectrum analyzer (BOSA-C, Aragón Photonics, resolution of 80 fm). Figure 19(c) shows the spectrum of the optical pulses shown in Fig. 19(b). The high resolution of the optical spectrum analyzer permits a direct observation of the individual cavity modes separated by 9 MHz, see the inset of Fig. 19(c). The measured 3-dB linewidth results in 2.8 pm, i.e. 360 MHz at 1530.5 nm. On the other side, the optical pulses have a temporal width of 780 ps, so according to the Fourier-transform-limited relation for a sech² pulse, time-bandwidth product of 0.315 (Wada et al., 2008), its bandwidth cannot be lower than 380 MHz. From the comparison between this last value and the linewidth measurement, we conclude that the optical pulses of our modelocked laser are transform-limited, i.e. un-chirped. A brief digression is in order here, the Fourier-transform-limited relation for Gaussian pulses with a time-bandwidth product of 0.44 (Kuizenga & Siegman, 1970; Wada et al., 2008), would result in a much higher linewidth (564 MHz), which is in contradiction with our spectral measurements. As a consequence, the fitting of the temporal pulses, see Fig. 19(b), the application of the Fourier-transform-limited relations, and dispersion measurements together with previous theoretical work, seems to corroborate a sech² nature for the emitted pulses.

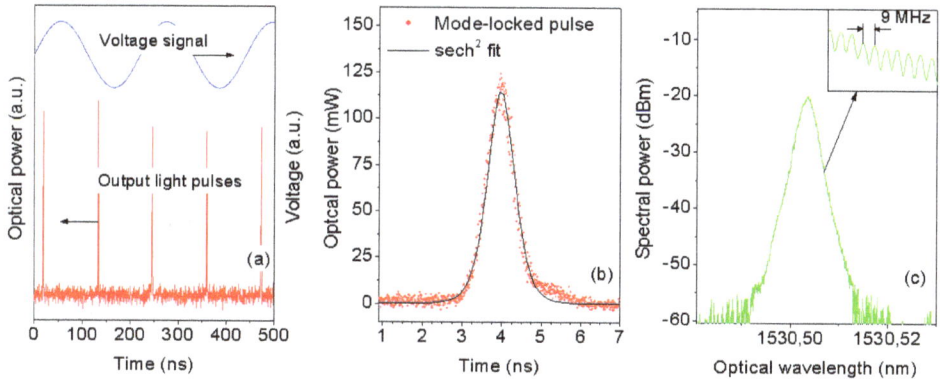

Fig. 19. (a) Voltage signal applied to the piezoelectric at 4.55 MHz and 16 V, and its corresponding modelocked train of pulses generated at twice this frequency with a pump power of 160 mW. (b) A single optical pulse and its corresponding fitting by a sech² function. (c) Optical spectrum of the output light pulses shown in (b); the inset shows the cavity modes distant by 9 MHz

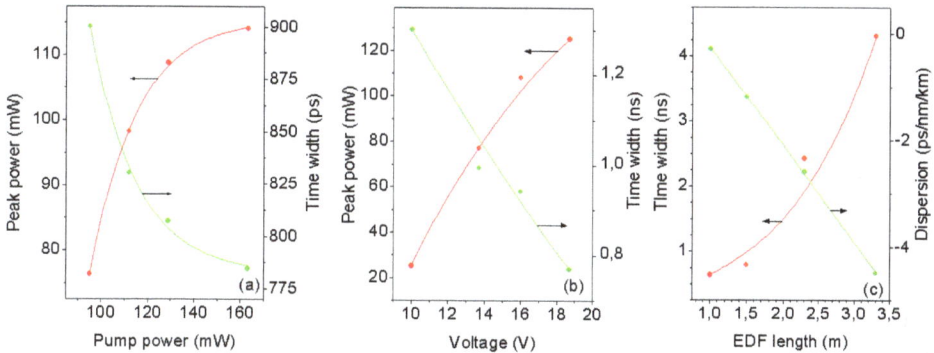

Fig. 20. (a) Peak power and pulse width of the optical pulses as a function of the pump power. (b) Peak power and pulse width of the output light pulses as a function of the voltage applied to the piezoelectric, for a pump power of 160 mW. (c) Pulse width of the output light pulses, as a function of the EDF length, for a fixed pump power of 160mW, additionally, the average dispersion of the cavity is also shown for each EDF length

Now we will show the variation in the pulse's parameter as typical variables are changed. First we analyze the optical pulse's peak power and temporal width as a function of the pump power, which is shown in Fig. 20(a). A smooth variation of the pulse width with pump power can be observed, before reaching the gain saturation. A higher available pump power would allow additional axial modes to contribute, increasing the spectral bandwidth, and thereby decreasing the pulse's width. Next, we analyzed the variation in the pulse's parameters as a function of the modulation voltage. The intensity of the reflection sidebands of the AOSLM can be dynamically controlled by varying the voltage applied to the piezoelectric as we explained before in subsection 3.2.1. This voltage controls the amplitude of the propagating acoustic wave and the amplitude of the standing wave, as a function of the reflection at the clamp. Thus, as opposed to mode-locking by bulk acousto-optic or electro-optic modulators, here the reflectivity and the modulation depth are intrinsically linked, i.e. when the applied voltage increases, both the reflectivity and the modulation amplitude increase. Figure 20(b) shows the peak power and time width, respectively, as a function of the voltage applied to the piezoelectric. At lower reflectivities and modulation amplitudes, the peak power diminishes, whereas the time width increases. According to Eq. (1), the pulse width and the modulation depth are inverse parameters, and in turn the later rises with the applied voltage. However, the pulses shorten very slowly with the increased modulation strength; hence this shortening procedure is generally discouraged (Kuizenga & Siegman, 1970). One can conclude that further increase of the voltage might improve the pulse parameters, but it is not likely to lead to a great enhancement in terms either of peak power or temporal width. We analyzed also the influence that dispersion has on the pulse's parameters, when the average dispersion of the cavity is changed. The dispersion for each type of optical fiber used in the setup was measured by the frequency-domain modulated-carrier method. The resulting average dispersion of this cavity is normal and its value is –1.2 ps/nm/km. Next, the optical fiber of the delay line –the Corning LEAF fiber– was replaced by a normal dispersion fiber –Fibercore SM980– leading to a lower (normal) overall dispersion of –5.8 ps/nm/km. With the new delay line, a higher pump reaches the EDF, since a better effective area compatibility is insured throughout the different fibers of the system –in fact, unlike LEAF fiber, the SM980 insures single-mode propagation of the pump power–. As a result, output pulses have higher peak power and shorter time width than in the previous configuration. In this way, optical pulses with temporal width and peak power of 640 ps and 160 mW, respectively, were obtained. These pulse parameters represent an improvement of 18% and 28% with respect to the previous configuration. Finally, in order to reverse the sign of the average dispersion of the cavity, we used an SMF28 optical fiber for the delay line, resulting in an average –anomalous– dispersion of 9.8 ps/nm/km. In this case, we did not observe mode-locking lasing. One reason for this could be motivated by the large average dispersion introduced within the cavity, since modelocked fiber lasers usually works with close-to-zero average dispersion cavities. Finally, the change on the pulse parameters was measured as a function of the EDF length –ranging from 0.5 to 3.3 m–, but keeping a constant cavity length of 11.4 m, by adjusting each time the length of LEAF fiber. The average dispersion remains normal regardless of the amount of EDF used, ranging from –0.3 ps/nm/km up to –4.5 ps/nm/km. The temporal pulse width as a function of the EDF length is shown in Fig. 20(c), for a fixed pump power of 160 mW. This figure includes the average dispersion of the cavity as a function of the EDF length. A direct relationship can be observed between the EDF length, the dispersion, and the time width of the pulses, with a minimum of 630 ps obtained for an EDF length of 1 m. This kind of interplay between gain and time width agrees with Eq. (1). Further narrowing of the optical pulses could not be reached by this trend, since no mode-locking lasing was observed with 0.5 m of EDF and the available pump power, as a result of insufficient cavity gain. Finally, we also developed the ytterbium-doped fiber version of this laser, with equally satisfactory results (Villegas et al., 2011).

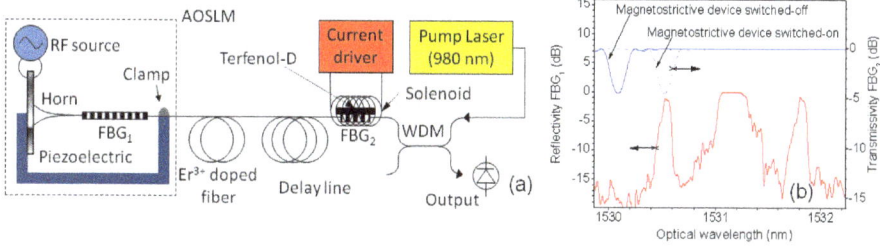

Fig. 21. (a) Q-switched mode-locked all-fiber laser setup when the AOSLM is driven by standing acoustic waves. (b) Reflection spectrum of the AOSLM when an RF signal of 4.55 MHz and 16 V is applied to the piezoelectric (left ordinate). Transmittance of FBG$_2$ (solid curve) when the magnetostrictive device is switched-off; the dotted curve shows the new spectral position of FBG$_2$ when the magnetostrictive device is switched-on (right ordinate)

Until now we have discussed the standard mode-locking operation. Now we demonstrate the possibility to simultaneously Q-switch and mode-lock, actively and independently. By combining both operations in a single laser, a superior performance is achieved with higher peak powers compared with ordinary mode-locked lasers, but almost the same pulse width as in a mode-locked laser is retained. Thus, the peak power of the central pulses of the mode-locked train, underneath the Q-switched envelope, can be greatly enhanced. The increased peak power provided by the Q-switching technique can be advantageous in applications such as wavelength conversion or super-continuum generation. It is worth to say, that any passive mechanism is used in this setup, neither for Q-switching nor for mode-locking. To this end, we design a solution for actively Q-switching, while simultaneously mode-locking, preserving the all-fiber configuration. This solution is based on fiber Bragg grating fast modulation using a magnetostrictive device. This feature provides a direct control of the repetition rate and a fully-modulated train of modelocked pulses. To the best of our knowledge, this was the first doubly active Q-switching mode-locking strictly all-fiber laser presented (Cuadrado-Laborde et al., 2009b). The setup of our doubly-active Q-switching mode-locking all-fiber laser is shown in Fig. 21(a). It is basically the same as it was described before for the standard mode-locking and shown in Fig. 18, with the same length of EDF and delay line, except by the magnetostrictive device controlling the FBG$_2$, see Fig. 21(a). It is composed of a 15 mm long (1 mm^2 cross-section) magnetostrictive rod of Terfenol-D bonded to the FBG$_2$. The rod and the fiber were placed inside a small coil driven by an electronic circuit designed to drive square current pulses with amplitudes up to 260 mA of any required duty cycle. Figure 21(b) shows the spectral positions of both FBGs –i.e. the FBG of the AOSLM and FBG$_2$, which helps understand the operation principle of this laser also. When the magnetic pulses generated in the solenoid stretch FBG$_2$, its central wavelength is brought to match the short-wavelength sideband of FBG$_1$ for a short period of time, which results in an increased Q-value. In this way by modulating the coil current with a given Q-switched frequency, the Q-factor is actively modulated at the same frequency. The magnetostriction has a low pass frequency response and, consequently, presents the advantage to permit a continuous tuning of both the Q-switched repetition rate and the duty cycle of the modulation pulses. The dependence of the wavelength shift with the coil current was measured by illuminating FBG$_2$ with a tunable laser source and detecting the reflected light. A quasi-linear behavior was observed, as a result of using moderate magnetic fields.

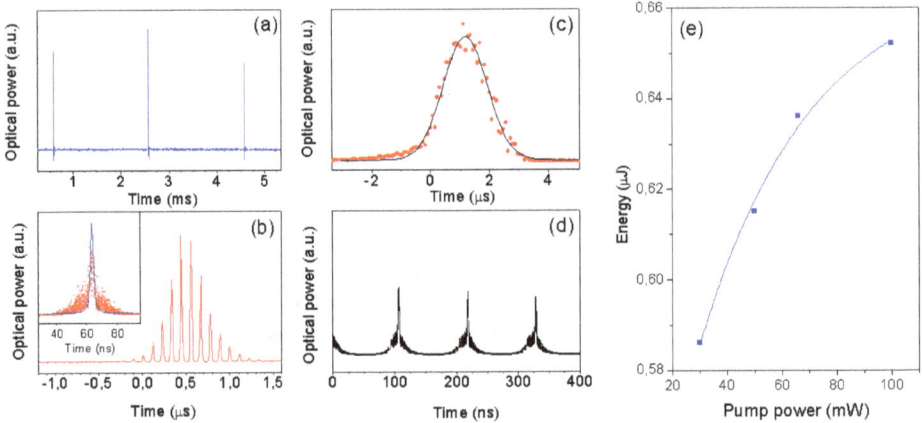

Fig. 22. (a) Q-switched mode-locked train of pulses at a Q-switching frequency of 500 Hz. (b) Single Q-switched pulse enveloping a train of 12–14 modelocked pulses; the inset shows a single modelocked pulse. (c) Single Q-switched pulse with no mode-locked. (d) Mode-locking operation of the laser without Q-switching. In all cases, the pump power was 70mW. (e) Energy of the Q-switched mode-locked pulses as a function of the pump power for a Q-switching repetition rate of 500 Hz

Now we discuss the Q-switching mode-locking laser operation. The laser emission wavelength was in 1530.55 nm, since the overlapping between the short-wavelength sideband of FBG$_1$ and the shifted position of FBG$_2$ takes place at this wavelength, see Fig. 21(b). Figure 22(a) exemplifies the Q-switched mode-locked laser behavior showing the train of optical pulses generated for a pump power of 70 mW. The optical Q-switched mode-locked pulses are 2 ms apart, since the repetition rate of the train of voltage pulses applied to the current driver was of 500 Hz. Figure 22(b) shows a single Q-switched envelope with a FWHM of 550 ns; it has between 12 and 14 fully-modulated modelocked pulses, with the expected temporal separation of 110 ns. However, the individual short pulses are broadened due to the limited bandwidth and sampling rate of the oscilloscope. In order to measure the FWHM of the individual short pulses, a 50 GHz sampling oscilloscope was used. In this case, the large difference between the Q-switching and mode-locking frequencies make difficult to trigger properly the oscilloscope. Even so, relatively good traces were recorded as the one depicted in the inset of Fig. 22(b). As we discussed before when we presented the standard mode-locking regime, the tuning sensitivity between the RF signal applied to the piezoelectric and the cavity round trip is not too critical in this setup. Only when the detuning is considerably –e.g. by a few kHz–, the pulse clearly deteriorates. Figure 22(c) shows the Q-switched pulse obtained by avoiding the mode-locking pulse formation, just by detuning the applied voltage to the piezoelectric by 10 kHz, in this way only the Q-switching operation is allowed within the cavity. As expected, the Q-switched pulse reproduces the waveform of the envelope of the Q-switched mode-locked pulses shown in Fig. 22(b), but with a much lower peak power (by a factor of ~4×10^{-3}) and an increased temporal width. The transition from the fully-modulated Q-switched mode-locked pulses to the pure Q-switching operation –i.e. from Fig. 22(b) to 22(c)– is progressive as we detuned the voltage applied to the piezoelectric. Finally, Fig. 22(d) shows the behavior of this laser when turning off the active Q-switching, i.e. standard mode-locking operation of the laser. To this end, a DC current was driven to the coil of the magnetostrictive device, in order to achieve a stationary overlap with the short wavelength sideband of the AOSLM. The trace of Fig. 22(d) was recorded with the 1 GHz oscilloscope. In this case, the thermal effects produced by the DC current made difficult an optimum and stable adjustment of FBG$_2$. The comparison of Figs. 22(b) and 22(d), illustrates the improved performance of the laser in terms of peak power when both Q-switching and mode-locking are operating. The energy of a Q-switched train of pulses, as that reported in Fig. 22(b), as a function of the pump power is shown in Fig. 22(e). This energy was measured directly with a pyroelectric detector. At high pump powers, the energy of the Q-switched mode-locked train of pulses reach gain saturation. Thus, a peak power higher than 250 W can be calculated for the central pulses of a train with energy of 0.65 µJ, assuming that the mode-locked pulses are 1 ns width. This result demonstrate a dramatic enhancement in comparison with the peak power achieved when the laser was operated in standard mode-locking regime (Cuadrado-Laborde et al., 2009a), since the ratio is higher than 2×10^3.

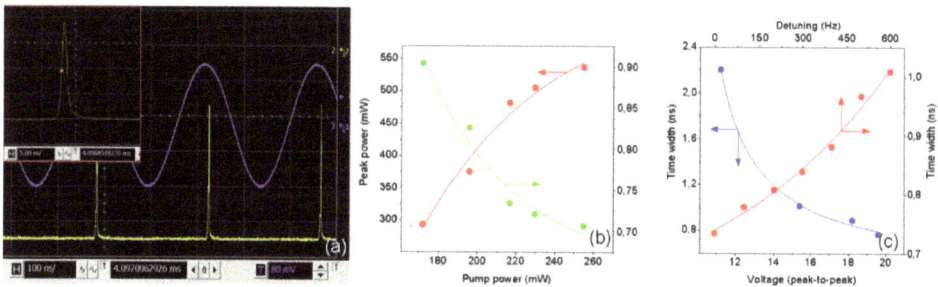

Fig. 23. (a) Oscilloscope traces of the sinusoidal electrical signal applied to the piezoelectric at 4.1 MHz repetition rate and 16 V (upper trace), and the train of optical pulses generated (lower trace); a pump power of 230 mW was used. The inset shows the detail of a single pulse of the optical train, with a FWHM of 720 ps and a peak power of 500 mW. (b) Peak power and pulse width (FWHM) of the optical pulses as a function of the pump power (solid and dotted curves, respectively). (c) Pulse width of the optical pulses when the applied voltage to the piezoelectric is varied, and when the frequency of the electrical signal is varied

3.2.3 All-fiber actively mode-locked laser with a fiber Bragg grating based acousto-optic modulator driven by travelling acoustic waves

The setup of the mode-locking laser driven by travelling acoustic waves is essentially the same described in Fig. 18, but with two important differences. First, the pigtailed laser diode emitting at 976 nm provides now a –higher– maximum pump power of 410 mW. Second, the modulation frequency is now at the same frequency of the acoustical signal (Cuadrado-Laborde et al., 2010b), since now we use travelling acoustic waves, see Subsection 3.2.1. Then, the delay line length is now much larger, because we must match the round-trip time with this frequency, which is lower. Since the selected piezoelectric operation point was of 4.1 MHz; then it results in a cavity length of 25 m –compare this with the 11.4 m of cavity length used before. Apart from the 1.4 m of EDF, the cavity was constructed entirely with Fibercore SM980 fiber, this result in an average dispersion of –4.9 ps/nm/km (i.e. normal). Figure 23(a) exemplifies the laser behavior showing the sinusoidal electrical signal applied to the piezoelectric at a frequency rate of 4.1 MHz and the train of optical pulses generated at the same frequency. The inset shows a detail of a single optical pulse with a temporal width (FWHM) of 710 ps. Once mode-locking was reached, the polarization controllers were adjusted in order to obtain the minimum pulse width.

The emission linewidth was measured using a classical heterodyne configuration (Galtarossa et al., 1993). To this end, a tunable laser was used as local oscillator, with a 100 kHz linewidth. The output of this laser was superimposed to the optical pulses of the modelocked laser, through a 1550 nm 50/50 coupler. The beat signal at the coupler output was detected with a 45 GHz bandwidth photodetector, and analyzed with a 2.5 GHz oscilloscope. This results in a linewidth for the mode-locked laser of 560 MHz for a 900 ps (FWHM) pulse, measured directly in the spectrum at –3 dB. Again, we found that the output pulses of this laser are best fitted by a $sech^2$ function rather than with a Gaussian function. From the Fourier-transform-limited relation for a $sech^2$ pulse, its time-bandwidth product cannot be lower than 0.315 (FWHM) (Wada et al., 2008). Since we obtain 560 MHz × 900 ps = 0.504, we conclude that the optical pulses of our modelocked laser could have some moderate degree of chirp.

According to Eq. (1), the pulse duration is limited by a variety of factors in an active modelocked laser. From these parameters, we believe the narrow spectral bandwidth of the AOSLM $\Delta\lambda$ plays a key role in this setup. As we explained before, these sidebands can be regarded as weak ghosts of the strong permanent Bragg grating, its FWHM bandwidth is that of a weak Bragg grating of the same length (Liu et al., 1997), namely $\Delta\lambda = 1.39\lambda^2/(\pi L n_0)$, where L is the fiber length and n_0 is the modal effective index. For the grating used in this chapter, with L = 120 mm, at λ = 1530 nm, this translates into $\Delta\lambda$ = 6 pm, which is equivalent to 770 MHz at the operation wavelength of this laser (1530.5 nm). As a rough estimation, since the cavity modes are distant 4.1 MHz, it is easily seen that this AOSLM only is able to lock a few percent of the axial modes available by the medium gain, namely 770 MHz /4.1 MHz = 192 modes (FWHM). If we assume for the output pulses of this laser a $sech^2$ envelope, then this parameter alone determines a lower limit for the pulse width around 0.315 /770 MHz = 410 ps. Therefore, we believe that narrower pulses can be reached by broadening the sidebands of the AOSLM in order to lock additional axial modes. The optical pulse's peak power and temporal width as a function of the pump power are shown in Fig. 23(b). A smooth variation of the pulse width with pump power can be observed, before reaching the gain saturation. Fig 23(c) shows the time width as a function of the frequency detuning. The behavior was asymmetric, i.e. it depends if the detuning was positive or negative. Once mode-locking is reached and the PC adjusted to get the minimum pulse width, a small detuning in either direction does not modify the pulse width –e.g. up to a few tens of Hertz, as expected for amplitude modulation mode-locking (Kuizenga & Siegman, 1970). However, when the detuning is considerably higher, a positive detuning continuously broadens the time width, whereas for a negative detuning, the mode-locking is rapidly lost, and the pulses drop out. We believe this asymmetry could be caused by the non-flat frequency response of the piezoelectric. We conclude this subsection by analyzing the behavior of this laser when the voltage applied to the piezoelectric changes. Figure 23(c) shows the influence on the time width of the optical pulses with the voltage applied to the piezoelectric. For higher voltages, the time width decreases; further narrowing by this trend appears to finish when the reflectivity of the sideband approaches the maximum (in addition, 20 V was the maximum voltage provided by the electrical amplifier used to drive the piezoelectric). Within the range of voltage available in our experiments, both the reflectivity and the modulation amplitude increase continuously. This is consistent with Eq. (1), by which the time width is inversely related with both, the modulation depth and the reflectivity.

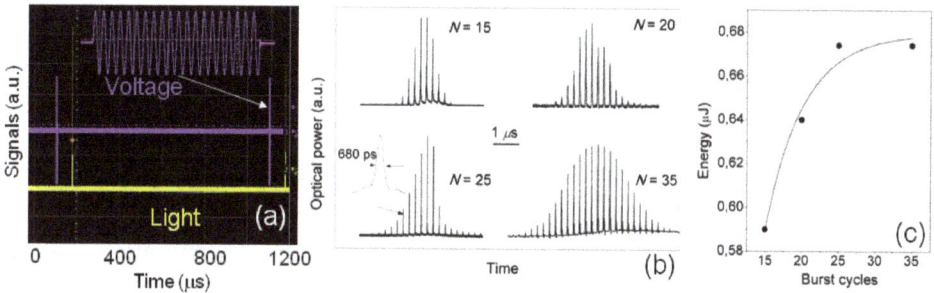

Fig. 24. (a) Voltage signal applied to the piezoelectric and Q-switched mode-locked train of pulses generated for a 1 kHz Q-switching repetition rate; the inset shows a detail of the burst signal. (b) Single Q-switched mode-locked pulses enveloping a train of modelocked pulses for different burst cycles; the inset shows a single-shot capture of a single modelocked pulse with a FWHM of 680 ps. (c) Energy of the Q-switched mode-locked pulses as a function of the cycles contained in the burst for a fixed Q-switching frequency of 500 Hz and a pump power of 270 mW

The most significant advantage of using travelling instead of standing acoustic waves is determined by the possibility to actively Q-switch this modelocked laser simply by launching to the grating a burst acoustical wave (Cuadrado-Laborde et al. 2010b). Therefore, the setup is exactly the same as the reported above for the standard mode-locking regime driven by travelling acoustic waves. The burst signal consists of an integer number of sinusoidal acoustical cycles N. Figure 24(a) shows the burst voltage signal applied to the piezoelectric together with the Q-switched mode-locked train of pulses generated, at a frequency rate of 1 kHz. Figure 24(b) shows several Q-switched mode-locked pulses as a function of N. When N is low, Q-switched mode-locked is not reached. For higher N, Q-switched mode-locked is reached and the emission is allowed. For example, the left bottom corner of Fig. 24(b) shows a single Q-switching envelope with a FWHM of 1.4 μs for $N = 25$; it has between 16 and 18 fully-modulated modelocked pulses, with the expected temporal separation given by 1/4.1 MHz = 244 ns. The inset shows a single-shot capture of a modelocked pulse within this train with a temporal width of 680 ps. As observed, the Q-switched mode-locked pulses of this laser have excellent inter-pulse characteristics. The energy of a Q-switched mode-locked train of pulses, as a function of N is shown in Fig. 24(c). This energy was measured directly with a pyroelectric detector. As expected, for a longer burst signal, the energy of the Q-switched mode-locked train of pulses increases, until it reaches saturation. Thus, a peak power higher than 200 W can be calculated for the central pulses of a train with energy of 0.68 μJ, assuming that the modelocked pulses are 680 ps width. This result clearly demonstrates the degree of enhancement in comparison with the peak power achieved when the laser was operated in the standard mode-locking regime, since the ratio is higher than 4×10^2.

3.3 Photonic true-time delay-line by acousto-optic interaction in a fiber Bragg grating in the short-wavelength regime

A photonic true-time-delay (PTTD) line is an optical device that permits to vary the group delay of an optical signal. The ideal PTTD line should exhibit fast tuning within a broad range of time delay values. PTTD lines attract much attention due to an increasing number of applications such as the control of phased array antennas (Liu et al., 2002; Perez-Millán et al., 2004; Italia et al., 2005;), tunable microwave band-pass filters (Capmany et al., 2005), buffering and packet synchronization (Li et al., 2007; Caucheteur et al., 2010). The advantages of PTTD lines are their broad bandwidth, high frequency operation and immunity to electromagnetic interference. In addition, fiber-based PTTD lines are readily compatible with fiber systems, robust, compact and lightweight, and they present low insertion loss.

Recently several types of fiber-based PTTD lines were proposed. For example, an optically controlled PTTD line based on stimulated Brillouin scattering demonstrates a fast time response within a relatively small range of the time delay (230 ps were reported using a fiber of 3.5 km length) (Zadok et al., 2007). Another example is the tunable PTTD line based on a uniform or chirped FBGs perturbed by heating or mechanical stress (Ortega et al., 2000; Liu et al., 2002; Perez-Millán et al., 2004), which permits simple experimental realization and easy tuning. However, such system has a very slow response (below 1 ms). In a number of PTTD lines tunable light sources are required to adjust the time delay (Liu et al., 2003; Blais et al., 2009), but this approach is not easily scalable when a large number of independent delays need to be adjusted.

Here we discuss a tunable PTTD line based on longitudinal ultra-sound modulation of a FBG written in a standard fiber similar to SMF-28. This device operates with a fixed optical carrier, demonstrates a wide time delay range (400 ps) and a fast response time [µs range, see Delgado-Pinar et al. (2006)].

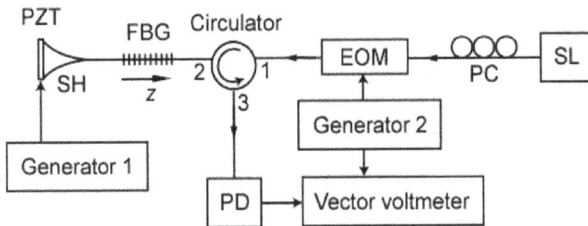

Fig. 25. Experimental setup. SL: semiconductor laser; PC: polarization controller; EOM: electro-optical amplitude modulator; FBG: fiber Bragg grating; PZT: piezoelectric transducer; SH: fused silica horn; PD: photodetector

The setup of the PTTD line is shown in Fig. 25. Light from a single-frequency semiconductor laser (SL) modulated by an electro-optical amplitude modulator (EOM) with a frequency of f_m = 400 MHz is sent through the optical circulator to a long uniform FBG written in a standard photosensitive fiber; the RF modulation depth is chosen to be comparatively low ($\sim$ 10%), which maintains a practically linear regime of light modulation. The FBG is 12 cm long, with a

Bragg wavelength of 1530.8 nm and reflectivity of 99.99% (40 dB transmission attenuation measured at the Bragg wavelength) that corresponds to a coupling coefficient $\kappa = 0.442$ cm^{-1}.

In the experimental arrangement, a longitudinal acoustic wave is generated by a piezoelectric (PZT) disk attached to a heat sink. This wave is launched along the single-mode optical fiber containing the uniform FBG using a fused silica horn (SH). The PZT is driven by Generator 1 at the PZT resonance frequency (2.08 MHz in our case). The light reflected by the FBG is registered by a photodetector connected to a vector voltmeter. AC voltage from Generator 2 drives the EOM; it is also used as a reference signal for the RF phase measurements.

The operation principle of such a PTTD line is based on the reflection of narrow-line light by one of the FBG side bands generated by the longitudinal acoustical wave. This wave modulates both the period and the refractive index of the grating producing thereby a sinusoidal chirp that moves along the fiber synchronously with the acoustic wave, and will be described by an effective phase modulation of the grating with amplitude $\Delta\Phi$. Under these conditions, the refractive index for the light mode propagating through the single-mode fiber core (so called LP_{01} mode) is written as follows:

$$n(z,t) = n_0 + \Delta n \cos\left[Kz - \Delta\Phi\cos(k_s z + \Omega t)\right] = n_0 + \Delta n J_0(\Delta\Phi)\cos(Kz)$$

$$+2\Delta n\cos(Kz)\sum_{m=1}^{\infty}(-1)^m J_{2m}(\Delta\Phi)\cos\left[2m(k_s z + \Omega t)\right] \tag{3}$$

$$+2\Delta n\sin(Kz)\sum_{m=1}^{\infty}(-1)^{m-1} J_{2m-1}(\Delta\Phi)\cos\left[(2m-1)(k_s z + \Omega t)\right],$$

where n_0 is the effective refractive index for LP_{01} mode, Δn is the grating amplitude, $K = 2\pi/\Lambda$ is the unperturbed grating wavenumber (Λ is the grating period), z is the distance along the fiber, $\Delta\Phi$ is the phase modulation amplitude, k_s is the acoustic wavenumber, Ω is the angular acoustic frequency, m is a natural number corresponding to the order of the FBG side band, and J_m is the Bessel function of the m-th order. From Eq. 3 one can see that the longitudinal acoustic wave produces a series of the traveling FBGs with amplitudes depending on $\Delta\Phi$.

Using the coupled-wave theory (Erdogan, 1997), one can obtain the following set of the coupled-wave equations for two contra-propagating optical waves (one probe wave, $A_{\pm l}$, and one reflected wave, $B_{\pm l}$) falling into the $\pm l$ sideband induced by the acoustic wave and interacting with the perturbed grating described by Eq. 3:

$$\frac{dA_{\pm l}}{dz} = -i\Delta\beta_{\pm l}A_{\pm l} + i^{(l-1)}\kappa_l B_{\pm l}, \tag{4}$$

$$\frac{dB_{\pm l}}{dz} = i\Delta\beta_{\pm l}B_{\pm l} + (-i)^{(l-1)}\kappa_l A_{\pm l}, \tag{5}$$

where $\pm l$ is the sideband number, $\Delta\beta_{\pm l} = \beta - (\beta_0 \mp lk_s/2) = 2\pi n_0(\lambda^{-1} - \lambda_0^{-1}) \pm l\pi/\lambda_s$ is the detuning from the $\pm l$-sideband Bragg wavelength, λ_0 is the Bragg wavelength of the unperturbed FBG, λ is the probe light wavelength, β and β_0 are the light wavenumbers that correspond to λ and λ_0, and $\kappa_l = \kappa_0 J_l(\Delta\Phi)$ is the coupling coefficient for the l-th sideband (κ_0 is the coupling coefficient of the unperturbed FBG).

The analysis based on the theory developed by Barmenkov et al. (2006, 2010) permits one to write the equations for the sideband reflectivity, R_l, and the sideband effective length, $L_{\pm l}^{eff}$, in the following form:

$$R_{\pm l} = \left(\tanh\left(\kappa_{\pm l}L\right)\right)^2,$$

(6)

$$L_{\pm l}^{eff} = \frac{1}{2\kappa_{\pm l}}\tanh\left(\kappa_{\pm l}L\right) = \frac{L\sqrt{R_{\pm l}}}{2\,\mathrm{atanh}\left(\sqrt{R_{\pm l}}\right)},$$

(7)

where L is the FBG length. The formula for the group delay is found as

$$\tau_{\pm l} = 2n_0 L_{\pm l}^{eff}/c.$$

(8)

From the last three equations one can conclude that the sideband diffraction efficiency, the effective length and the group delay may be controlled by the amplitude of the phase modulation of the grating induced by the acoustic wave, which, in turn, depends on the acoustic wave magnitude. The sideband Bragg wavelength can be tuned by adjusting the acoustical frequency within the PZT resonance in the case of a slight tuning, or by replacing the PZT with another one having the necessary resonance frequency.

The diffraction efficiency of the FBG sidebands can be measured using a wide-spectrum LED connected directly to the circulator port 1, and an optical spectrum analyzer (OSA) connected, in turn, to the port 3 instead of a photodetector (see Fig. 25); The Generator 2 is switched off. Fig. 26(a) shows the FBG spectra at different AC voltages applied to the PZT. At low voltages, the ±1 sidebands have amplitudes much higher than that of the high-order sidebands. Reflectivity of the ±1 sidebands did not reach 100% because the OSA resolution was not enough to resolve comparatively narrow reflection peaks (3 dB spectrum width is about 40 pm). Note that a proper FBG design based on the grating apodization and chirping should permit increasing the sideband width.

As it is seen from Fig. 26 (b), the dependence of the +1 sideband efficiency on the voltage applied to the PZT is in good agreement with Eq. (6), assuming that the amplitude of the phase modulation of the grating is proportional to the PZT voltage; a slight difference observed at high voltage amplitudes (> 4V) could be explained by a small heating of the PZT, which slightly changes the PZT electromechanical coupling factor. Note that the ±1 sideband allows operation of AOSLM at low PZT voltages in comparison with high-order sidebands, the −1 sideband has the same properties as the +1 sideband.

The RF modulation envelope phase, Ψ_{+1}, is related directly to the +1 sideband group delay τ_{+1} and the effective length L_{+1}^{eff}:

$$\Psi_{+1} = 2\pi\tau_{+1}f_m = 4\pi n_0 L_{+1}^{eff} f_m/c$$

(9)

Thus, the dependence of Ψ_{+1} on +1 sideband efficiency is the basic feature that permits to implement a dynamic control of the group delay and the effective length of the grating by means of the AC voltage. The same is for the −1 sideband.

Fig. 27(a) shows the experimental relationship between Ψ_{+1} and the amplitude of the AC voltage applied to the PZT. One can see that the phase decreases as voltage amplitude increases, which is explained by decreasing of the group delay and the effective length. Figure 27(b) plots these two parameters versus the +1 sideband diffraction efficiency. The experimental data show a good agreement with Eqs. (7) and (8). Since the optical path between the EOM and the photodetector was relatively long (the fiber length was ~ 2 m), producing an additional phase shift into the measured Ψ_{+1} values, the experimental data were corrected by a phase offset (the right scale in Fig. 27(a) that permits to compensate the fiber length excess. The correcting parameter was chosen so that the +1 sideband effective length is equal to a half of FBG physical length at low sideband efficiency (low PZT voltage), as the Eq. (7) predicts.

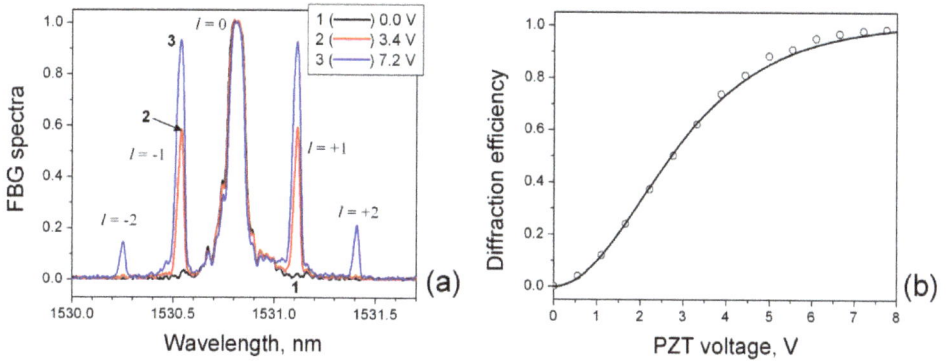

Fig. 26. (a) Spectra of FBG perturbed by ultra-sound wave at different PZT voltages (the voltage values are shown in the upper right corner); l indicates the sideband number. The OSA resolution is 20 pm. (b) Dependence of the +1 sideband efficiency on PZT voltage. Circles: experimental data; solid line: fitting

From Fig. 27(b) one can conclude that the group delay and the effective length for the +1 sideband depend on its efficiency that is controlled electrically by the AC voltage applied to the PZT. In the experiments, the group delay was adjustable from 150 ps to 550 ps, i.e., covering a range of 400 ps. Within the 3 dB range of the sideband efficiency, the group delay could be continuously adjusted from 150 to 450 ps (a range of 300 ps). The implementation of an automatic SL power control would permit to adjust the group delay at constant optical power, compensating for the reflectivity changes of the sideband. The response time of this PTTD line is approximately 20 µs and is determined by the acoustic wave speed and the grating length.

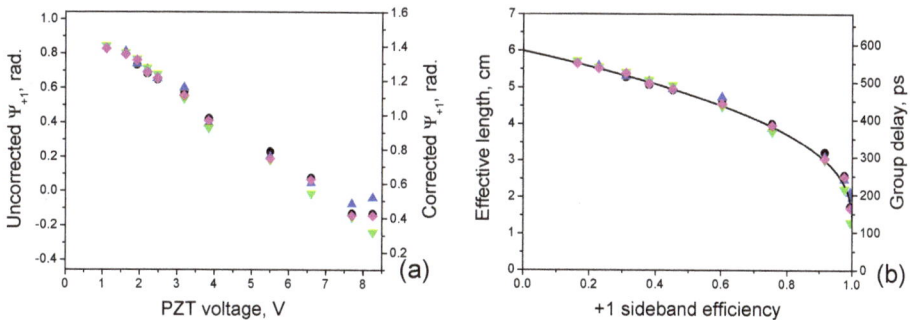

Fig. 27. (a) Dependence of the RF modulation envelope phase of light reflected from the FBG +1 sideband on amplitude of voltage applied to PZT. The left scale: uncorrected values, the right scale: corrected values. (b) FBG +1 sideband effective length and group delay versus the diffraction efficiency. Symbols: experimental data; solid line: curve calculated by Eq. (7). In both figures different symbols correspond to different experimental series

In summary, the experimental and theoretical study of the phase and group delay response of a FBG modulated by a longitudinal acoustic wave permits the implementation of electrically-tuned PTTD line controlled by the AC voltage applied to the piezo-electric transducer that generates the acoustic wave. The proposed PTTD line permits to vary the group delay in the range of 400 ps.

3.4 Q-switching in a distributed feedback fiber laser

When the acoustic perturbation is not a harmonic wave but a single pulse, its passage through the FBG creates a defect which can be used to control the Q-factor in a DFB all-fiber laser. Single-mode narrow-linewidth lasers are of great interest as they have a wide applicability in fields such as high-resolution interferometry, distributed Brillouin sensing, optical coherent communications, etc. DFB lasers based on semiconductor technology have linewidths typically in the MHz domain; due to the short cavity length and low thermal dissipation its output spectrum is typically broad. Fiber lasers usually operate in a multi-mode regime and have a broad spectrum; to make efficient single-mode narrow-linewidth fiber lasers different approaches have been implemented. Fiber rings lasers fulfill the requirement and produce narrow-linewidth outputs; most importantly, they can be made widely tunable. However, they are more complex and –because of their long cavities– susceptible to mode hopping. Distributed Bragg reflector (DBR) fiber lasers is another option, however, temperature stabilization is still required to prevent mode hopping and, in the short laser cavity, the pump wave absorption is low, as a consequence the DBR fiber laser is not efficient and external optical amplifiers are generally needed (Babin et al., 2007). Distributed feedback fiber lasers can overcome part of these problems. They have the simplest, robust and compact design providing operation without mode hopping. Its fabrication is relatively simple, and involves the writing of a grating structure –i.e. a fiber Bragg grating– with ultraviolet light into an active fiber. Single-mode pump leads to an alignment free resonator with optimum overlap of pump and signal light. For these FBG-based DFB lasers, the distributed reflection occurs in the grating when a phase shift has been generated within it. A number of techniques have been proposed for this; however, statics phase shifts only allow CW operation, which translates into low power emission. Recently, some approaches have been reported to obtain single-frequency pulsed all-fiber lasers, based on active Q-switching of DFB fiber cavities. The pulsed operation of a DFB laser is interesting because for certain applications (e.g. Brillouin sensing), it is not only required a narrow linewidth but also a peak power about a minimum threshold, which otherwise could not be reached by a CW operation (Cuadrado-Laborde et al., 2008).

Figure 28(a) shows the scheme of the proposed DFB all-fiber laser. The FBG was 100 mm long, and was written in a 1500 ppm erbium hydrogen-loaded fiber (codoped with germanium and aluminum) of the same length using a doubled argon laser and a uniform period mask. The FBG shows more than 30 dB attenuation at the Bragg wavelength of 1532.45 nm and a 3 dB bandwidth of 88 pm. The FBG was pumped through a 980/1550 nm WDM with a 980 nm semiconductor laser, providing a maximum pump power of 130 mW. A square shape rod of a magnetostrictive material (Terfenol-D, 15 mm long and 1 mm^2 section) was bonded outside the FBG to a free section of fiber at 88 mm from the center of the grating, and placed inside a small coil, see Fig. 28(a).

The acoustic pulse generated by using a magnetostrictive device has a superior performance as compared with piezoelectric devices which have a frequency-dependent response characterized by strong mechanical resonances. On the contrary, a magnetostrictive device has a lower frequency range but its frequency response is basically flat. The magnetostrictive rod can be bonded directly over the FBG, but this has some detrimental consequences. First, any external actuator bonded directly on the fiber is likely to exhibit long-term instabilities (Andrés et al., 2008). Second, once the magnetostrictive rod is fixed to the fiber, any temperature change generates a differential expansion between the fiber and the magnetostrictive rod, producing a local static perturbation of the FBG that may cause CW emission. Because of this, we avoided these drawbacks by attaching he magnetostrictive rod outside the FBG and generating a dynamic defect in the FBG through an acoustic axially-propagating pulse. Thereby, when a pulse of electric current is applied to the coil, the small rod lengthens and stretches the section of fiber attached to it, generating in this way a longitudinal acoustic pulse that propagates toward the FBG. The pulse propagating along the grating generates a phase shift opening a transmission peak within the reflection band of the grating; as a consequence a high Q resonance is produced and a laser pulse is emitted (Pérez-Millán et al., 2005b; Delgado-Pinar et al., 2007). Otherwise, if no perturbation is present within the FBG, there is no efficient feedback for the optical signal, and the laser emission is not allowed. Since the defect is not fixed, but induced by a travelling acoustic pulse with a time-varying position along the FGB, one could think that this might have important consequences on the spectral position of the transmission peak. However, this is not the case, as it was recently demonstrated (Andrés et al., 2008). The spectral position of the resonance is constant, no matter the actual spatial location of the acoustic pulse, although some short transients are produced when the pulse overlaps the extremes of the FBG. This property insures that the laser will emit with a narrow linewidth, preserving one of the most attractive properties of DFB fiber lasers. The transmission properties of the passive FBG –i.e. without pumping– interacting with the acoustic pulses were investigated by illuminating the FBG with a tunable laser at the Bragg wavelength and detecting the reflected signal. When the coil current is zero, the reflectance is maximal; but when a current pulse of 220 mA and 5 μs temporal width is applied to the coil, a transmission peak opens the reflection band, being as narrow as 2 μs (FWHM), see Fig. 28(b). Figures 29(a) to 29(c) shows the voltage pulses applied to the current driver, the backward output train of the DFB laser, a detail of the optical pulse, and a single current pulse applied to the coil, respectively. The observed delay time between the voltage signal and the optical pulses is mainly due to the distance that pulses have to travel from the magnetostrictive device to the FBG, see Fig. 28(a).

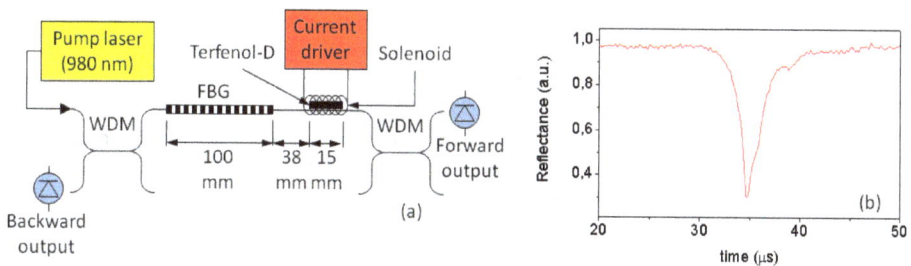

Fig. 28. (a) Q-switched distributed feedback-fiber laser setup. (b) Reflection at the Bragg wavelength when an acoustic pulse travels along the FBG

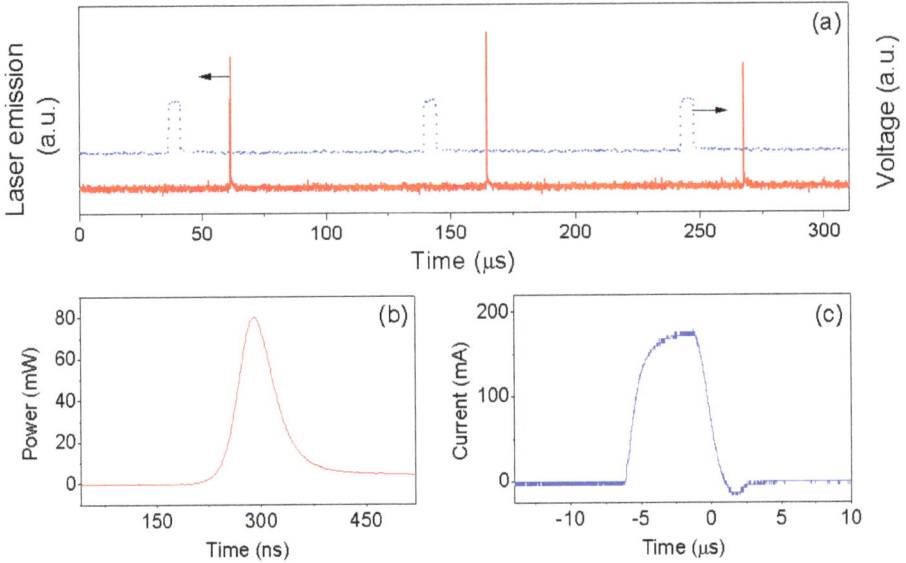

Fig. 29. Q-switched DFB behavior at 10 kHz repetition rate and 80 mW of pump power. (a) Emitted optical train pulses and voltage signal applied to the current driver, (b) detail of a single optical pulse of the train, and (c) detail of a single current pulse applied to the coil

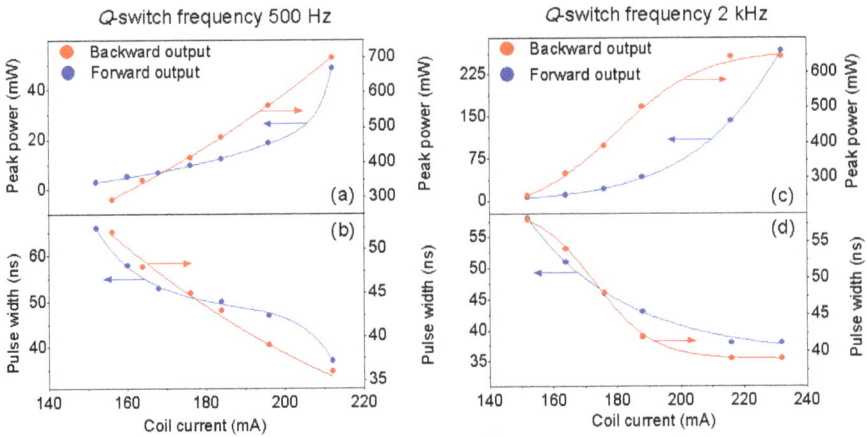

Fig. 30. Peak power and pulse width for backward and forward outputs as a function of the coil current for two different Q-switched repetition rates: 500 Hz (a, b) and 2 kHz (c, d), respectively, and 55mW of pump power

One key characteristic in this setup is its versatility, which is given by the possibility of selecting a variety of peak powers and time widths just by varying the coil current. Figure 30 shows the peak power and pulse width for backward and forward outputs as a function of the coil current for two different repetition rates: 500 Hz and 2 kHz, Figs. 30(a-b) and 30(cd), respectively. There is a low threshold value for the coil current (I), bellow this value; there is not laser emission, since the Q value is not high enough (i.e. I < 150 mA). Above this value, the laser emission is allowed and one optical pulse per cycle is emitted; at higher electric currents (i.e., I > 230 mA) there are two optical pulses per current pulse. At different repetition rates the peak power and time width values change but the general trend is preserved. The effect of pump power on the optical pulses, for repetition rates ranging from 200 Hz to 20 kHz, is shown in Figs. 31(a,b) and 31(c,d), for the backward and forward outputs, respectively. A fixed coil current of 200 mA amplitude was used in all cases. At each different frequency, there is a corresponding pump power threshold. Above threshold, the peak power increases with pump power and there is a corresponding reduction of the pulse width. Pulses of 800 mW peak power and 32 ns time width were obtained at 500 Hz repetition rate and 46 mW pump power for the backward output. If pump power is high enough, the laser emits more than one pulse per cycle, defining in this way an upper limit also. It can be observed the differences in peak powers between backward and forward outputs, being higher in the first case. Temporal widths and peak power jitters were measured to be below 5 %.

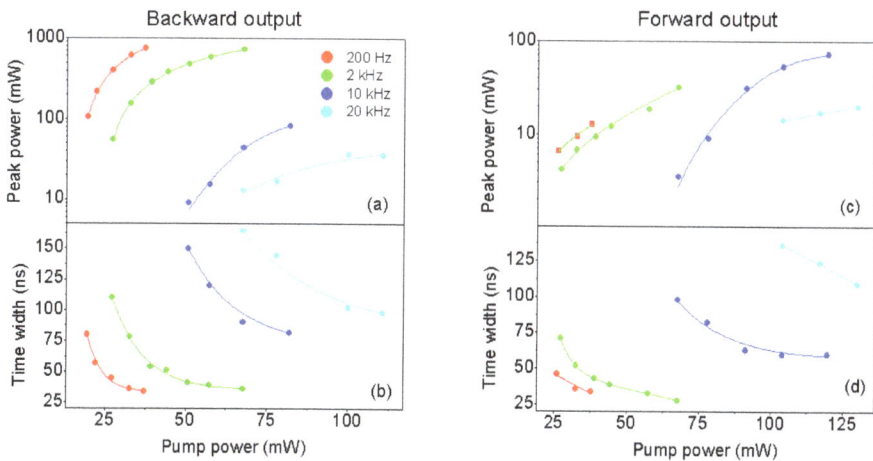

Fig. 31. Peak power and pulse width for the backward (a, b) and forward (c, d) laser outputs, as a function of the pump power, for several repetition rates

The emission linewidth was measured, using a classical heterodyne configuration (Galtarossa et al., 1993). A tunable laser was used as local oscillator with spectral linewidth of 100 kHz. The output of this laser was superimposed to the optical pulses of the DFB laser (backward output), through a 1550 nm 20 dB coupler. The beat signal at the coupler output was detected with a 1 GHz bandwidth photodetector, and analyzed with a 500 MHz bandwidth oscilloscope. From the spectrum of the beating signals, resulted a DFB laser linewidth of 6 MHz. The optical pulses from the DFB all-fiber laser had a temporal width of 80 ns, so according to the time-frequency uncertainty principle (Agrawal, 2001), its bandwidth cannot be lower than 5.5 MHz. From the comparison between this last value and the linewidth measurement, we conclude that the optical pulses of our DFB fiber laser are transform-limited. Finally, as an example of application of this compact laser source, we demonstrate the possibility to generate Brillouin scattering for sensing purposes. Brillouin scattering essentially refers to the scattering of a light wave by an acoustic wave (Agrawal, 2001). When a coherent pulse of light propagates through a medium, part of its energy is backscattered due to a non elastic interaction with the acoustic phonons. This back-scattered light is composed of a frequency down-shifted Stokes light and an up-shifted anti-Stokes light, whose spectral positions are dependent on temperature and strain of the fiber, in this way allowing its use as a sensing mechanism (Culverhouse et al., 1989a, 1989b; Bao et al., 1995; Parker et al., 1998; Y. Li et al., 2003). Figure 32(b) shows the backscattered light spectrum after illuminating a 10.5 km length optical fiber spool (Corning SMF-28) with the backward output of our DFB all-fiber laser (4 kHz repetition rate and 74 mW pump power). The Brillouin spectrum was registered with an optical spectral analyzer (resolution of 20 pm). The extreme of the fiber optic spool was terminated with a matching refractive index liquid (n = 1.46). The central (highest) peak corresponds to the proper laser beam reflections after successive connections and splices together with Rayleigh scattering. Peaks symmetrically positioned at both sides correspond to the Brillouin backscattering by Stokes and anti-Stokes processes (Parker et al., 1998). The measured Brillouin shift results in 88 pm (i.e. 11.24 GHz at 1532.4 nm), which correspond to the expected value in this fiber (Agrawal, 2001). Fiber optics distributed temperature, and/or strain, sensors have becoming very attractive for applications requiring sensing lengths of many kilometers, principally due to its inexpensiveness and availability. Optical fiber based distributed sensor systems normally make use of the principle of optical time domain reflectometry (Parker et al., 1998). Therefore, an optical pulse is launched into one end of the fiber system and the variation of the scattered light is detected as a function of time, giving in this way information of temperature or strain as a function of distance. A key requirement in this measurement system is a stable light source with a narrow enough spectral linewidth. In addition, for time domain reflectometry applications, the sensor spatial resolution proportionally depends on the optical pulse width, so it must be considered also. In order to fulfill all these requirements, solid state lasers with external cavities –plus amplifiers and amplitude modulators– are currently used in these systems (Parker et al., 1998). Here we propose using this compact all-fiber pulsed light source as a relatively simpler alternative.

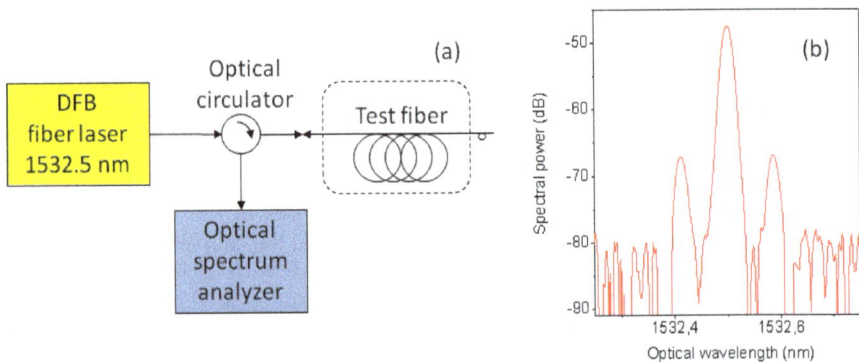

Fig. 32. Schematic diagram of the setup used for the Brillouin backscattering measurements. Brillouin spectra at room temperature for a 10.5-km length Corning SMF-28 optical fiber spool, for a backward output at 4 kHz repetition rate and 74 mW pump power

4. Conclusions

In this chapter we have discussed the use of acoustic waves to control all-fiber devices for different applications. In Section 2 we focused on novel applications of acousto-optic fiber devices based on flexural acoustic waves. As an example of application, we described in subsection 2.1 an actively Q-switched ytterbium-doped strictly all-fiber laser. Q-switching modulation was achieved by intermodal modulation induced by flexural acoustic waves travelling in a tapered optical fiber. Q-switched light pulses at 1064.1 nm were successfully obtained at repetition rates in the range 1-10 kHz, with pump powers between 59 mW and 88 mW. Best results were for laser pulses of 118 mW peak power, 1.8 µs of time width, with a pump power of 79 mW, at 7 kHz repetition rate.

Next, in subsection 2.2 we described an actively mode-locked fiber ring laser. As mode-locker was proposed an acousto-optic modulator driven by standing flexural acoustic waves, which couples core-to-cladding modes in a standard single-mode optical fiber. Among the remarkable features of the modulator, we mention its high modulation depth (72%), broad bandwidth (187 GHz), easy tunability in the optical wavelength, and low insertion losses (0.7 dB). The narrowest optical pulses obtained were of 95 ps time width, 21 mW peak power, repetition rate of 4.758 MHz, and 110 mW of pump power. We also characterized this laser as a function of the RF voltage that controls the modulator, the length of the active fiber, and the optical bandwidth of an intracavity filter implemented with a fiber Bragg grating. Best results were for output light pulses of 34 ps time width, and 1.4 W peak power.

In Section 3 we focused on novel applications of acousto-optic devices based on the interaction of longitudinal acoustic waves with fiber Bragg gratings. From the comparison of the acoustic wavelength with the periodicity of the FBG, two well-different situations could be distinguished: the long-wavelength and the short-wavelength regimes. As an example of application of the former, we described in subsection 3.1 an in-fiber resonant acousto-optic modulator suitable for Q-switching applications. The modulator consists of a short-length FBG modulated by a long-wavelength standing longitudinal acoustic wave. The acoustic wave shifts periodically in time the reflection band of the grating along a given wavelength range; by using this modulator, we demonstrated an actively Q-switched all-fiber laser. Output light pulses of 1.6 W peak power and 172 ns time width were obtained at 18 kHz repetition rate.

Next, in section 3.2 and 3.3 we focused on applications of acousto-optic devices based on the interaction of longitudinal acoustic waves with fiber Bragg gratings in the short-wavelength regime. Thus, in subsection 3.2 we proposed the use of an acousto-optic supper-lattice modulator as mode-locker. Since an acousto-optic supper-lattice modulator can be driven in two different ways either by standing or traveling acoustic waves; we discussed first the construction of a mode-locked laser driven by standing acoustic waves. In this configuration we obtained transform-limited optical light pulses of up to 120 mW peak power and 780 ps pulse width generated at a fixed repetition rate of 9 MHz, with an emission linewidth of 2.8 pm at 1530.5 nm. We also study the influence of different parameters on the mode-locking process such as: frequency detuning, EDF length, amplitude modulation, and dispersion. In this case, narrower pulses were obtained at higher modulation depths, normal dispersion, and shorter lengths of active fiber. Best results were reached when the laser was optimized according to these variables (160mW peak power and 630 ps pulse width). On the other hand, we showed that by slightly modifying the cavity, it is possible to operate the laser in a double-active Q-switching mode-locking regime. It is worth to mention that this approach was unique, being this double-active all-fiber laser the first of its kind. To this end, we attached a magnetostrictive rod to the output FBG to modulate the Q factor of the Fabry–

Perot cavity. Fully modulated Q-switched mode-locked trains of optical pulses were obtained for a wide range of pump powers and repetition rates. For a Q-switched repetition rate of 500 Hz and a pump power of 100 mW, the laser generates trains of 12–14 modelocked pulses of about 1 ns each, within an envelope of 550 ns, an overall energy of 0.65 μJ, and a peak power higher than 250 W for the central pulses of the train. Then, we discussed the construction of a mode-locked laser when the mode-locker is driven by travelling acoustic waves. In this case, the modulation frequency is half the frequency obtained when standing acoustic waves are used. Optical pulses were obtained of 530 mW peak power, 700 ps pulse width, at a repetition rate of 4.1 MHz. The variation of the pulses parameters under frequency detuning and applied voltage was also studied. Finally, we demonstrated that it is not necessary to modify the setup in order to reach double active Q-switching and mode-locking, when travelling acoustic waves were used to drive the mode-locker. In this case the commutation between mode-locking and Q-switching mode-locking is remarkably simple; it just needs use of a different electrical signal to drive the piezoelectric of the mode-locker, i.e. from a sinusoidal to a burst-sinusoidal electrical signal. In this case, fully modulated 10–25 modelocked pulses around 700 ps each within a Q-switching envelope around 1 μs and a maximum overall energy of 0.68 μJ were obtained.

In subsection 3.3 we shown another example of application of acousto-optic devices based on the interaction of longitudinal acoustic waves with fiber Bragg gratings in the short-wavelength regime. Thus, we carried out an experimental and theoretical study of the phase and group delay response of the acousto-optic supper-lattice modulator. The phase properties of the first sidebands permit the implementation of electrically-tuned photonic true-time delay line controlled by the AC voltage applied to the piezoelectric transducer that generates the acoustic wave. The proposed photonic true-time delay line permits to vary the group delay up to 400 ps.

Finally, when the acoustic perturbation is not a harmonic wave but a single pulse, its passage through the FBG creates a defect which can be used to control the Q-factor in a DFB all-fiber laser, subsection 3.4. Thus, we shown a single-mode, transform-limited, actively Q-switched distributed-feedback fiber laser. Optical pulses of 800 mW peak power, 32 ns temporal width, and up to 20 kHz repetition rates were obtained. The measured linewidth demonstrates that these pulses are transform limited: 6 MHz for a train of pulses of 10 kHz repetition rate, 80 ns temporal width, and 60 mW peak power. Efficient excitation of spontaneous Brillouin scattering was demonstrated.

In summary, photonic devices can benefit highly of a strictly all-fiber configuration which provides them a series of attractive advantages. Among all the proposed in-fiber solutions, devices controlled by acoustic waves have been by far the most employed, especially in mode-locked lasers, providing a broad range of alternatives. The recent advances in acoustically controlled photonic systems positioned them as a promising candidate for commercially available systems in the near future.

5. Acknowledgments

This work has been financially supported by the *Ministerio de Ciencia e Innovación* and the *Generalitat Valenciana* of Spain (projects TEC2008-05490 and PROMETEO/2009/077, respectively). C. Cuadrado-Laborde acknowledges the *Secretaría de Estado de Universidades e Investigación del Ministerio de Ciencia e Innovación* (Spain) and ANPCyT (project PICT 2008-1506, Argentina).

6. References

Agrawal, G. P. (2001). *Nonlinear Fiber Optics*, Academic Press, New York, 2001.

Álvarez-Chávez, J. A.; Offerhaus, H. L.; Nilsson, J.; Turner, P. W.; Clarkson, W. A.; Richardson, D. J. (2000). High-energy, high-power ytterbium-doped Q-switched fiber laser. *Optics Letters*, 25 (1): 37-39 Jan 1 2000.

Andersen, T. V.; Pérez-Millán, P.; Keiding, S. R.; Agger, S.; Duchowicz, R.; Andrés, M. V. (2006). All-fiber actively Q-switched Yb-doped laser. *Optics Communications*, 260 (1): 251-256 Apr 1 2006.

Andres, M. V.; Cruz, J. L.; Díez, A.; Pérez-Millán, P.; Delgado-Pinar, M. (2008). Actively Q-switched all-fiber lasers. *Laser Physics Letters*, 5 (2): 93-99 Feb 2008.

Babin, S. A.; Churkin, D. V.; Ismagulov, A. E.; Kablukov, S. I.; Nikulin, M. A. (2007). Single frequency single polarization DFB fiber laser. *Laser Physics Letters*, 4 (6): 428-432 Jun 2007.

Bao, X.; Dhliwayo, J.; Heron, N.; Webb, D. J.; Jackson, D. A. (1995). Experimental and theoretical-studies on a distributed temperature sensor-based on Brillouin-scattering. *Journal of Lightwave Technology*, 13 (7): 1340-1348 Jul 1995.

Barmenkov, Y. O.; Zalvidea, D.; Torres-Peiró, S.; Cruz, J. L.; Andrés, M. V. (2006). Effective length of short Fabry-Perot cavity formed by uniform fiber Bragg gratings. *Optics Express*, 14 (14): 6394-6399 Jul 10 2006.

Barmenkov, Y. O.; Cruz, J. L.; Díez, A.; Andrés, M. V. (2010). Electrically tunable photonic true-time-delay line. *Optics Express*, 18 (17): 17859-17864 Aug 16 2010.

Belanger, P. A. (2005). On the profile of pulses generated by fiber lasers. *Optics Express*, 13 (20): 8089-8096 Oct 3 2005.

Bello-Jiménez, M.; Cuadrado-Laborde, C.; Sáez-Rodriguez, D.; Díez, A.; Cruz, J. L.; Andrés, M. V. (2010). Actively mode-locked fiber ring laser by intermodal acousto-optic modulation. *Optics Letters*, 35 (22): 3781-3783 Nov 15 2010.

Bello-Jiménez, M.; Cuadrado-Laborde, C.; Díez, A.; Cruz, J. L.; Andrés, M. V. (2011). Experimental study of an actively mode-locked fiber ring laser based on in-fiber amplitude modulation. *Applied Physics B*, in press (2011).

Birks, T. A.; Russell, P. S. J.; Pannell, C. N. (1994). Low-power acoustooptic device based on a tapered single-mode fiber. *IEEE Photonics Technology Letters*, 6 (6): 725-727 Jun 1994.

Blais, S.; Yao, J. P. (2009). Photonic true-time delay beamforming based on superstructured fiber Bragg gratings with linearly increasing equivalent chirps. *Journal of Lightwave Technology*, 27 (9): 1147-1154 May 1 2009.

Bonadeo, N. H.; Knox, W. H.; Roth, J. M.; Bergman, K. (2000). Passive harmonic mode-locked soliton fiber laser stabilized by an optically pumped saturable Bragg reflector. *Optics Letters*, 25 (19): 1421-1423 Oct 1 2000.

Capmany, J.; Ortega, B.; Pastor, D.; Sales, S. (2005). Discrete-time optical processing of microwave signals. *Journal of Lightwave Technology*, 23 (2): 702-723 Feb 2005.

Caucheteur, C.; Mussot, A.; Bette, S.; Kudlinski, A.; Douay, M.; Louvergneaux, E.; Megret, P.; Taki, M.; Gonzalez-Herraez, M. (2010). All-fiber tunable optical delay line. *Optics Express*, 18 (3): 3093-3100 Feb 1 2010.

Costantini, D. M.; Limberger, H. G.; Lasser, T.; Muller, C. A. P.; Zellmer, H.; Riedel, P.; Tunnermann, A. (2000). Actively mode-locked visible upconversion fiber laser. *Optics Letters*, 25 (19): 1445-1447 Oct 1 2000.

Cuadrado-Laborde, C.; Delgado-Pinar, M.; Torres-Peiró, S.; Díez, A.; Andrés, M. V. (2007). Q-switched all-fibre laser using a fibre-optic resonant acousto-optic modulator. *Optics Communications*, 274 (2): 407-411 Jun 15 2007.

Cuadrado-Laborde, C.; Perez-Millán, P.; Andres, M. V.; Díez, A.; Cruz, J. L.; Barmenkov, Y. O. (2008). Transform-limited pulses generated by an actively Q-switched distributed fiber laser. *Optics Letters*, 33 (22): 2590-2592 Nov 15 2008.

Cuadrado-Laborde, C.; Díez, A.; Delgado-Pinar, M.; Cruz, J. L.; Andres, M. V. (2009a). Mode locking of an all-fiber laser by acousto-optic superlattice modulation. *Optics Letters*, 34 (7): 1111-1113 Apr 1 2009.

Cuadrado-Laborde, C.; Diez, A.; Cruz, J. L.; Andres, M. V. (2009b). Doubly active Q switching and mode locking of an all-fiber laser. *Optics Letters*, 34 (18): 2709-2711 Sep 15 2009.

Cuadrado-Laborde, C.; Diez, A.; Cruz, J. L.; Andres, M. V. (2010a). Experimental study of an all-fiber laser actively mode-locked by standing-wave acousto-optic modulation. *Applied Physics B-Lasers and Optics*, 99 (1-2): 95-99 Apr 2010.

Cuadrado-Laborde, C.; Diez, A.; Cruz, J. L.; Andres, M. V. (2010b). Actively Q-switched and modelocked all-fiber lasers. *Laser Physics Letters*, 7 (12): 870-875 Dec 2010.

Culverhouse, D. O.; Farahi, F.; Pannell, C. N.; Jackson, D. A. (1989a). Potential of stimulated Brillouin-scattering as sensing mechanism for distributed temperature sensors. *Electronics Letters*, 25 (14): 913-915 Jul 6 1989.

Culverhouse, D. O.; Farahi, F.; Pannell, C. N.; Jackson, D. A. (1989b). Stimulated Brillouin-scattering - a means to realize tunable microwave generator or distributed temperature sensor. *Electronics Letters*, 25 (14): 915-916 Jul 6 1989.

Culverhouse, D. O.; Richardson, D. J.; Birks, T. A.; Russell, P. S. J. (1995). All-fiber sliding-frequency Er^{3+}/Yb^{3+} soliton laser. *Optics Letters*, 20 (23): 2381-2383 Dec 1 1995.

Delgado-Pinar, M.; Zalvidea, D.; Diez, A.; Perez-Millan, P.; Andres, M. V. (2006). Q-switching of an all-fiber laser by acousto-optic modulation of a fiber Bragg grating. *Optics Express*, 14 (3): 1106-1112 Feb 6 2006.

Delgado-Pinar, M.; Diez, A.; Cruz, J. L.; Andres, M. V. (2007). Single-frequency active Q-switched distributed fiber laser using acoustic waves. *Applied Physics Letters*, 90 (17): Art. No. 171110 Apr 23 2007.

Erdogan, T. (1997). Fiber crating spectra. *Journal of Lightwave Technology*, 15 (8): 1277-1294 Aug 1997.

French, P. M. W. (1995). The generation of ultrashort laser pulses. *Reports on Progress in Physics* 58 (2): 169-267 Feb 1995.

Galtarossa, A.; Nava, E.; Valentini, G. (1993). *Single-Mode Optical Fiber Measurement: Characterization and Sensing*, Ed. G. Cancellieri, Artech Pub. 1993.

Geister, G.; Ulrich, R. (1988). Neodymium-fiber laser with integrated-optic mode locker. *Optics Communications*, 68 (3): 187-189 Oct 1 1988.

Haus, H. A. (2000). Mode-locking of lasers. *IEEE Journal of Selected Topics in Quantum Electronics*, 6 (6): 1173-1185 Nov-Dec 2000.

Huang, D. W.; Liu, W. F.; Yang, C. C. (2000). Q-switched all-fiber laser with an acoustically modulated fiber attenuator. *IEEE Photonics Technology Letters*, 12 (9): 1153-1155 Sep 2000.

Hudson, D. D.; Holman, K. W.; Jones, R. J.; Cundiff, S. T.; Ye, J.; Jones, D. J. (2005). Mode-locked fiber laser frequency-controlled with an intracavity electro-optic modulator. *Optics Letters*, 30 (21): 2948-2950 Nov 1 2005.

Imai, T.; Komukai, T.; Yamamoto, T.; Nakazawa, M. (1997). A wavelength tunable Q-switched fiber laser using fiber Bragg gratings. *Electronics and Communications in Japan Part II-Electronics*, 80 (11): 12-21 Nov 1997.

Italia, V.; Pisco, M.; Campopiano, S.; Cusano, A.; Cutolo, A. (2005). Chirped fiber Bragg gratings for electrically tunable time delay lines. *IEEE Journal of Selected Topics in Quantum Electronics*, 11 (2): 408-416 Mar-Apr 2005.

Jeon, M. Y.; Lee, H. K.; Kim, K. H.; Lee, E. H.; Oh, W. Y.; Kim, B. Y.; Lee, H. W.; Koh, Y. W. (1998). Harmonically mode-locked fiber laser with an acousto-optic modulator in a Sagnac loop and Faraday rotating mirror cavity. *Optics Communications*, 149 (4-6): 312-316 Apr 15 1998.

Kee, H. H.; Lees, G. P.; Newson, T. P. (1998). Narrow linewidth CW and Q-switched erbium-doped fibre loop laser. *Electronics Letters*, 34 (13): 1318-1319 Jun 25 1998.

Kim, B. Y.; Blake, J. N.; Engan, H. E.; Shaw, H. J. (1986). All-fiber acoustooptic frequency shifter. *Optics Letters*, 11 (6): 389-391 Jun 1986.

Kim, H. S.; Yun, S. H.; Kwang, I. K.; Kim, B. Y. (1997). All-fiber acousto-optic tunable notch filter with electronically controllable spectral profile. *Optics Letters*, 22 (19): 1476-1478 Oct 1 1997.

Kuizenga, D. J.; Siegman, A. E. (1970). FM and AM mode locking of homogeneous laser .1. Theory. *IEEE Journal of Quantum Electronics*, QE 6 (11): 694-& 1970.

Li, Y. H.; Lou, C. Y.; Han, M.; Gao, Y. Z. (2001). Detuning characteristics of the AM mode-locked fiber laser. *Optical and Quantum Electronics*, 33 (6): 589-597 Jun 2001.

Li, Y. Q.; Zhang, F. C.; Yoshino, T. (2003). Wide-range temperature dependence of Brillouin shift in a dispersion-shifted fiber and its annealing effect. *Journal of Lightwave Technology*, 21 (7): 1663-1667 Jul 2003.

Li, X. W.; Peng, L. M.; Wang, S. B.; Kim, Y. C.; Chen, J. P. (2007). A novel kind of programmable 3(n) feed-forward optical fiber true delay line based on SOA. *Optics Express*, 15 (25): 16760-16766 Dec 10 2007.

Liu, W. F.; Russell, P. S. J.; Dong, L. (1997). Acousto-optic superlattice modulator using a fiber Bragg grating. *Optics Letters*, 22 (19): 1515-1517 Oct 1 1997.

Liu, W. F.; Russell, P. S. J.; Dong, L. (1998). 100% efficient narrow-band acoustooptic tunable reflector using fiber Bragg grating. *Journal of Lightwave Technology*, 16 (11): 2006-2009 Nov 1998.

Liu, Y. Q.; Yang, J. L.; Yao, J. P. (2002). Continuous true-time-delay beamforming for phased array antenna using a tunable chirped fiber grating delay line. *IEEE Photonics Technology Letters*, 14 (8): 1172-1174 Aug 2002.

Liu, Y. Q.; Yao, J. P.; Yang, J. L. (2003). Wideband true-time-delay beam former that employs a tunable chirped fiber grating prism. *Applied Optics*, 42 (13): 2273-2277 May 1 2003.

Myren, N.; Margulis, W. (2005). All-fiber electrooptical mode-locking and tuning. *IEEE Photonics Technology Letters*, 17 (10): 2047-2049 Oct 2005.

Ortega, B.; Cruz, J. L.; Capmany, J.; Andres, M. V.; Pastor, D. (2000). Analysis of a microwave time delay line based on a perturbed uniform fiber Bragg grating operating at constant wavelength. *Journal of Lightwave Technology*, 18 (3): 430-436 Mar 2000.

Parker, T. R.; Farhadiroushan, M.; Feced, R.; Handerek, V. A.; Rogers, A. J. (1998). Simultaneous distributed measurement of strain and temperature from noise-initiated Brillouin scattering in optical fibers. *IEEE Journal of Quantum Electronics*, 34 (4): 645-659 Apr 1998.

Pérez-Millán, P.; Torres-Peiró, S.; Mora, J.; Díez, A.; Cruz, J. L.; Andres, M. V. (2004). Electronic tuning of delay lines based on chirped fiber gratings for phased arrays powered by a single optical carrier. *Optics Communications*, 238 (4-6): 277-280 Aug 15 2004.

Pérez-Millán, P.; Díez, A.; Andres, M. V.; Zalvidea, D.; Duchowicz, R. (2005a). Q-switched all-fiber laser based on magnetostriction modulation of a Bragg grating. *Optics Express*, 13 (13): 5046-5051 Jun 27 2005.

Pérez-Millán, P.; Cruz, J. L.; Andres, M. V. (2005b). Active Q-switched distributed feedback erbium-doped fiber lasers. *Applied Physics Letters*, 87 (1): Art. No. 011104 Jul 4 2005.

Philippov, V. N.; Kiryanov, A. V.; Unger, S. (2004). Advanced configuration of erbium fiber passively Q-switched laser with Co^{2+}: ZnSe saturable absorber. *Ieee Photonics Technology Letters*, 16 (1): 57-59 Jan 2004.

Phillips, M. W.; Ferguson, A. I.; Hanna, D. C. (1989a). Frequency-modulation mode-locking of a Nd^{3+}-doped fiber laser. *Optics Letters*, 14 (4): 219-221 Feb 15 1989.

Phillips, M. W.; Ferguson, A. I.; Kino, G. S.; Patterson, D. B. (1989b). Mode-locked fiber laser with a fiber phase modulator. *Optics Letters*, 14 (13): 680-682 Jul 1 1989.

Russell, P. S. J.; Liu, W. F. (2000). Acousto-optic superlattice modulation in fiber Bragg gratings. *Journal of the Optical Society of America A-Optics Image Science and Vision*, 17 (8): 1421-1429 Aug 2000.

Russo, N. A.; Duchowicz, R.; Mora, J.; Cruz, J. L.; Andres, M. V. (2002). High-efficiency Q-switched erbium fiber laser using a Bragg grating-based modulator. *Optics Communications*, 210 (3-6): 361-366 Sep 15 2002.

Schaffer, C. B.; Garcia, J. F.; Mazur, E. (2003). Bulk heating of transparent materials using a high-repetition-rate femtosecond laser. *Applied Physics A-Materials Science & Processing*, 76 (3): 351-354 Mar 2003.

Siegman, E. (1986). *Lasers*, University Science, Mill Valley, CA, 1986.

Tuchin, V. V. (1993). Lasers and fiber optics in biomedicine. *Laser Physics*, 3 (4): 767-820 1993.

Vicente, S. G. C.; Gámez, M. A. M.; Kiryanov, A. V.; Barmenkov, Y. O.; Andrés, M. V. (2004). Diode-pumped self-Q-switched erbium-doped all-fibre laser. *Quantum Electronics*, 34 (4): 310-314 Apr 2004.

Villegas, I. L.; Cuadrado-Laborde, C.; Abreu-Afonso, J.; Díez, A.; Cruz, J. L.; Martinez-Gámez, M. A.; Andrés, M. V. (2011a). Mode-locked Yb-doped all-fiber laser based on in-fiber acoustooptic modulation. *Laser Physics Letters*, 8 (3): 227-231 Mar 2011.

Villegas, I. L.; Cuadrado-Laborde, C.; Díez, A.; Cruz, J. L.; Martinez-Gámez, M. A.; Andrés, M. V. (2011b). Yb-doped strictly all-fiber laser actively Q-switched by intermodal acousto-optic modulation. *Laser Physics, in press* 2011.

Wada, K.; Fujita, J.; Yamada, J.; Matsuyama, T.; Horinaka, H. (2008). Simple method for estimating shape functions of optical spectra. *Optics Communications*, 281 (3): 368-373 Feb 1 2008.

Yu, C. X.; Haus, H. A.; Ippen, E. P.; Wong, W. S.; Sysoliatin, A. (2000). Gigahertz-repetition-rate mode-locked fiber laser for continuum generation. *Optics Letters*, 25 (19): 1418-1420 Oct 1 2000.

Zadok, A.; Raz, O.; Eyal, A.; Tur, M. (2007). Optically controlled low-distortion delay of GHz-wide radio-frequency signals using slow light in fibers. *IEEE Photonics Technology Letters*, 19 (5-8): 462-464 Mar-Apr 2007.

Zalvidea, D.; Russo, N.A.; Duchowicz, R.; Delgado-Pinar, M.; Díez, A.; Cruz, J. L.; Andrés, M. V. (2005). High-repetition rate acoustic-induced Q-switched all-fiber laser. *Optics Communications*, 244 (1-6): 315-319 Jan 3 2005.

14

Surface Acoustic Waves and Nano–Electromechanical Systems

Dustin J. Kreft and Robert H. Blick
University of Wisconsin – Madison
Department of Electrical and Computer Engineering
U.S.A.

1. Introduction

Surface acoustic waves (SAW) follow the industrial trend of reducing the size, enhancing the speed, while enhancing the efficiency of energy coupling. Integrating this with micro-electromechanical systems (MEMS) and nano-electromechanical systems (NEMS) offers a wide variety of applications such as touch screens, gas and biological sensors, and embedded RFID devices. With modern lithographic techniques, allowing the fabrication of smaller SAW devices, we now use SAWs to probe the mechanical interactions of nano structures. In particular, SAWs can be used to actuate NEMS which gives rise to many interesting phenomena including anomalous acoustoelectric currents, shock waves in suspended devices, and few electron transport, to only name a few, (Beil et al., 2008; Talyanskii et al., 1997). Today, SAWs are also used to generate a quantized current for use as a current standard. In practice two counter-propagating SAWs are used to observe a quantized acoustoelectric current. This leads to population and depopulation of discrete states (Kataoka et al., 2007). In the following we want to give an overview of the state of the art of applying SAWs to nanomechanical devices. We will also give a brief introduction to recent nanoelectromechanical systems with integrated low-dimensional electron gases, which have the potential to reveal insights into quantized acoustoelectric states.

2. Fabrication

The focus for generating SAWs in this chapter will involve the fabrication of interdigitated transducers (IDT). An IDT is simple in concept but can be very involved when fine tuning a structure for engineering applications. Such topics as electronic impedance matching to RF lines, effects of bulk waves in contrast to SAWs, and increasing bandwidth will not be covered; though, this is simply a shortened list of things to consider when designing a proper IDT for engineering applications, they do fall outside the scope of this chapter. Nevertheless, another fabrication step we will consider is the use of acoustic waveguides for acoustic impedance matching of IDTs to nanomechanical devices.

2.1 Interdigitated Transducer Design
The main equation to consider when designing an IDT is:

$$v = \lambda f \qquad\qquad (1)$$

Where v is the velocity of sound in the material, λ is the SAW wavelength or pitch of the IDT, and f is the frequency of the propagating SAW. The pitch of the IDT fingers is the same as the SAW wavelength, λ, which will propagate across the sample. Fig. 1 shows a simple IDT schematic along with a scanning electron micrograph (SEM) image and optical image.

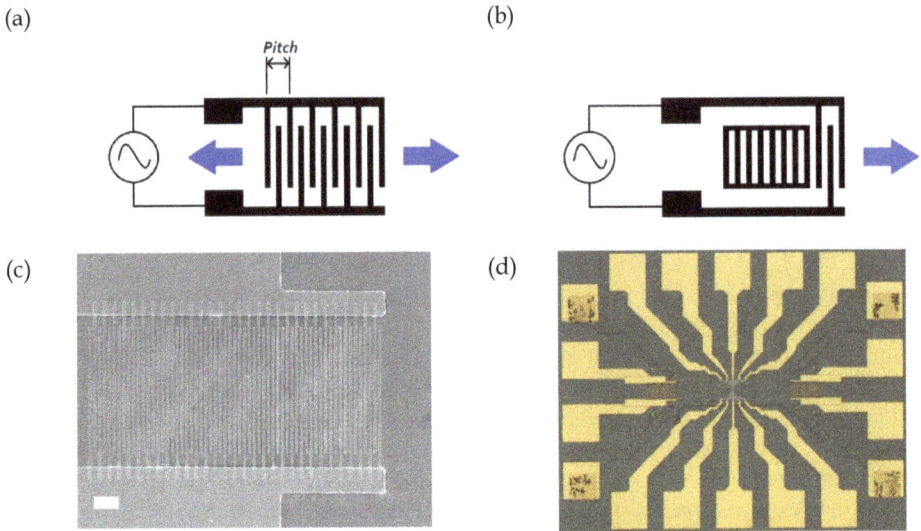

(a)   (b)

(c)   (d)

Fig. 1. Schematics and scanning electron microscope images of IDTs. (a) Schematic of bidirectional IDT, (b) schematic of unidirectional IDT through the use of a reflection grating, (c) SEM image of a bidirectional IDT on GaAs with a pitch of 4 µm and a center frequency of ~715 MHz. The scale bar is 10 µm. (d) Optical image of device with an IDT on both ends and a nanomechanical device placed in the center

In Fig. 1 the arrows indicate the SAW propagation directions. It can be easily seen that the SAW will propagate in both directions away from the IDT. A singly propagating SAW direction can be achieved by placing a reflection grating on one end of the SAW. The reflection grating will have the same geometry as the IDT; that is, the same finger spacing and width. The grating distance from the IDT is $\sim \lambda/2$. The IDT fingers are typically chosen with evenly spaced fingers, where the spacing between the fingers is equal to the finger width. The finger width and finger spacing is $\lambda/4$ in this scenario and gives a metallization ratio of $\eta = 0.5$, which will generate only odd harmonics with no response of the third harmonic (Campbell, 1998). The bandwidth of an IDT is defined as $BW\% = f/N_{Pairs}$. Where f, again, is the center frequency and N_{Pairs} is the number of finger pairs of the IDT. The IDT in Fig. 1a has five pairs of fingers and is symmetric about a center point along the axis of SAW propagation. It is typical to fabricate two IDTs, one on each end of the device, so that their propagating waves can interfere either constructively or destructively across the center region of the device; this is typically the region where a Quantum Point Contact (QPC) or other structures resides. This can be seen in Fig. 1d where in the center a double quantum dot (DQD) is placed and the IDTs are placed to the left and right.

2.2 Acoustic Waveguide Design

Another detail of fabrication is the use of waveguides for the SAW. A waveguide for SAWs is a pattern that is fabricated onto the device so the SAWs can be guided into a certain region allowing a stronger SAW signal or amplitude; it is analogous to a coplanar waveguide for RF signals. When fabricating IDTs, and a device as a whole, using waveguides does not have much of a use in the areas of RFID, cellular delay lines, sensors, and other non-region specific devices. However, this chapter will focus on NEMS and the use of SAWs to interact with these devices. Since the NEMS device is orders of magnitude smaller than the IDT aperture very little SAW power can interact with the NEMS structure; it can be beneficial to include a waveguide to focus the SAW onto the nanostructure.

Waveguides can be modeled using standard acoustic, or horn, waveguides with some minor tuning. There are several shapes which can be used and each shape offers its own benefits regarding a particular application. We will start by first mentioning the base equation for hyperbolic horns used in speaker systems, these horns can also be referred to as hyperbolic-exponential horns or hypex horns (Salmon, 1946).

$$A = A_t[\mathrm{Cosh}(kx) + T \cdot \mathrm{Sinh}(kx)]^2 \tag{2}$$

$$z = \frac{\rho_0^v}{A_t}\left(\sqrt{1 - \left(\frac{f_c}{f}\right)^2} + i\frac{f_c T}{f}\right)\left(1 - \frac{f_c^2(1 - T^2)}{f^2}\right)^{-1} \tag{3}$$

Eq. 2 gives the wave front area expansion of the wave as propagating through the waveguide or horn. A_t is the area of the throat, or base, of the waveguide, T is a factor describing the shape of waveguide; for $T = 1$ the waveguide becomes exponential in shape, and becomes canonical as $T \rightarrow \infty$, k is given as $(2\pi f_c)/v$; where v is the velocity of sound, and f_c is the cutoff frequency of the waveguide. For hyperbolic and exponential waveguides there is no transmission below the cutoff frequency, f_c, since at this point the impedance is purely reactive. Eq. 3 gives the impedance of the waveguide; all variables are defined previously with the addition of ρ_0 which is the density of the material.

Fig. 2 contains three SEM images of an exponential waveguide with a tube waveguide in the center. A tube waveguide is simply a small or narrow SAW delay line. The material is a GaAs/AlGaAs heterostructure containing a two-dimensional electron gas (2DEG) and a sacrificial layer. We use a 2DEG because the high electron mobility makes this an ideal candidate for nanostructures, such as quantum dot (QD) and DQD systems. The waveguide was defined using electron-beam lithography to open up an etch region through the PMMA. The material was then wet etched with H_3PO_4:H_2O_2:H_2O into the sacrificial layer. Then diluted HCl was used to remove the sacrificial layer which resulted in a suspended region in the center and around the waveguide. This suspended region is visible as a darkened shadow indicating a height difference due to strain relaxation of the material. The suspended region forms the actual nanomechanical device. Since this device contains a DQD in the center region of the structure it is beneficial to use a waveguide to ensure a higher SAW power density coupled into that region. Since the IDT aperture of this device is 50 μm and the DQD region is roughly 600 nm at the widest QPCs.

(a)

(b)

(c)

(d)

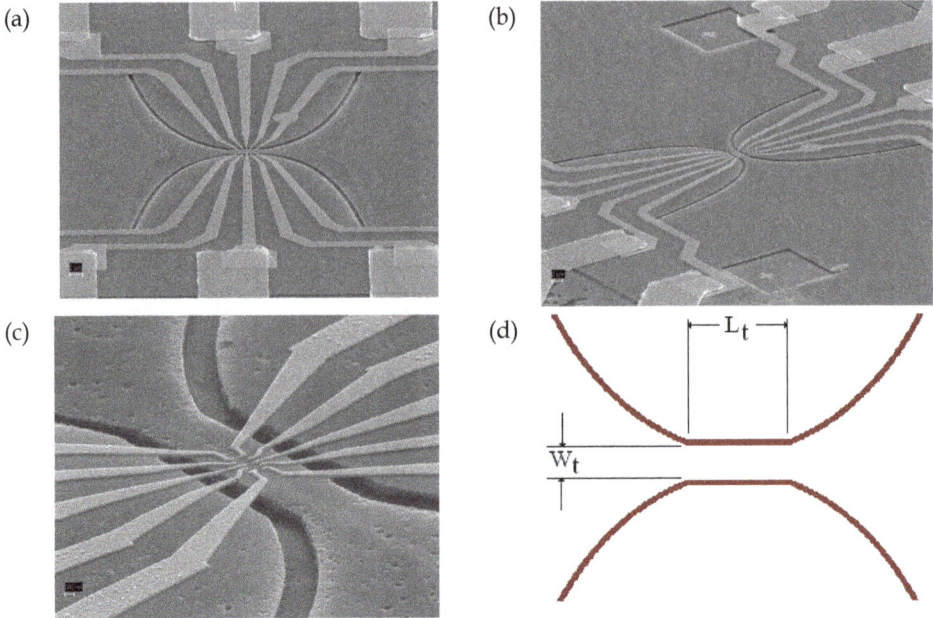

Fig. 2. SEM images of an exponential waveguide with a cutoff frequency of 75 MHz and throat width of ~1.5 µm. The darker region around the waveguide shows that the area is suspended, which was achieved by an HCl etch to remove the sacrificial layer below the 2DEG, (a), (b) and (c). (d) Schematic of waveguide, center region is the throat (pipe waveguide) with width W_t and length L_t. The outer curves are the exponential portion of the waveguide

3. Surface Acoustic Waves in Two Dimensional Electron Gases

3.1 Surface Acoustic Wave Basics

When applying an RF signal to the IDTs an alternating electric potential is seen by the electrodes, this will create an electric field distribution within the piezoelectric substrate, see Fig. 3a. This field distribution, in turn, will cause a mechanical deformation of the material through the inverse piezoelectric effect. As stated before, this mechanical deformation will propagate away from the IDTs and continue to move along the surface of the device, see Fig. 3d. Here we neglect other forms of waves being produced such as bulk waves.

Once the RF signal of the IDT is coupled into the piezoelectric substrate a Rayleigh-wave is produced. The elliptical wave propagating (Fig. 3b, d) along the surface can be described quite accurately by a traveling wave:

$$U = |U| e^{i(\omega t - kz)} e^{-k|y|} \tag{4}$$

The electric potential being created by the propagating wave can also be described in a like manner:

$$\Phi = |\Phi| e^{i(\omega t - kz)} e^{-k|y|} \tag{5}$$

In both Eqs. 4 and 5, ω is the angular frequency and k is the phase constant. The SAW penetrates into the depth of the material by about one wavelength, $\sim\lambda$. This value is different in the suspended region. Hence, the actuation of nanomechanical resonators is considerably enhanced. Another property of the SAW is that the electric field created from the induced electric potential does not terminate at the surface of the material but can extend beyond by λ.

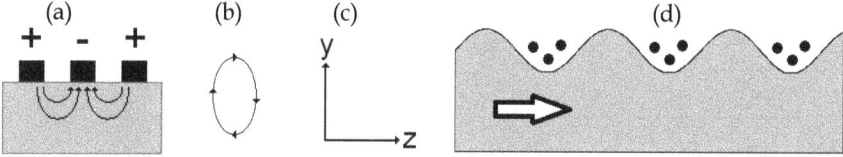

Fig. 3. (a) Side view of IDT with an applied voltage, the electric field couples into the piezoelectric substrate causing a deformation. (b) The elliptical motion of the Rayleigh-wave. (c) Axis used for reference in equations as it applies to the orientation of the piezoelectric substrate. (d) Side view of piezoelectric material as it is deformed causing a SAW to propagate

The set of base equations used for describing SAW phenomena are listed.

$$E_i = -\frac{\partial \Phi}{\partial x_i} \tag{6}$$

$$T_{ij} = c_{ijkl}S_{kl} + e_{nij}E_n \tag{7}$$

$$S_{kl} = \frac{1}{2}\frac{\partial u_k}{\partial x_l} + \frac{\partial u_l}{\partial x_k} \tag{8}$$

Eq. 6 is the electric field intensity that is produced from the deformed piezoelectric material from the SAW. Eq. 7 is the piezoelectric mechanical stress and Eq. 8 is the linear strain displacement.

3.1.1 Attenuation
SAW attenuation can be described by the following equations (Wixworth et al., 1989):

$$\Gamma = k\frac{K^2_{eff}}{2}\frac{\sigma_s / \sigma_M}{1 + (\sigma_s / \sigma_M)^2} \tag{9}$$

$$\frac{\Delta v}{v} = k\frac{K^2_{eff}}{2}\frac{1}{1 + (\sigma_s / \sigma_M)^2} \tag{10}$$

Here the attenuation occurs because part of the longitudinal electric field of the propagating wave couples into the electrons of the 2DEG. This not only causes a current to flow but pulls power from the SAW due to ohmic losses. This attenuation is described by Eq. 9. A SAW velocity shift is also observed due to the piezoelectric stiffening of the substrate, see Eq. 10 (Wixworth et al., 1989). Below are the recreated graphs from (Wixworth et al., 1989) to show the relationship of the attenuation and sound velocity shift due to a change in conductivity.

(b)

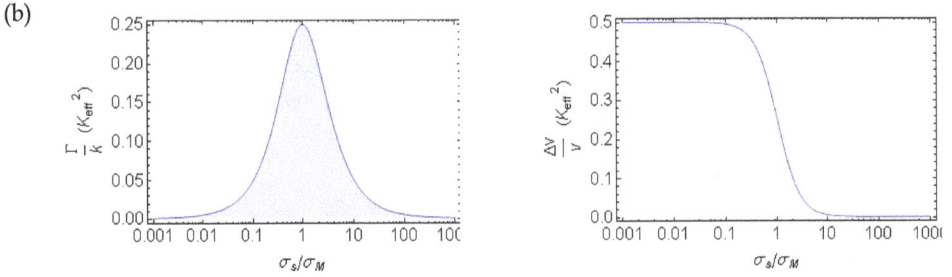

Fig. 4. (a) The SAW attenuation in units of K_{eff}^2, see Eq. 9. (b) The change in SAW velocity in units of K_{eff}^2, see Eq. 10

Where k is the SAW wave vector, K_{eff}^2 is the effective piezoelectric coupling coefficient, σ_s is the 2DEG sheet conductivity, and $\sigma_M = v_0(\varepsilon_1 + \varepsilon_2)$. Again, where v_0 is the sound velocity and ε_1 and ε_2 are the dielectric constants of the piezoelectric substrate and half space above it.

3.2 Acoustoelectric Current

As the SAW propagates across the material it creates two types of currents, one is the normal acoustoelectric current and the other being the anomalous acoustoelectric current. The normal acoustoelectric current is created by electrons being "dragged" across the material and can be described by Eq. 11 where I is the current, n is the number of electrons, e is the charge of an electron, and f is the frequency of the SAW or RF signal to the IDT. This current always flows in the direction of the SAW and is produced as a DC current despite an oscillating RF signal being applied to the IDTs. Eq. 11 shows that at higher frequencies the normal acoustoelectric current becomes quantized.

$$I = nef \tag{11}$$

$$j(z) = \frac{a\omega}{2\pi} \int_0^{2\pi/\omega} \sigma_{zz}(z,t)E_z(z,t)dt \tag{12}$$

Once this deformation occurs the energy bands in the material bend as well causing the electrons to fall into the created quantum wells and are dragged along with the SAW, see Fig. 3d. As the frequency increases the wavelength, and pitch of the IDT, decreases causing fewer energy states to be available within the wells. This idea is also implemented to generate QDs within the SAW, (Barnes et al., 2000).

The anomalous acoustoelectric current is produced from the deformation of the material and flows as a DC current. The difference is that the anomalous current always flows in one direction regardless of which IDT, left or right, produced the SAW. The current is smaller than the normal acoustoelectric current and is detected by different methods. The anomalous acoustoelectric current can be obtained by sweeping the RF signal of the IDTs and typically appears at an off-center frequency. Since the normal acoustoelectric current is much smaller, the anomalous current can be more easily detected, as shown in Fig. 5b. One such method for direct detection of the anomalous acoustoelectric current is to apply an RF signal to both the left and right IDTs, while phase locking the two signals (Beil et al., 2008). By phase locking the IDT signals a standing SAW can be created. Thus, a surface deformation is produced without a net direction of propagation. As seen in Fig. 5a the phase is shifted between the two IDTs as the current is measured. At a phase difference of 180° the current tends towards zero.

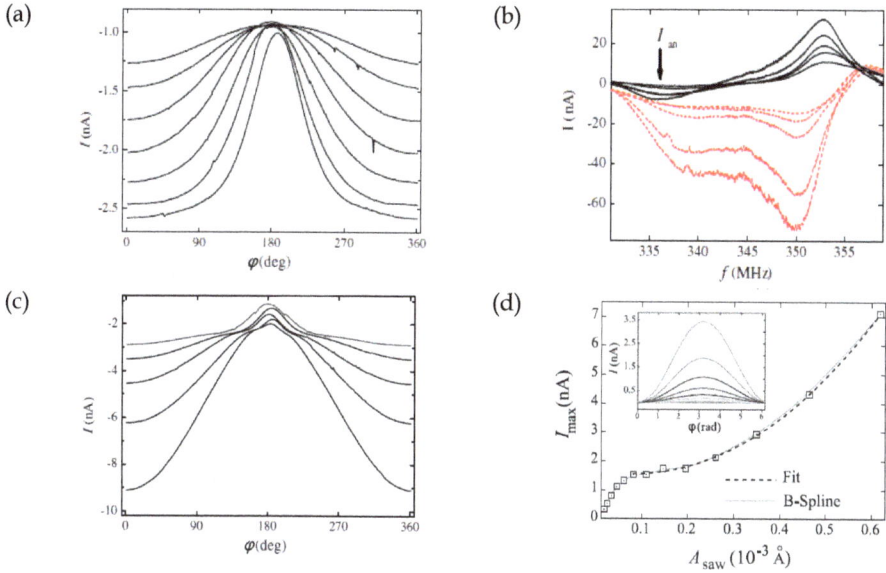

Fig. 5. Results for a suspended beam that is 1.2 μm long and 300 nm wide. (a) Anomalous acoustoelectric current as a result of phase adjusting two counter propagating SAWs. (b) Acoustoelectric current vs. applied frequency of the IDTs. There is always a negative current value to the left and is indicative of the anomalous current. (c) Formation of a shock wave seen by its Sinc(x) shape. The device's response jumps to a higher order mode and returns to Sin(x). (d) Derived anomalous acoustoelectric current amplitude as a result of the calculated SAW amplitude. All images are taken from Beil et al., 2008, Copyright (2008) by The American Physical Society

This zero current occurs because the two SAWs interfere destructively and create a nearly smooth surface with no deformation. At 0°, and also 360°, the current is at a maximum since the two SAWs interfere constructively allowing a maximum in the surface deformation.
The anomalous acoustoelectric current can be used to probe the SAW amplitude of a suspended beam or nanostructure. The accompanying graphs show the acoustoelectric current from the device. It has been shown that the anomalous acoustoelectric current through the device in relation to the SAW amplitude can be described by Eqs. 12-14; where Eq. 12 is the general equation for an anomalous acoustoelectric current, 13 is the derived equation, and 14 is the strain equation of the beam.

$$j_{an}(L/2) = -\frac{\sigma_0 e_{z4}}{2\varepsilon_{GaAs}}(\Pi_{zzzy}S_{zy}(L/2)^2) \tag{13}$$

$$S_{zy}(z,t) = 2A_t\left(\left(\frac{6z^2}{L^3} - \frac{6z}{L^2}\right)\right)(Cos(\varphi/2) - Cos(k_{SAW}L + \varphi/2)) \tag{14}$$

Where σ_0 is the unperturbed conductivity, ε_{GaAs} is the permittivity of GaAs, Π is a tensor which describes the effect of strain S, and A_t is the transverse component of the SAW.

An interesting phenomenon that occurs is the formation of shock waves in the suspended structure. From Figs. 5a and 5c it is seen that the device exhibits a Sin(x) type behavior in the anomalous acoustoelectric current at low RF powers. Once the RF power starts to increase it can be seen that the shape of the current starts to exhibit more of a Sinc(x) shape. This transition from a linear to a non-linear response indicates the formation of a shock wave in the suspended beam. As the power increases, the beam jumps into the next higher order mode and the current trace returns back to a linear response again. The shock wave formation is an indication that the beam will transition from one mode to another.

3.3 Surface Acoustic Waves in Magnetic Fields

We can use SAWs to probe the characteristics of a 2DEG under the presence of a magnetic field. Since a high mobility 2DEG is subject to Shubnikov-de Haas (SdH) oscillations, which creates changes in the conductivity, and SAWs are more sensitive to a conducting plane or surface, and the conductance of that plane, makes this combination an ideal candidate to investigate quantum effects of the 2DEG. The real interesting features are seen when integral Landau level filling factors are observed which causes a drop in conductivity of the 2DEG. At these quantized values the SAW responds strongly to the conductivity, σ.

In Fig. 6 it can be seen that the SdH oscillations and the acoustoelectric current oscillations are nearly identical. The SdH measurement was measured using a standard four-point lock-in technique, where as the acoustoelectric current was created by a SAW and was taken from two ohmic contacts on opposite sides of the sample. The peaks and valleys of the two measurements line up, for the most part, but there is a small offset. This offset is due to the fact that the SAW attenuation is not a linear with respect to the 2DEG conductivity, or magneto conductivity in this case, see Eq. 9.

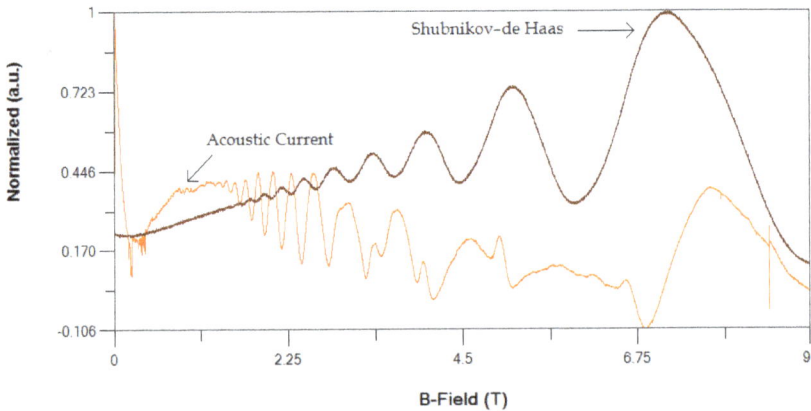

Fig. 6. A magnetic field, of maximum value 9 T, was applied perpendicular to the surface of the device. The device is shown in figures 1d and 2a-2d and contains a 2DEG 40 nm below the surface. Upper trace is the Shubnikov-de Haas oscillations measured using standard lock-in techniques and normalized to 713.4 Ω. The lower trace is the acoustoelectric current generated from a SAW with -12 dBm applied to the right IDT at 1.488 GHz and is normalized to 10.47 nA

Special attention has to be given to the acoustic current trace of Fig. 6 around 3 – 4 Tesla. There is a splitting of the peaks that occurs. This information does not appear in the SdH oscillation and is only isolated to a SAW effect. During this splitting the conductivity of the 2DEG is very low and σ_s becomes much smaller than σ_M, see Eq. 9. When this happens a maximum in SAW attenuation will occur which results in a reduced SAW amplitude. The center of a SAW current split is a minimum in the 2DEG conductivity; further details are discussed in Wixworth et al., 1989.

From Fig. 6 there are two observations that need to be explained. One is the spike-like feature or discontinuity around the 8.5 T mark of the acoustic trace. This is caused from SAW reflections on the sample. This spike-like feature can be seen on other data plots as well with SAW currents. The final feature is the negative acoustic current that was measured. Since at this point the SAW was highly attenuated, causing little or no current to flow, caused the sample itself to heat. The negative current may be a combination of a thermal current and a small offset in the measuring equipment.

4. Surface Acoustic Waves in Quantum Electronics

SAWs are used to produce a current in low dimensional electronic systems and NEMS. This can be used in various applications and for numerous different designs. The information presented in this section will deal with piezoelectric materials with an embedded two dimensional electron gas (2DEG); mainly GaAs/AlGaAs heterostructures unless otherwise stated. These measurements are carried out at liquid helium temperatures or lower, ≤ 4.2 K.

4.1 Quantum Point Contacts and Low Dimensional Channels

SAWs can be used to create a quantized current based upon Eq. 7. The advantage of the SAW inducing a quantized current is that this process can be used to populate and depopulate QDs and DQDs at higher frequencies then what can be used by applying an oscillating signal to the source-drain of the 2DEG. When using a SAW as the current source the acoustoelectric current can be pinched off in a Coulomb blockade just like a standard source-drain current can be. This pinch off is done via a quantum point contact (QPC) or a set of QPCs, which can be used to form QDs; see Fig. 7.

4.1.1 Quantum Point Contact Fabrication

Fabricating QPCs is done with the same methods as fabricating IDTs, see Sec. 2. Since QPCs are have small dimensions it is most common to use Electron Beam Lithography, or e-beam lithography, to create the structures. The exact dimensions of the QPC pair depends on what works best for the user and there is no set rules for design like there is for IDTs. When viewing Fig. 7 it can be seen that there are five sets of QPCs. A single QPC is seen as an electrode with another electrode opposite its position. All of the QPCs shown have a few common features; the tip of the QPC is small when compared to the rest of the electrode and the gap between the electrode tips is small as well. The tip is small so the electric field being emitted from the QPC is very localized, and the majority of the electrode is made wider so that it covers a wider portion of the 2DEG so the electrons are repelled. The gap between electrodes is small so pinch-off can be achieved with small voltages, more on this in Sec. 4.1.2.

4.1.2 Quantum Point Contact Operation

QPCs work by applying a negative voltage to them, this negative voltage produces an electric field that penetrates into the 2DEG causing electrons to be repelled at or around the QPC region; this can be seen if Fig 7(c). Here the black region represents the metallic electrode while the blue region represents the electric field and the green region represents the 2DEG or substrate. It can be taken as a good estimate that the electric field penetrates into the 2DEG from the QPC electrode at a 45° angle. As a larger negative voltage is applied to the QPC the effective screening electric field becomes larger causing the electron path within the 2DEG to become more constricted until single electron transport is achieved; this leads to a Coulomb Blockade, see Sec. 4.2. Then finally, the current is completely pinched off, see Figs 9 and 10.

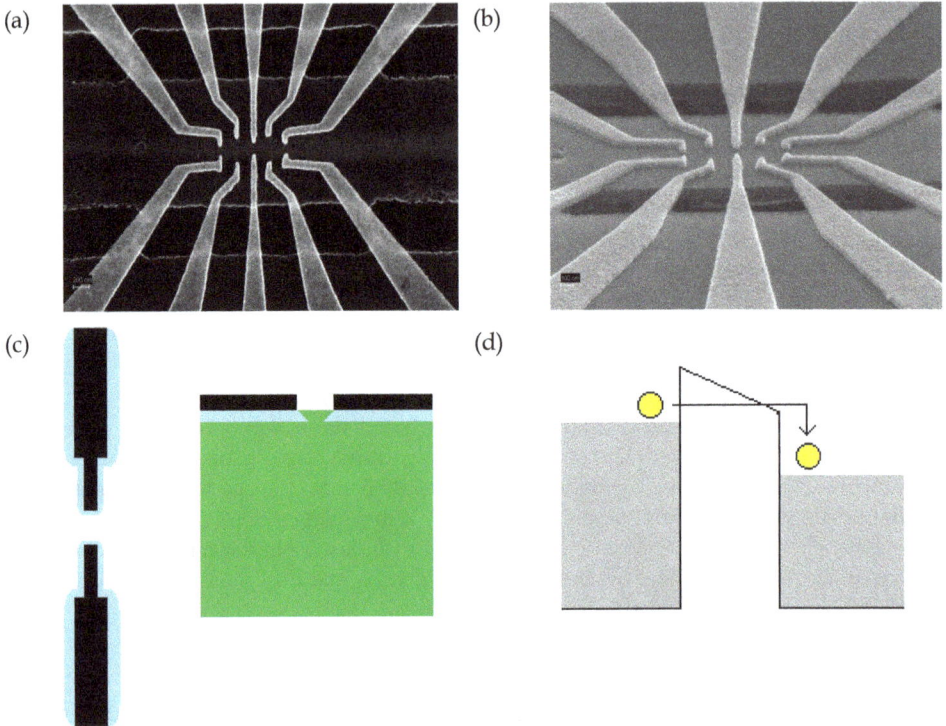

Fig. 7. Two SEM images of the same DQD structure on top of a suspended tube section of an acoustic waveguide. The DQD is made up of five pairs of QPCs. The material is a GaAs/AlGaAs heterostructure with an embedded 2DEG about 40 nm below the surface. (a) Top view with a scale bar of 200 nm, (b) angled view with a scale bar of 100 nm. (c) schematic view of QPC with applied voltage, the blue area represents the electric field surrounding the QPC and penetrating into the 2DEG (green). (d) Tunneling of electrons across potential barrier created by applied voltage from the QPC

The QPC creates a tunneling barrier which separates the source and drain regions of the sample. As the voltage magnitude becomes larger the strength, or height, or the tunneling barrier is increased and the width of the barrier becomes larger. When plotted a step like feature can be seen which corresponds to a single conductance step which has a value of G = $2e^2/h$, where h is Plank's constant and e is the charge of an electron. Now the temperature must be low so the thermal energy, E = k_bT, of the background is smaller than the tunneling energy needed for the electrons to "jump" across the barrier, where k_b is the Boltzmann constant. As seen in Figs 9 and 10 this step like feature can be seen by doing an I-V measurement. By changing the temperature of the system the phonon energy is increased and causes scattering events to increase, or increase the electron-phonon interaction, and the steepening or smoothing of the step like feature is a direct measure of this.

4.1.3 Usage of Quantum Point Contacts and Surface Acoustic Waves
The use of QPCs offers a benefit of determining which SAW mode(s) are propagating in the sample. Different SAW modes, such as bulk, longitudinal, and transverse with propagate at different frequencies due to the fact that they have different sound velocities, see Eq. 1. Another factor, which affects the sound velocity, is the propagation direction of the SAW with respect to the crystal orientation of the material. In Fig. 8 a QPC had an applied voltage of -0.8 V, which puts the QPC into pinch-off mode. Since it is in pinch-off higher RF power is required to create a sufficiently strong SAW that will overcome the potential barrier. As the power is increased from -18 dBm to -10 dBm, three peaks emerge as transferring current through the tunnel barrier. From Eq. 7 we can calculate the electron count to be 6, 3, and 2 for RF powers of -10 dBm, -12 dBm, and -14 dBm, respectively (some rounding is taken into account, due to thermal errors in measurement).

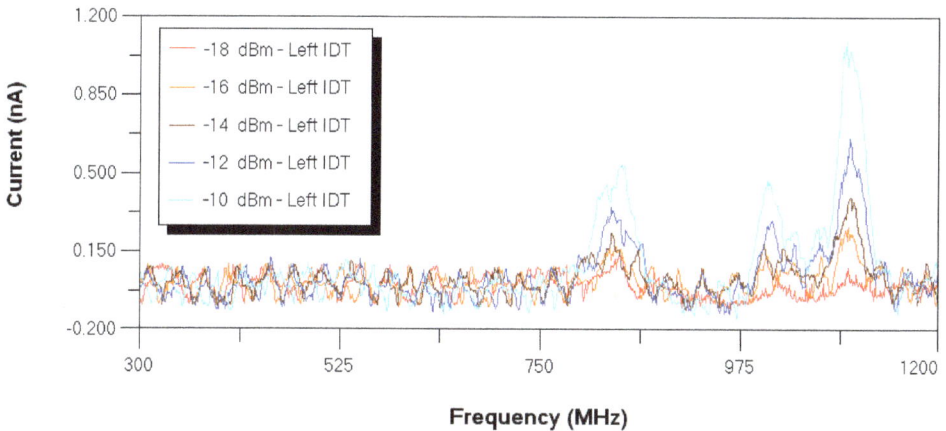

Fig. 8. A frequency sweep of varying RF powers while the center QPC of the sample in Fig. 4 is held at -0.8 V. The first peak is at 840 MHz with a current of 540 pA and a velocity of 3,368 m/s, the second peak at 1.005 GHz with a current of 472 pA and a velocity of 4,020 m/s, the third peak is at 1.095 GHz with a current of 1.098 nA and a velocity of 4,380 m/s

Now the three peaks represent different SAW modes. The highest frequency peak of Fig. 8 of 1.095 GHz represents a longitudinal wave with an acoustic velocity of 4,380 m/s and an angle of about 10° off from the (110) direction (Kuok et al., 2001). This small angle variation is due to a small misalignment during the lithography process. When viewing the lower peak of 840 MHz at a velocity of 3,368 m/s, this coincides with a fast transverse wave with, again, a 10° difference between the SAW direction and the (110) GaAs crystal orientation (Kuok et al., 2001).

4.2 Coulomb Blockade of Acoustoelectric Current

When looking over Fig. 9, we see that single steps are observed just before the total acoustoelectric current becomes zero. Focusing on Fig. 9a, we see that the last step is at about 0.5 nA. When solving with Eq. 11 the first step yields an answer of n = 1, i.e. a single electron is being transferred. We can look at all of the remaining steps and solve for them as well which will reveal integer numbers for n and will increase by 1 for each step, as is expected. Fig. 9a also shows that an applied voltage across the source and drain contacts, or ohmics, has very little effect on the final quantized acoustoelectric current. We do see, however, that a larger, or smaller, gate voltage is required for final pinch-off in the system but this is due to the small offset in the Fermi energy because of the applied bias. The only real difference is the shift from negative current to a positive current which is easy explained by the fact that the source-drain bias is producing a current that is opposite in direction, when V_{ds} < 0 mV, to that of the acoustoelectric current. Another aspect is the RF power dependence on the quantized current. When viewing Fig. 9b there is a similar trend to that of Fig. 9a. The small change in RF power has only a small effect on the acoustoelectric current. This is because the RF power is directly proportional to the SAW amplitude and we do not really identify any significant difference in the number of electrons being transferred, or Coulomb steps, until a drop of about 2 dBm.

So far an acoustoelectric current behaves in the same way as a standard source-drain bias current when a QPC is near pinch off. There is, however, another effect which can arise. As seen in Fig 9c a negative current arises and still exhibits the step like behavior. This negative current is the negative anomalous acoustoelectric current. This is said to be produced as an effect of SAW reflections, this is mostly seen in a two IDT system. The second IDT acts as a reflector much as the same way a reflection gradient is used for a unidirectional SAW, see Fig 1b. This can cause a standing wave to occur in the sample in such a way that the reflections effectively reduce the potential of the SAW and cause fewer electrons to be transported.

With slight phase shifts added the standing wave with respect to the QPC a net negative potential in the energy landscape can exist on one side of the QPC which will cause the current to change direction. These steps are best observed for QPCs that are long when compared to the SAW wavelength. This makes it possible to observe the acoustoelectric step current without applying a magnetic field (Shilton et al, 1996). As mentioned earlier, the high mobility of the 2DEG will screen the acoustoelectric current but this effect is not as prevalent when inside of a long QPC channel. The current screening is reduced around the QPC region and thus electrons can be transported through the channel. A long channel can allow ballistic transport of the electron causing it to shuttle from one side of the QPC to another. Since the screening is minimal, the electron will be "dragged" through the channel and the current steps of Fig. 9 can be observed.

Another phenomenon is oscillations in the acoustoelectric current once approaching the QPC pinch-off limit. As described in Shilton et al, 1996, and Maaø & Galperin, 1997, the acoustoelectric current oscillations occur at the same positions as the Coulomb steps. This oscillation is described as interference effects near the QPC at high frequencies which may be attributed to impurities in the 2DEG channel. The theory presented suggests that these oscillations are due to state transitions, both propagating and reflecting (Maaø & Galperin, 1997). The SAW may reflect from the QPC and create interference patterns; these patterns will affect the electron transmission probability through the QPC simply by changing the potential landscape. This mechanism is also sensitive to scattering events near the QPC, where energy quanta are emitted and absorbed between two waves. Others have contributed to the theory and more is being added.

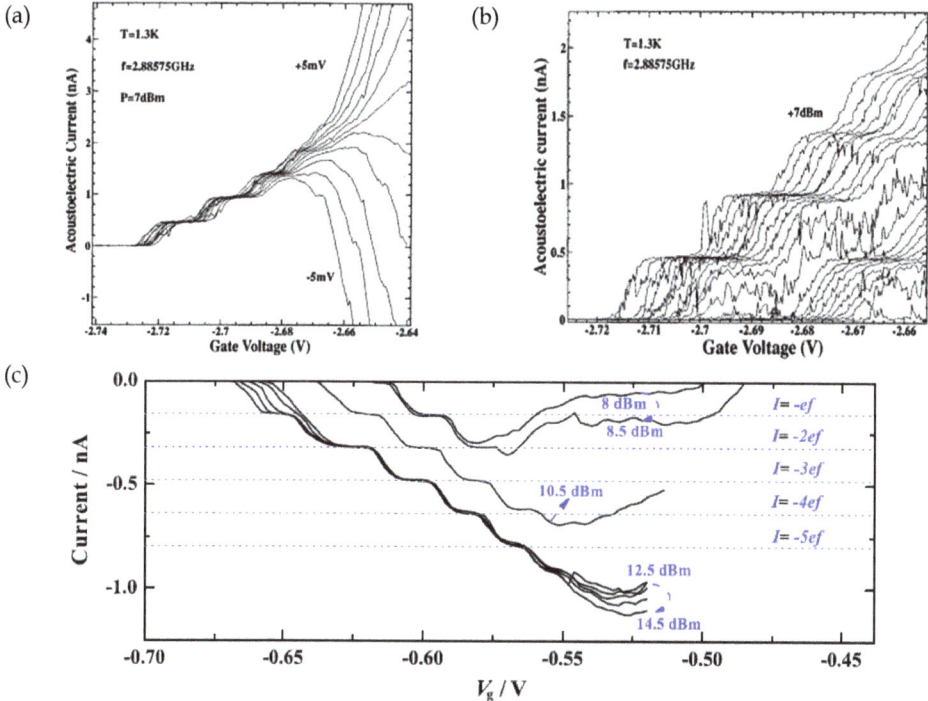

Fig. 9. QPC pinch-off of acoustoelectric current: (a) pinch-off with varying source-drain voltages applied at the SAW, changes are in 1 mV increments (Talyanskii et al., 1997) Copyright (1997) by The American Physical Society. (b) Pinch-off with different SAW power levels, the left most trace is 7 dBm and decrements by 0.2 dBm (Talyanskii et al., 1997) Copyright (1997) by The American Physical Society. (c) A negative anomalous acoustoelectric current for different gate voltages and SAW powers; $f = 1007.426$ MHz, $V_{ds} = 0$ V, T = 1.7 K, (Song et al., 2010)

Fig. 10 shows the oscillations of the acoustoelectric current as it is being driven through a QPC (orange trace) as it approaches pinch-off. It can be seen that the step like features observed in Fig. 9 are have been replaced by oscillations. These oscillations are negative, or on the lower cycle, when the current from the source-drain bias is flat or non-changing. We then see that as the current starts to decrease the acoustoelectric current has a positive value, or is on the upper cycle of the oscillation. The inset graph of Fig. 10 shows the entire sweep of the QPC gate voltage. As one conduction channel pinches off at the $V_g = -0.65$ V we see that there is a large, nearly single, oscillation in the acoustoelectric current. Since SAWs are very sensitive to 2DEGs we get what seems to be an amplified signal when compared to a source-drain bias. This can be used to identify information that may otherwise have been too weak or nearly washed out from thermal effects.

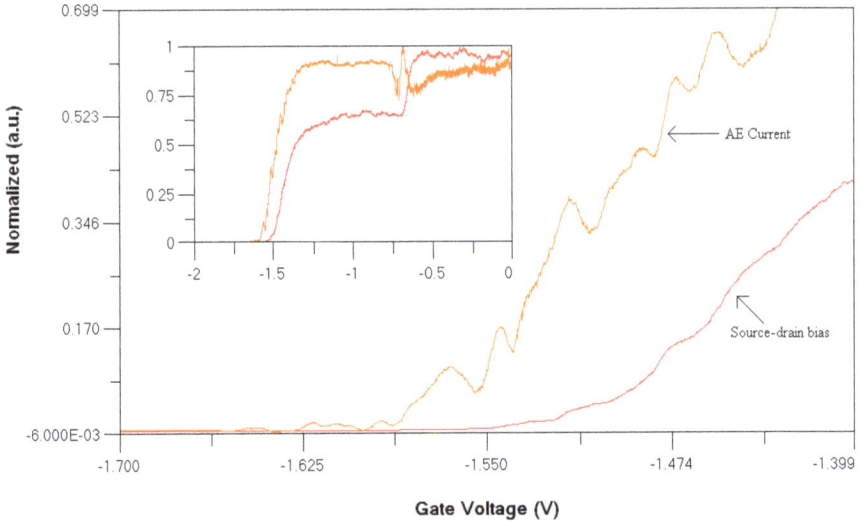

Fig. 10. Acoustoelectric current oscillations as the QPC nears pinch-off. The data plot has been normalized to show the relationship between the acoustoelectric current oscillations (orange trace) and the pinch-off when using a voltage bias across the source-drain electrodes (red trace). The acoustoelectric current trace has been normalized to -8.832 nA and the source-drain bias trace has been normalized to 4.692 nA. Measurements were taken at 4.2 K with the sample shown in Fig. 7. Inset plot shows entire gate voltage sweep range

4.3 Population and Depopulation of Quantum Dots
SAWs have been used widely as a mechanism to control the population and depopulation of QDs and DQDs. The high speed, or frequency, of operation combined with quantized current production make SAWs suitable for use with quantum systems. In a traditional QD system the gate and source-drain voltages are changed to allow single electron population and depopulation. This has several drawbacks; one being it is hard to decouple the dot from the surrounding environment due to the precise control of the voltages. There is always some noise either in the system or from the electronics. Another problem is if one gate is changed then the others must be changed to ensure a constant electron count in each dot, if applicable.

A SAW can be used as an alternative and has gotten more attention lately for the use in a QD or DQD system. The biggest reason for the attention is that SAWs can be used to interact with QD systems at much higher frequencies then what has been previously achieved by pulsing the gate electrodes. Since the acoustoelectric current can be well defined a QD system can use very high gate voltages to ensure no leakage current. We will take a look at the work done in Kataoka et al., 2007. This work describes very well the use of SAWs as a way to populate and depopulate a QD.

When the QD needs to be populated, the center plunger gate voltage value is lowered; causing the Fermi energy to lower below the Fermi energy of the surrounding 2DEG. The gate electrodes are still held at a high voltage value so the electrons cannot tunnel in or out of the QD, see Fig. 11e. The IDT is pulsed and as the SAW enters the QD region it changes the tunneling barrier created by the electrodes. The barrier is changed just enough that a single electron can tunnel into the dot, this is population; see Figs. 11a-b. The IDT is pulsed many times since the tunneling barrier is so high that the additional pulses are needed to increase the tunneling probability of the electron.

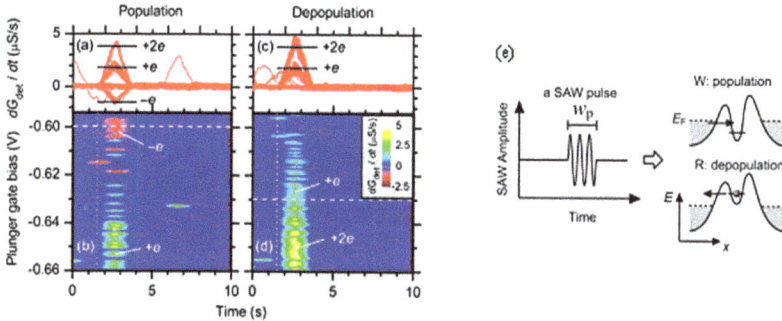

Fig. 11. Results from the paper Kataoka et al., 2007, Copyright (2007) by The American Physical Society. (a) and (b) The population of the QD, the –e line is an electron entering the dot. (c) and (d) Depopulation of the QD. (e) Schematic of SAW for population and depopulation of the QD

Likewise, as shown in Figs. 11c-d, the QD is depopulated by a similar mechanism. Here the plunger gate voltage is raised, made to be more negative, which causes the Fermi energy of the QD to be raised in comparison to the 2DEG Fermi energy outside of the dot, Fig. 11e. At this time the IDT is pulsed and the SAW enters the QD region. The SAW alters the tunneling barrier created by the gate electrodes when the potential of the SAW is superimposed onto the barrier, see Eq. 5. When the barrier is raised nothing happens since this will simply ensure the electron stays in the system, but when the negative cycle of the SAW superimposes with the barrier it decreases it which causes the electron to tunnel out.

This method of operation has proven, quite nicely, that SAWs and high potential QPCs offer a more robust method of single electron population and depopulation. As shown in Figs. 11a-d the QD can maintain the electron confinement for a long period of time which can be difficult in a traditional setup due to noise reduction of the measurement electronics. The IDTs can be accessed from an outside system such as an antenna, for example, and interact with the QD allowing the system another degree of freedom from the traditional closed electronics. However, this may come at the cost of additional noise reduction equipment and filtering.

5. Conclusion

This chapter has shown all of the basic properties and uses of SAWs in nano-structures and nano-systems. The reader is shown the parameters required for fabrication, theory of operation, real results, and application. The use of SAWs is quite limitless in the area of

NEMS. As people continue to refine the research through experiment and by continuing to add to the theory more applications will emerge.

6. Acknowledgment

We like to thank the National Science Foundation for support under a NIRT grant (No. ECCS-0708759) and the Air Force Office for Scientific Research for support under a MURI grant (No. FA9550-08-1-0337).

7. References

Barnes, C. H. W., Shilton, J. M., & Robinson, A. M. (2000). Quantum computing using electrons trapped by surface acoustic waves. *Physical Review B*, Vol. 62, No. 12, (September 2000), pp. 8410-8419, ISSN 0163-1829

Beil, F. W., Wixforth, A., Wegscheider, W., Schuh, D., Bichler, M., & Blick, R. H. (2008). Shock Waves in Nanomechanical Resonators. *Physical Review Letters*, Vol. 100, No. 2, (January 2008), pp. 026801 1-4, ISSN 0031-9007

Campbell, C. (1998). *Surface Acoustic Wave Devices for Mobile and Wireless Communications*, Academic Press, Inc, ISBN 0-12-157340-0, San Diego, CA, USA

Kataoka, M., Schneble, R. J., Thorn, A. L., Barnes, C. H. W., Ford, C. J. B., Anderson, D., Jones, G. A. C., Farrer, I., Ritchie, D. A., & Pepper, M. (2007). Single-Electron Population and Depopulation of an Isolated Quantum Dot Using a Surface-Acoustic-Wave Pulse. *Physical Review Letters*, Vol. 98, No. 4, (January 2007), pp. 46801-1 – 46801-4, ISSN 0031-9007

Kuok, M., Ng, S. C., & Zhang, V. L. (2001). Angular dispersion of surface acoustic waves on (001), (110), and (111) GaAs. *Journal of Applied Physics*, Vol. 89, No. 12, (June 2001), pp. 7899-7902, ISSN 0021-8979

Maaø, F. A. & Galperin, Y. (1997). Acoustoelectric effects in quantum constrictions. *Physical Review B*, Vol. 56, No. 7, (August 1997), pp. 4028-4036, ISSN 0163-1829

Salmon, V. (1946). A New Family of Horns. *Jounal of the Acoustical Society of America*, Vol. 17, No. 3, (January 1946), pp. 212-218, ISSN 0001-4966

Shilton, J. M., Mace, D. R., Talyanskii, V. I., Galperin, Y., Simmons, M. Y., Pepper, M., & Ritchie, D. A. (1996). On the acoustoelectric current in a one-dimensional channel. *Journal of Condensed Matter Physics*, Vol. 8, (March 1996), pp. L337-L343, ISSN 0953-8984

Song, L., Chen, S. W., He, J. H., Zhang, C. Y., Lu, C., & Gao, J. (2010). The anomalous negative acoustoelectric current in single-electron transport. *Solid State Communications*, Vol. 150, (2010), pp. 292-296, ISSN 0038-1098

Talyanskii, V. I., Shilton, J. M., Pepper, M., Smith, C. G., Ford, C. J. B., Linfield, E. H., Ritchie, D. A., & Jones, G. A. C. (1997). Single-electron transport in a one-dimensional channel by high-frequency surface acoustic waves. *Physical Review B*, Vol. 56, No. 23, (December 1997), pp. 15180-15184, ISSN 0163-1829

Wixforth, A., Scriba, J., Wassermeier, M., Kothaus, J. P., Weimann, G., & Schlapp, W. (1989). Surface acoustic waves on GaAs/$Al_xGa_{1-x}As$ heterostructures. *Physical Review B*, Vol. 40, No. 11, (October 1989), pp. 7874-7887

Permissions

The contributors of this book come from diverse backgrounds, making this book a truly international effort. This book will bring forth new frontiers with its revolutionizing research information and detailed analysis of the nascent developments around the world.

We would like to thank Marco G. Beghi, for lending his expertise to make the book truly unique. He has played a crucial role in the development of this book. Without his invaluable contribution this book wouldn't have been possible. He has made vital efforts to compile up to date information on the varied aspects of this subject to make this book a valuable addition to the collection of many professionals and students.

This book was conceptualized with the vision of imparting up-to-date information and advanced data in this field. To ensure the same, a matchless editorial board was set up. Every individual on the board went through rigorous rounds of assessment to prove their worth. After which they invested a large part of their time researching and compiling the most relevant data for our readers. Conferences and sessions were held from time to time between the editorial board and the contributing authors to present the data in the most comprehensible form. The editorial team has worked tirelessly to provide valuable and valid information to help people across the globe.

Every chapter published in this book has been scrutinized by our experts. Their significance has been extensively debated. The topics covered herein carry significant findings which will fuel the growth of the discipline. They may even be implemented as practical applications or may be referred to as a beginning point for another development. Chapters in this book were first published by InTech; hereby published with permission under the Creative Commons Attribution License or equivalent.

The editorial board has been involved in producing this book since its inception. They have spent rigorous hours researching and exploring the diverse topics which have resulted in the successful publishing of this book. They have passed on their knowledge of decades through this book. To expedite this challenging task, the publisher supported the team at every step. A small team of assistant editors was also appointed to further simplify the editing procedure and attain best results for the readers.

Our editorial team has been hand-picked from every corner of the world. Their multi-ethnicity adds dynamic inputs to the discussions which result in innovative outcomes. These outcomes are then further discussed with the researchers and contributors who give their valuable feedback and opinion regarding the same. The feedback is then collaborated with the researches and they are edited in a comprehensive manner to aid the understanding of the subject.

Apart from the editorial board, the designing team has also invested a significant amount of their time in understanding the subject and creating the most relevant covers. They scrutinized every image to scout for the most suitable representation of the subject and create an appropriate cover for the book.

The publishing team has been involved in this book since its early stages. They were actively engaged in every process, be it collecting the data, connecting with the contributors or procuring relevant information. The team has been an ardent support to the editorial, designing and production team. Their endless efforts to recruit the best for this project, has resulted in the accomplishment of this book. They are a veteran in the field of academics and their pool of knowledge is as vast as their experience in printing. Their expertise and guidance has proved useful at every step. Their uncompromising quality standards have made this book an exceptional effort. Their encouragement from time to time has been an inspiration for everyone.

The publisher and the editorial board hope that this book will prove to be a valuable piece of knowledge for researchers, students, practitioners and scholars across the globe.

List of Contributors

Alexander Zinovev, Igor Veryovkin and Michael Pellin
Argonne National Laboratory, USA

Josef Foldyna
Institute of Geonics of the ASCR, v. v. i., Ostrava, Czech Republic

N.E. Bykovsky and Yu.V. Senatsky
Lebedev Physical Institute, Russian Academy of Sciences, Moscow, Russia

Adam Brański
Rzeszow University of Technology, Poland

Natalya Naumenko
Moscow Steel and Alloys Institute (Technological University), Russia

Takahiko Yanagitani
Nagoya Institute of Technology, Japan

Trang Hoang
Faculty of Electrical-Electronics Engineering, University of Technology, HoChiMinh City, Vietnam

Eduard Rocas and Carlos Collado
Universitat Politècnica de Catalunya (UPC), Barcelona, Spain

Nicolas Woehrl and Volker Buck
University Duisburg-Essen and CeNIDE, Duisburg, Germany

Jyoti Prakash Kar
Department of Electronics Engineering, University of Tor Vergata, Rome, Italy

Gouranga Bose
Department of Applied Electronics and Instrumentation Engineering, Institute of Technical Education and Research, Bhubaneswar, Orissa, India

Ivan D. Avramov
Georgi Nadjakov Institute of Solid State Physics, Sofia, Bulgaria

Cinzia Caliendo
Istituto dei Sistemi Complessi, ISC-CNR, Area della Ricerca Roma 2, Rome, Italy

I. L. Villegas and Y. O. Barmenkov
Departamento de Física Aplicada y Electromagnetismo, ICMUV, Universidad de Valencia, Burjassot, Valencia, Spain
Centro de Investigaciones en Óptica, León, Guanajuato, Mexico

A. Díez, M. V. Andrés, J. L. Cruz and M. Bello-Jimenez
Departamento de Física Aplicada y Electromagnetismo, ICMUV, Universidad de Valencia, Burjassot, Valencia, Spain

A. Martínez-Gámez
Centro de Investigaciones en Óptica, León, Guanajuato, Mexico

C. Cuadrado-Laborde
Departamento de Física Aplicada y Electromagnetismo, ICMUV, Universidad de Valencia, Burjassot, Valencia, Spain
CONICET La Plata, Buenos Aires, Argentina

Dustin J. Kreft and Robert H. Blick
University of Wisconsin – Madison, Department of Electrical and Computer Engineering, USA